New Frontiers of Educational Research

More information about this series at http://www.springer.com/series/10795

Yiming Cao · Frederick K.S. Leung
Editors

The 21st Century Mathematics Education in China

Springer

Editors
Yiming Cao
Beijing Normal University
Beijing
China

Frederick K.S. Leung
Faculty of Education
The University of Hong Kong
Hong Kong
Hong Kong

ISSN 2195-3473 ISSN 2195-349X (electronic)
New Frontiers of Educational Research
ISBN 978-3-662-55779-2 ISBN 978-3-662-55781-5 (eBook)
https://doi.org/10.1007/978-3-662-55781-5

Library of Congress Control Number: 2017952020

Printed on acid-free paper

This Springer imprint is published by Springer Nature
The registered company is Springer-Verlag GmbH Germany
The registered company address is: Heidelberger Platz 3, 14197 Berlin, Germany

Preface

The major characteristics of traditional Chinese mathematics are its emphasis on algorithms and the application of mathematics in solving practical problems; mathematics theories are embodied in the combination of these. In contrast, ancient Greek mathematics is an established axiomatic system consisting of a series of theorems derived from given axioms (or postulates) and definitions through deductive reasoning. Chinese education and cultural tradition place much emphasis on the humanities and social sciences, and science and technology are relatively less emphasized. In terms of classroom teaching, concise exposition, ample practice and careful reflection are stressed, and for a long time this has been considered by some as a traditional and conservative teaching method. The literature has pointed out a number of shortcomings in Chinese classroom teaching, such as a uniformed didactic approach, teacher-dominated instruction and passive student reception, and large class sizes. On this phenomenon, Biggs[1] and other western scholars put forward the famous "Chinese Learners' Paradox", that Chinese learners, despite learning mathematics in a seemingly unsatisfactory environment, performed better than their western counterparts in international studies of mathematics achievement.

Since the 1980s, phenomenal changes have taken place in China's society and economy. Faced with the new challenges of the time and new problems in classroom teaching arising out of curriculum reforms, the Ministry of Education conducted a survey from June 1996 to 1997 on the implementation status of the compulsory national curriculum, including the mathematics curriculum. The survey results showed that, despite significant achievements such as students' acquisition of solid foundations of basic knowledge and basic skills, major problems have remained in the implementation of the curriculum, for example, the contents being too complex, difficult, distorted and obsolete; students suffering from the pressures of memorization and rote learning; teachers' time taken up setting excessive

[1] Biggs, J.B. & Watkins, D.A. (2001). Insight into teaching the Chinese Learner. In D.A. Watkins & J.B. Biggs, Teaching the Chinese Leaner: Psychological and pedagogical perspectives: 277–300.

practice exercises; and over-emphasis on achievement scores in the assessment of student learning. Under this traditional and unified evaluation system, substantial numbers of students have been subjected to immense pressure in learning mathematics, which has had the negative impact of lowering their confidence greatly, even leading to a hatred of mathematics and its learning.

Under the influence of the worldwide trend of curriculum reform in mathematics, China launched a new round of reforms at the basic education stage, based on relevant research. At the level of design and implementation, it aimed to develop a new system of Chinese mathematics education based on an orientation towards student development, with a curriculum structure emphasizing student choice. The curriculum is organized around the major theme of unifying various mathematics contents, with an integrative and modular structure, thus extending the traditional stress on basic knowledge, basic skills and basic ability.

The mathematics curriculum reform since the beginning of the twenty-first century has been based on student-centred approaches, with emphasis on mathematics application and the development of students' abilities. Different aspects of mathematics education, including the curriculum standards, teaching materials and teaching practice, as well as examination and evaluation, have undergone different degrees of changes, some of which were very radical ones. With changes, there have been controversies as well, and amid these changes and controversies, Shanghai's participation in PISA for the first time in 2009 and its first-place rankings in mathematics, science and reading caught attention around the world. Shanghai students' superior performances came as a surprise because developing students' mathematics literacy has not been a traditional strength of education in China. Yet the PISA results have shown that the education system in Shanghai is able to inculcate students who can perform so well in PISA, which focuses on assessing mathematics literacy and applications. The latest release of the PISA 2015 assessment results, in December 2016, gives a slightly more balanced picture. Whereas Shanghai was the only participating city in mainland China in PISA 2009 and 2012, three additional Chinese provinces, namely Beijing, Jiangsu and Guangdong, also participated in PISA 2015. In addition, some changes were made in the PISA 2015 assessment, including the introduction of computer tests and assessment of capabilities in cooperative problem solving. So it could be argued that the results of PISA 2015 reflect more accurately the performance of students in the more developed regions in mainland China.

The PISA results have invoked much reflection nationally, as well having aroused interest in the mathematics education community internationally. After more than a decade of curriculum reform, what changes in mathematics education have occurred in China? Are the "two basics" still the only concern? Are there any special characteristics of Chinese mathematics education which have led to the outstanding performances of Chinese students in PISA? What are the characteristics of students' mathematics learning and teachers' instructional practices and professional development in China? What are the main problems faced by the education system in the country?

Such international attention on mathematics education in China has prompted us to carry out an in-depth study into the current status of mathematics education in China since the beginning of the twenty-first century, including its strengths and weaknesses. At a meeting in Beijing Normal University in March 2014, a group of mathematics educators expressed a common interest in conducting a study of different aspects of mathematics education in China. In June of the same year, Prof. Paul Cobb of Vanderbilt University and Prof. David Clarke of Melbourne University visited Beijing Normal University, and they also expressed support for such a project. In July 2014, during the Fourteenth Session of the National Institute of Mathematical Education, scholars in mathematics education from both home and abroad were consulted on what aspects of mathematics education in China are deemed to be of interest to the international community. Views and opinions about the project were also collected from various stake holders through interviews and questionnaires. Through these efforts, the preliminary framework of the book was determined.

This book aims at summarizing different aspects of the status and achievements of mathematics education in China from the perspective of the "insiders" in order to share with the international mathematics education community our experiences and the lessons learned. At the same time, it is hoped that mathematics instruction in China will also be enhanced and improved through our work in the project. The book consists of twenty-three chapters organized around six parts, covering almost all aspects of mathematics education in China. The six parts include: an overview of mathematics education in China which introduces the traditions, the examination system, the curriculum reform and family education; mathematics curriculum and teaching materials, covering the curriculum and teaching materials at primary and secondary school levels and the characteristic features of the mathematics teaching materials in the new century; classroom instructional practices, including different types of mathematics teaching, their characteristics, task design, mathematics teaching objectives and the integration of information technology into mathematics education; professional development of teachers, which covers pre-service education, school-based professional development of in-service teachers, teachers' beliefs about mathematics teaching and teachers' knowledge in teaching. In addition, the book also includes topics on learning and evaluation in mathematics education.

This book is the result of a concerted effort of authors including university professors who have been engaged in research in mathematics education for a long time; young scholars who research in specific areas of mathematics education; and experienced researchers, curriculum developers and textbook writers. The editors and authors have different areas of expertise. Frederick Leung and Yiming Cao have long been interested in comparative studies of mathematics education and the influence of culture on mathematics teaching and learning; Dai Qin, Liu Jian, Zhu Yan and Wang Lidong have been engaged in research on mathematics curriculum reform and educational policy; Lv Shihu, Li Haidong and Ye Beibei are experts in the study of the mathematics curriculum and textbooks; the main research interests of Wang Guangming, Hu Dianshun and Shao Zhenhong's have been on mathematics classroom practices; Guo Yufeng has been engaged in research on students' mathematics learning; Cai Jinfa, Yang Xinrong, He Xiaoya and Han Jiwei have been involved in research on mathematics teachers' professional development; and Tu Rongbao, Ning Lianhua and Zhang Chunli have rich research experience in the evaluation of mathematics education. The authors work in teacher education institutions, comprehensive universities, secondary schools and publishing houses from virtually all parts of China, including the University of Hong Kong, Beijing Normal University, East China Normal University, Northeast Normal University, Nanjing Normal University, South China Normal University, Southwestern University, Inner Mongolia Normal University, Northwest Normal University, the High School Affiliated to Renmin University of China, Beijing Jingshan Middle School and the People's Education Press. This representative group of scholars is best qualified to produce a comprehensive picture of different aspects of mathematics education research and practices in China since the twenty-first century.

The book has pulled together the commitments and efforts of different authors, taking more than two years to complete, from the design of the framework in 2014, to reviews of the abstracts of the chapters by the editors, peer reviews of all the chapters, final reviews by the editors, and proof reading of the final Chinese and English versions, to its completion in 2017. Making use of attendance at mathematics education conferences, the editorial team met numerous times in Chongqing, Wuhan, Beijing and other places, in order to ensure the quality of the book.

We are also indebted to other scholars in mathematics education at home and abroad for their help and support in the preparation of the book. Professor David Clarke of the University of Melbourne, Prof. Paul Cobb of Vanderbilt University, Dr. Zsolt Lavicza of the University of Cambridge and others have rendered insightful comments and suggestions. Li Haidong, Director of the People's Education Press, has invested great effort in the production of the book, and Dr. Li Xinlian has helped prepare the manuscripts in her capacity as project assistant. To all of these people, we express our thanks and appreciation.

Research and practice in mathematics education in China have attracted and will continue to attract the attention of researchers and practitioners at home and abroad. We sincerely hope that this book will enhance the development of research and practice in mathematics education in this country, as well as promote cooperation and interchange of ideas between China and the rest of the world.

Prof. Yiming Cao
School of Mathematical Sciences
Beijing Normal University

Prof. Frederick K.S. Leung
Kintoy Professor in Mathematics Education,
The University of Hong Kong
Southwest University

Contents

Part I
A General Introduction of Chinese Mathematics Education

Chapter 1
Mathematics Education of Chinese Communities from the Perspective of International Studies of Mathematics Achievement

Frederick K.S. Leung

Abstract International studies of mathematics achievement have been gaining importance in the past decades in the mathematics education community and have drawn much attention from policy makers around the world. In this chapter, results of some international studies in terms of mathematics achievements, student attitudes and classroom teaching are reviewed in order to gain a better understanding of mathematics education in the Chinese communities. It was found that Chinese students did very well in mathematics as measured by these international studies, however they did not do correspondingly well in terms of the affective outcomes of mathematics education. Superficially, mathematics teaching in the Chinese systems looked very traditional and backward, but a more fine-grained analysis of the data in these international studies of classroom teaching showed that the quality of teaching was high. It is argued that the underlying cultural values of these Chinese communities may be a plausible explanation of these findings. Chinese communities should be proud of their students' achievement, but there is certainly no scope for complacency.

1.1 Introduction: The Surge of International Studies

International studies of mathematics achievement have been gaining importance in the past decades in the mathematics education community and have drawn much attention from policy makers around the world. Most noticeable of these studies are the Trends in International Mathematics and Science Study (TIMSS), under the auspices of the International Association for the Evaluation of Educational

---

F.K.S. Leung (✉)
The University of Hong Kong, Hong Kong SAR, China
e-mail: frederickleung@hku.hk

Y. Cao and F.K.S. Leung (eds.), *The 21st Century Mathematics Education in China*, New Frontiers of Educational Research,
https://doi.org/10.1007/978-3-662-55781-5_1

Achievement (IEA) and the Programme for International Student Assessment (PISA) conducted by the Organisation for Economic Co-operation and Development (OECD). These studies evaluate mathematics achievements of students in different countries and purport to identify factors that explain student achievement. But policy makers and the general public are usually more interested in the ranking of countries according to the scores of these studies, and, to some, TIMSS and PISA have become the "Olympics of Education" (Scardino, 2008). Notwithstanding these distorted uses of their results, if we look beyond the scores and rankings, these studies provide an informative lens for us to understand the education system. They provide invaluable data for understanding an education system in the light of comparison with other countries and afford insights which cannot be gained through studies within one's own education system.

In this Chapter, we intend to gain some knowledge and insight of mathematics education in China and other Chinese communities[1] through this lens of international studies of mathematics education. The first such international study that China participated in was a series of studies conducted at the University of Michigan in the early 1980s, and China was one of the countries which participated in the International Assessment of Educational Progress (IAEP) study in the early 1990s. Unfortunately, China did not participate fully in the more recent large-scale international studies such as TIMSS and PISA (it withdrew from participation in TIMSS after joining the initial preparation stage of the project in the early 1990s). Shanghai participated in PISA 2009 and 2012, and a number of provinces and municipals in China, namely Beijing, Shanghai, Jiangsu and Guangdong (B-S-J-G) participated in PISA 2015. Beyond mainland China, Taiwan was in the University of Michigan Study and IAEP, and it has been participating in TIMSS since 1999. Hong Kong joined TIMSS from the outset in 1995 and has been participating since; it has also been participating in PISA since its beginning in 2000. Taiwan and Macao started to participate in PISA in 2006 and have been in the study since. So as far as mathematics education in mainland China is concerned, we only have limited data from these international studies. Notwithstanding this limitation, together with the information on the Chinese communities of Hong Kong and Macao and Taiwan, results from international studies have provided us with an informative and interesting picture about mathematics education in China.

[1]The full name for China is the People's Republic of China. There are other Chinese communities or systems outside the Chinese mainland, and they include Hong Kong and Macao (which are Special Administrative Regions of China), and Taiwan (referred to as "Chinese Taipei" in these international studies). In this Chapter, when the Chinese mainland is referred to in contrast to these systems, the term "mainland China" is used. When mainland China and these systems are referred to collectively, the term "Chinese communities" is used. International studies sometimes use the term "countries" to refer to these different communities beyond mainland China, but we know they are referring to these political entities rather than "countries".

1.2 Findings of International Studies: Mathematics Achievement

1.2.1 The University of Michigan Studies

Prior to the bloom of international studies triggered by the release of the results of TIMSS in the mid-1990s, there was a series of studies based at the University of Michigan comparing academic achievements of children in Japan, Taiwan, mainland China and the USA conducted in the late 1970s and early 1980s. The first study took place in 1979–80, and Kindergarten, Grade 1 and Grade 5 pupils were sampled in the cities of Sendai, Japan; Taipei, Taiwan and the Minneapolis metropolitan area of USA. Data for the second study were collected in 1985–86 (the Chicago metropolitan area in the USA was studied instead in the second study), and the data included those collected through interviews with parents and teachers, classroom observations and cognitive tests. For the mathematics tests in the first study, computation, for year one and year five students, and geometry, for year five only, were covered (see Table 1.1). For the second study, a series of tests including word problems, operations, visualization, graphing, mental calculation, number concepts, estimation and mental image transformation were administered.

1.2.1.1 Results

For the second study, as shown in Table 1.2, year one and year five students in Beijing outperformed their Chicago counterparts in all areas of mathematics except for *Graphs and Tables* and *Visualization* in year five (Stevenson et al., 1990, p. 1058).

Table 1.1 Mean mathematics scores for kindergarten, Grade 1 and Grade 5 children of USA, Taiwan and Japan in the first University of Michigan Study

Grade	USA	Taiwan	Japan
K	37.5 ± 5.6	37.8 ± 7.4	42.2 ± 5.1
1	17.1 ± 5.3	21.2 ± 5.5	20.1 ± 5.2
5	44.4 ± 6.2	50.8 ± 5.7	53.3 ± 7.5

Table 1.2 Mean mathematics scores for Grade 1 and Grade 5 children of USA and Beijing in the second University of Michigan Study

	Asian American	White	Black	Hispanic	Beijing
Grade 1					
Computation	14.3	14.0	11.2	11.5	18.1
Grade 5					
Computation	51.3	46.3	44.9	43.7	57.5
Geometry	5.6	4.6	3.7	3.8	10.3

1.2.2 IAEP

Another international study, before the emergence of TIMSS and PISA, in which China participated was the International Assessment of Educational Progress (IAEP) study conducted by the Educational Testing Service (ETS) in the USA. The purpose of IAEP was to collect and report data on what students knew and could do, on educational and cultural factors associated with achievement, and on students' attitudes (Mead, 1995). The first IAEP study was conducted in 1988, but the Chinese systems were not among the six countries involved in the study. The second IAEP study was conducted in 1990/1991 to assess the proficiency of 9- and 13-year-old students from 20 countries. About 1650 students from 20 provinces/municipal cities in mainland China participated in the 13-year-old test, and about the same number from Taiwan took part in both the 9-year-old and 13-year-old tests (Lapointe, Mead, & Askew, 1992).

1.2.2.1 Results

The results of the second IAEP study showed that Chinese students performed very well in the mathematics tests (see Tables 1.3 and 1.4). Their scores were well above the IAEP averages, with averages of 68 and 73% for the 9-year-olds and 13-year-olds, respectively, from Taiwan, and 80% for the 13-year-olds from China; China's 13-year-olds' performance was the highest among the participating countries.

Table 1.3 Mathematics, age 9—distribution of percent correct scores by population

Population	Average percent correct
Korea	75
Hungary	68
Taiwan	**68**
Emilia-Romagna, Italy[a]	68
Soviet Union	66
Scotland*	66
Israel	64
Spain	62
Ireland	60
Canada	60
England[a]	59
USA	58
Slovenia	56
Portugal[a]	55

[a]Populations with exclusions or low participation
Chinese communities in bold letters

Table 1.4 Mathematics, age 13—distribution of percent correct scores by population

Population	Average percent correct
China[a]	**80**
Korea	73
Taiwan	**73**
Switzerland	71
Soviet Union	70
Hungary	68
France	64
Emilia-Romagna, Italy	64
Israel	63
Canada	62
Scotland	61
Ireland	61
England[a]	61
Slovenia	57
Spain	55
USA	55
Portugal[a]	48
Jordan	40
Sao Paulo, Brazil[a]	37
Fortaleza, Brazil[a]	32
Mozambique cities[a]	28

[a]Populations with exclusions or low participation
Chinese communties in bold letters

IAEP also investigated three process aspects of mathematics achievement: Conceptual Understanding, Procedural Understanding and Problem-Solving (Lapointe et al., 1992). For 13-year-olds, "students from Mainland Chinese ranked first in all three aspects, whereas those from Taiwan also performed quite well, ranking either second or third. However, at the 9-year-old age level, the Chinese students (from Taiwan) did not do well on Problem-Solving tasks, and their average percentage correct (55.7) was lower than the international average (58.5), ranking tenth among the 14 education systems" (Fan & Zhu, 2004, p. 16).

1.2.3 TIMSS

TIMSS is a large-scale international study of mathematics and science achievement conducted under the auspices of IEA; TIMSS originally stood for the Third International Mathematics and Science Study. One of the first studies conducted by IEA was the First International Mathematics Study (FIMS), which took place in the 1960s, but none of the Chinese communities participated in this. The Second

International Mathematics Study (SIMS) took place in the 1980s, and Hong Kong was the first of the Chinese communities to join such an international study. Meanwhile, a First International Science Study (FISS) and a Second International Science Study took place and, by the early 1990s when IEA considered the third study, it was decided that the mathematics and science studies be combined. The first National Project Coordinators (NPC) (later renamed National Research Coordinators, NRC) meeting of the **Third** Mathematics International Mathematics and Science Study (TIMSS) took place in 1991, and it was decided that students at three grade levels (namely Grades 3 and 4, Grades 7 and 8 and the last Grade of secondary education) were to be sampled for study. Data collection was administered in 1995, and the results were released in late 1996.

The 1995 TIMSS results came as a surprise for many. Countries generally considered as strong in mathematics, such as Germany, Hungary and USA, did not do as well as expected. On the other hand, students from a cluster of countries in East Asia, namely Singapore, Korea, Japan and Hong Kong, did very well. Partly because of this surprising result, the IEA decided to conduct a follow-up study of TIMSS in 1999 (known as TIMSS-Repeat or TIMSS-R) for Grade 8 children, catching the cohort of Grade 4 students in 1995.

Hong Kong also participated in TIMSS-R, and Taiwan joined the study as well. The results of TIMSS-R were released in 2000, and the findings were no less surprising. The four East Asian countries that did well in 1995 continued to surpass other countries, and Taiwan became a fifth member of this group of high-performing countries.

Because the results of TIMSS and TIMSS-R caught the attention of educators and the public globally, IEA decided that, instead conducting fourth and fifth studies and so on, the TIMSS studies would be conducted in four-year cycles thereafter. Grade 4 and Grade 8 students are tested, so that the Grade 8 students in each cycle correspond to the Grade 4 students in the previous cycle, making it a quasi-longitudinal study. The series of studies has been renamed **Trends** in International Mathematics and Science Study, keeping to the acronym TIMSS. The subsequent cycles of TIMSS took place in 2003, 2007 and 2011, with the latest cycle taking place in 2015.

1.2.3.1 Results

As shown in Table 1.5, East Asian countries have continued to outperform their counterparts elsewhere. In particular, the Chinese systems of Hong Kong and Taiwan have never fallen out of the first five positions (Beaton et al., 1996; Mullis et al., 1997, 2000; Mullis, Martin, & Foy, 2008; Mullis, Martin, Foy, & Arora, 2012; Mullis, Martin, Foy, & Hooper, 2016; Mullis, Martin, Gonzalez, & Chrostowski, 2004). Moreover, not only did students in these Chinese systems do very well in mathematics on average, these systems also have been able to nurture top performers without sacrificing catering for students with low performances. From 2003 onward, TIMSS included a measure of the percentages of students in each country who attained various international benchmarks of achievement. Students attaining the advanced benchmark are those who are the very best in their cohorts in terms of mathematics achievement, while those attaining the low benchmark have reached the minimum standard expected of a student of that grade. As shown in Table 1.6, the Chinese systems have a large proportion of students who attained the advanced benchmark internationally; at the same time, only a

Table 1.5 Summary of TIMSS results of Chinese systems for various years (ranking of system in bracket)

TIMSS	Chinese systems	Grade 4	Grade 8
1995	Hong Kong	587 (4)	588 (4)
1999	Hong Kong	–	582 (4)
	Chinese Taipei	–	585 (3)
2003	Hong Kong	575 (2)	586 (3)
	Chinese Taipei	564 (4)	585 (4)
2007	Hong Kong	607 (1)	572 (4)
	Chinese Taipei	576 (3)	598 (1)
2011	Hong Kong	602 (3)	584 (4)
	Chinese Taipei	591 (4)	609 (3)
2015	Hong Kong	615 (2)	594 (4)
	Chinese Taipei	597 (4)	599 (3)

Table 1.6 Percentages of students in Chinese systems attaining advanced and low benchmarks in various cycles of TIMSS

TIMSS	Chinese systems	Grade 4		Grade 8	
		Advanced benchmark	Low benchmark	Advanced benchmark	Low benchmark
2003	Hong Kong	22	99	31	98
	Chinese Taipei	16	99	38	96
	Int. average	9	82	7	74
2007	Hong Kong	40	100	31	94
	Chinese Taipei	24	99	45	95
	Int. average	5	90	2	75
2011	Hong Kong	37	99	34	97
	Chinese Taipei	34	99	49	96
	Int. average	4	90	3	75
2015	Hong Kong	45	100	37	98
	Chinese Taipei	35	100	44	97
	Int. average	6	93	5	84

negligible proportion of their students fell below the low benchmark. This shows that, in the Chinese systems, the high achievements of the high flyers are not achieved at the expense of taking care of the low achievers.

1.2.4 PISA

As mentioned at the beginning of this chapter, PISA is an OECD project which runs every three years. PISA tests the "literacy" of students in the three areas of mathematics, science and reading, with one of the three being the focus of study in each cycle. PISA is meant to complement TIMSS (and PIRLS), which is curriculum based. For mathematics literacy, PISA measures "an individual's capacity to identify and understand the role that mathematics plays in the world, to make well-founded judgements and to use and engage with mathematics in ways that meet the needs of that individual's life as a constructive, concerned and reflective citizen …Rather than being limited to the curriculum content students have learned, the assessments focus on determining if students can use what they have learned in the situations they are likely to encounter in their daily lives" (OECD, 2003, p. 24).

PISA uses an age-based instead of a grade-based sample, as it "seeks to assess how well 15-year olds are prepared for life's challenges" (OECD, 2004, p. 20). The first PISA took place in 2000, with the focus area being reading. Mathematics was the focus area in 2003 and 2012. Hong Kong joined PISA in 2000, Taiwan and Macau joined in 2006, Shanghai joined in 2009 and Beijing-Shanghai-Jiangsu-Guangdong (B-S-J-G) joined in 2015.

1.2.4.1 Results

Some might have thought that students in the Chinese systems did well in TIMSS because it is a curriculum-based test and the Chinese systems are known to place a lot of emphasis on the curriculum content and its examination. Since PISA measures mathematics literacy instead of curriculum knowledge, the Chinese systems are not expected to perform as well. Contrary to this expectation, the Chinese systems also did very well in mathematics in PISA, as shown in Table 1.7. And similar to the results of TIMSS, the Chinese communities have a large proportion of students who are top performers, and a relatively small proportion who are low achievers (OECD, 2001, 2003, 2004, 2007, 2010, 2014, 2016).

One observation, however, is that, relatively speaking, the Chinese communities did not do as well in the area of "problem-solving"—in PISA 2012 and 2015, there was a measure of problem-solving competence. Taking the results of PISA 2012 as an example, notwithstanding the fact that the Chinese communities, together with the other East Asian countries of Singapore, Korea and Japan, topped the rest of the

Table 1.7 Summary PISA results of Chinese systems for various years (ranking of system in bracket)

PISA	Chinese systems	Achievement
2000	Hong Kong	560 (1)
2003	Hong Kong	550 (1)
	Macao	527 (9)
2006	Hong Kong	547 (3)
	Macao	525 (8)
	Chinese Taipei	549 (1)
2009	Hong Kong	555 (3)
	Macao	525 (12)
	Chinese Taipei	543 (5)
	Shanghai	600 (1)
2012	Hong Kong	561 (3)
	Macao	538 (6)
	Chinese Taipei	560 (4)
	Shanghai	613 (1)
2015	Hong Kong	548 (2)
	Macao	544 (3)
	Chinese Taipei	542 (4)
	BSJG	531 (6)

world in terms of achievement scores, the performances of Chinese students in problem-solving lagged behind the other East Asian countries—their scores were at least 12 score points behind the other East Asian countries. This seems to be consistent with the findings of a study by Cai (2000) that, although Chinese students outperformed US students on process-constrained tasks, US students outperformed Chinese students on process-open tasks.

1.2.5 Summary

It can be seen from the section above that Chinese students have been doing extremely well in mathematics achievement as measured by international comparative studies, when compared with their counterparts elsewhere. Does this mean that mathematics education in the Chinese systems is superior to that of systems in the West? This actually is not the whole story. Achievement is only one of the benchmarks of a good education system. What about the affective outcomes of mathematics education of the system? Most of the international studies discussed in the last section included surveys of students' attitudes. Do Chinese students hold positive attitudes that are comparable to their superior mathematics achievement?

1.3 Findings of International Studies: Attitudes/Affective Outcomes of Chinese Students

1.3.1 The University of Michigan Studies

In the second University of Michigan Study, children were asked how well they liked mathematics. The results are summarized in Table 1.8 (Stevenson et al., 1990).

1.3.2 IAEP

In the IAEP Study, a student questionnaire on their family background, classroom experiences and the schools they attended was administered, and schools were also invited to complete a school questionnaire. The students who took the tests were also asked to respond to a questionnaire on their attitudes towards mathematics. As shown in Table 1.9, in contrast to their superior performances in the mathematics tests, China and Taiwan had fewer students (79%) who expressed positive attitudes towards mathematics compared with other countries in the study (the percentage for Canada, e.g. was 94%).

1.3.3 TIMSS

Again, in contrast to their superior mathematics achievements in TIMSS, the questionnaire results showed that Chinese students held rather negative attitudes towards mathematics. From Table 1.10, it can be seen that the Chinese students did not enjoy mathematics much, they did not think of it as being very important, and they consistently had low confidence in their mathematics abilities.

Table 1.8 Attitudes of students in the second University of Michigan Study

	Beijing	Chicago
Like mathematics (%)	85	72
Want to do mathematics (%)	8	22
Mathematics is hard (%)	20	8
Do mathematics well (%)	39	52
Optimistic about future performance (Grade 1) (%)	50	75
Optimistic about future performance (Grade 5) (%)	29	58
Meet parents' expectation (%)	55	89

Table 1.9 Percent of age 13 students who expressed positive attitudes towards mathematics in IAEP

Population	Average percent of students who expressed positive attitudes towards mathematics
Canada	94
Scotland	91
England[a]	91
Israel	90
USA	90
Spain	89
Ireland	88
Mozambique cities[a]	88
Emilia-Romagna, Italy	86
Fortaleza, Brazil[a]	86
Switzerland	85
Hungary	85
Portugal[a]	84
Slovenia	83
Sao Paulo, Brazil[a]	83
France	81
Taiwan	**79**
China[a]	**79**
Jordan	77
Soviet Union	76
Korea	71

[a]Populations with exclusions or low participation
Chinese communties in bold letters

1.3.4 PISA

As pointed out above, 2003 and 2012 were the years when mathematics was the focus year for PISA. As shown in Table 1.11, the PISA 2003 and 2012 results for students' attitudes towards mathematics were rather similar to those of TIMSS.

1.3.5 Summary

As can be seen from the discussion in this section, the attitudes of Chinese students towards mathematics and mathematics learning were rather negative, notwithstanding their superior mathematics achievement. Could the negative student attitudes, perhaps as well as their superior student achievement, be attributed to the kind of teaching that is going on in Chinese classrooms? Do Chinese students do

Table 1.10 Attitudes of students in various cycles of TIMSS

TIMSS	Chinese systems	Grade 4 (%)	Grade 8 (%)
1995 Like learning mathematics	Hong Kong	36	17
	International average	49	19
Positive self-concept in mathematics	Hong Kong	17	5
	International average	37	23
1999 Like learning mathematics	Chinese Taipei	–	15
	Hong Kong	–	22
	International average	–	24
Positive self-concept in mathematics	Chinese Taipei	–	11
	Hong Kong	–	14
	International average	–	18
2003 Enjoy learning mathematics	Chinese Taipei	31	13
	Hong Kong	30	15
	International average	50	29
Value mathematics	Chinese Taipei	–	25
	Hong Kong	–	35
	International average	–	55
Confidence in mathematics	Chinese Taipei	41	26
	Hong Kong	40	30
	International average	55	40
2007 Positive affect towards mathematics	Chinese Taipei	50	37
	Hong Kong	67	47
	International average	72	54
Value mathematics	Chinese Taipei	–	45
	Hong Kong	–	60
	International average	–	78
Confidence in mathematics	Chinese Taipei	36	27
	Hong Kong	46	30
	International average	57	43
2011 Like learning mathematics	Chinese Taipei	34	14
	Hong Kong	47	19
	International average	48	26
Value mathematics	Chinese Taipei	–	13
	Hong Kong	–	26
	International average	–	46
Confidence in mathematics	Chinese Taipei	20	7
	Hong Kong	24	7
	International average	34	14
2015 Like learning mathematics	Chinese Taipei	23	11
	Hong Kong	35	15
	International average	46	22

(continued)

Table 1.10 (continued)

TIMSS	Chinese systems	Grade 4 (%)	Grade 8 (%)
Value mathematics	Chinese Taipei	–	10
	Hong Kong	–	19
	International average	–	42
Confidence in mathematics	Chinese Taipei	15	9
	Hong Kong	19	10
	International average	32	14

Table 1.11 Summary of PISA results for various years

PISA	Chinese systems	Index
2000 Interest in mathematics	Hong Kong	0.59
Self-concept in mathematics	Hong Kong	0.6 (country average: 2.4)
2003 Interest in and enjoyment of mathematics	Macao	0.13
	Hong Kong	0.22
Instrumental motivation in mathematics	Macao	−0.03
	Hong Kong	−0.12
Self-concept in mathematics	Macao	−0.20
	Hong Kong	−0.26
Anxiety in mathematics	Macao	0.24
	Hong Kong	0.23
2012 Intrinsic motivation to learn mathematics	Shanghai	0.43
	Chinese Taipei	0.07
	Macao	0.15
	Hong Kong	0.30
Instrumental motivation to learn mathematics	Shanghai	0.01
	Chinese Taipei	−0.33
	Macao	−0.26
	Hong Kong	−0.23
Self-concept in mathematics	Shanghai	−0.05
	Chinese Taipei	−0.45
	Macao	−0.19
	Hong Kong	−0.16
Mathematics anxiety	Shanghai	0.03
	Chinese Taipei	0.31
	Macao	0.19
	Hong Kong	0.11
Mathematics intentions	Shanghai	0.03
	Chinese Taipei	−0.18
	Macao	−0.17
	Hong Kong	−0.31

well in mathematics because their teachers teach mathematics well? What do the results of international studies tell us about classroom teaching in Chinese communities? Classroom teaching is, of course, one aspect of mathematics education, and international studies of mathematics classrooms may provide us with some insight into this important aspect of mathematics education in the Chinese communities.

1.4 Findings of International Studies: Classroom Teaching

In the past 30 years, there have been three major international studies of mathematics classrooms in which the Chinese communities were involved: the University of Michigan Studies, the TIMSS Video Study and the Learners' Perspective Study (LPS). What do these studies tell us about mathematics teaching in China?

1.4.1 The University of Michigan Studies

There was a classroom observation component in both the first and the second University of Michigan Studies. The first study employed a systematic time-sampling method for data collection "in which observers coded the presence or absence of predetermined categories of (pupil and teacher) behaviour" (Stigler, Lee, & Stevenson, 1987). The target (either teacher or student) was observed for 10 s, and then the observer spent the next 10 s for coding the presence or absence of behaviour from a checklist of 49 categories. The second study was designed "specifically at understanding the cross-cultural differences in mathematics learning". In addition to the "objective" coding system, narrative descriptions were recorded by a second observer (Stigler et al., 1987).

In the first study, the results showed that the Taiwanese (and Japanese) children spent "significantly more time learning mathematics" and "engaged [more] in academic activities than did U.S. children" (Stigler & Perry, 1988, p. 207), who spent more time "engaged in inappropriate, off-task activities" (Stigler & Perry, 1988, p. 212). Classrooms in Taiwan (and Japan) were also found "highly organized and orderly; those in the United States more disorganized and disorderly" (Stigler & Perry, 1988, p. 209), and because of the differences in organization, "U.S. students experience being taught by the teacher a much smaller percentage of time than do the Asian students, even though U.S. classes contain roughly half the number of students" (Stigler & Perry, 1988, p. 211). These may have resulted from the fact that Taiwanese (and Japanese) teachers spent most of their time working with the whole class, "imparting information about mathematics to their students"

(Stigler et al., 1987, p. 1281) rather than letting students work in small groups or individually, which was more prevalent in the USA.

The results of the second study were similar to those found in the first. In addition, it was found that "both Chinese and Japanese classrooms provided more opportunities than U.S. classrooms for students to construct a coherent representation of the sequence of events that make up a typical mathematics class and to understand the goals of the activities in which they are engaged" (Stigler & Perry, 1988, p. 213). Also, Taiwanese classrooms were found to emphasize performance (Japanese classrooms were found to emphasize reflection and verbalization), while the US classrooms "appear confused in this regard, and accomplish neither goal well".

Furthermore, it was found that "Students are actively involved in learning tasks. They have opportunities to think mathematically. Teachers also applied their own teaching strategies to lead students to construct mathematical concepts. Although the stereotypical Asian education system places a strong emphasis on drilling procedural skills, the data illustrated that East Asian students also had frequent classroom experiences that facilitated their conceptual understanding of mathematics" (Gu, Huang, & Marton, 2004, p. 312; Lee, 1998; Stevenson & Lee, 1995).

1.4.2 TIMSS Video Study

The TIMSS Video Study was conducted as part of the TIMSS-R study in 1999, and seven countries or systems were involved: Australia, Czech Republic, Hong Kong, Japan, the Netherlands, Switzerland and the USA. The main goal of the study was to describe and compare eighth-grade mathematics teaching in the seven countries/systems (Hiebert et al., 2003, pp. 1–2). A national probability sample of eighth-grade mathematics lessons was videotaped, and the achieved sample was 638 videotaped lessons. An international team of bilingual researchers developed and applied codes to analyse the video data quantitatively. In addition, a specialist group of mathematics educators and mathematicians (known as the Mathematics Quality Analysis Group) reviewed a randomly selected subset of the lessons based on the "expanded lesson tables" of the lessons (which include details about the classroom activities) created by the international video coding team and evaluated the lessons qualitatively "country-blind", with all indicators that might reveal the country removed.

Hiebert et al. (2003) found peculiar features of classroom practices in each of the seven countries, which they called "national scripts". Hong Kong was the only Chinese system which participated in the study. Compared with the other six systems, the following instructional practices, as portrayed by the quantitative analysis of the codes, were found (Leung, 2005):

(1) The was a dominance of teacher talk,
(2) Students had more opportunities to learn new content,

(3) The mathematics problems worked on by students were more complex,
(4) The problems were mainly set up using mathematical language, and
(5) The problems involved more proof.

On the other hand, the instructional practices, as portrayed by the qualitative analysis of the Mathematics Quality Analysis Group, showed that in the Hong Kong lessons:

(1) More advanced content was covered,
(2) The content of the lessons was more coherent,
(3) The presentation of the lessons was more fully developed,
(4) Students were more likely to be engaged in the lessons, and
(5) The overall quality of the lessons was consistently high.

The findings seem to have confirmed the stereotype of the Chinese classroom that the teaching was teacher-dominated and rather traditional and outdated. But the findings also showed that Hong Kong students learned more new, complex and advanced mathematics content than their Western counterparts. The quality of the lessons was also judged by experts to be high.

1.4.3 Learners' Perspective Study

To complement more quantitative studies such as the TIMSS Video Study, based on probability samples, the Learner's Perspective Study (LPS) documented sequences of (10) competently taught eighth-grade lessons for three schools in each of 16 countries "in a more integrated and comprehensive fashion". The idea of "national script" proposed by the TIMSS Video Study was considered too simplistic, since teachers in the same country differ in their instructional practices. Individual teachers also differ in their teaching from one lesson to another. Post-lesson video-stimulated interviews of students and teachers were conducted to study the meanings that the classroom activities held for the participants, especially the students (hence, the name **Learner's** Perspective Study).

In an analysis of the LPS data from Shanghai, Mok (2006) found that the mathematics lessons in Shanghai were "teacher-dominated", but with "a clear rationale" (Mok, 2006, p. 95). Students were given opportunities for discussion and were welcome to express ideas in their own words, but the teacher controlled their activities by choosing tasks with limited options and corrected students' language to the "standardised language" (Mok, 2006, p. 96).

In another analysis of the same dataset, Huang, Mok, and Leung (2006) found that the Chinese teachers "paid much attention to practicing with variation" (as opposed to "repetitive practicing"), and "students seemed to be offered many more opportunities to … recognize the underpinning principles … and mathematical thinking" (Huang et al., 2006, p. 270).

1.4.4 Summary

From the findings reported in this section, mixed pictures of the Chinese mathematics classroom have been portrayed. On the one hand, the findings of some studies have confirmed the stereotype of the Chinese classroom being very conservative and traditional, especially when the studies relied on quantitative methods in studying the classroom activities. On the other hand, a more qualitative and in-depth analysis of the Chinese classroom has shown that it is of high quality.

1.5 Discussion

Synthesizing the results of the three kinds of international studies reported above, What do the findings tell us about different aspects of mathematics education in the Chinese communities?

1.5.1 Mathematics Achievement

First, notwithstanding all the limitations of these large-scale studies, they all point to the clear conclusion that Chinese students did very well in mathematics, as measured by these international studies of mathematics achievement. It seems that the education systems in these Chinese communities are able to nurture students to be very competent in mathematics. The education in these systems is not only able to prepare a general student population who is doing very well in mathematics on average, it is also able to nurture high flyers in mathematics, while at the same time leaving very few students falling behind the minimum competency.

Looking closer into the different aspects of the mathematics achievement, generally speaking, we can say that in terms of basic knowledge and skills, Chinese students' performance is very strong. On the other hand, relatively speaking, they do not do as well in reasoning, or in problem-solving. In the world of today where high-order thinking and problem-solving capability are far more important than familiarity with facts, we should reflect on whether our education is preparing the right kind of personnel and expertise for the modern world. There should be a stronger stress on problem-solving and deductive reasoning rather than simple recall of facts in our curriculum (and the assessment) so as to prepare our future population with stronger higher-order thinking and skills.

This is particularly important when we compare the achievements of the Chinese communities with neighbouring countries such as Japan, Korea and Singapore, because these neighbouring East Asian countries are the main economic competitors. Are the Chinese communities doing less well than these neighbouring countries in terms of inculcating higher-order thinking skills, and in terms of nurturing the most talented in mathematics?

1.5.2 Attitudes Towards Mathematics and Mathematics Learning

No matter how well Chinese students may have excelled in mathematics achievement, one finding from the international studies is clear: the Chinese students do not do correspondingly well in terms of the affective outcomes of mathematics education. Students in general do not like mathematics much. They do not attach a lot of importance to it, and their confidence in their own ability is very low.

In the literature, it has been established that there is a strong correlation between students' achievement and their attitudes towards mathematics and its learning. For better mathematics achievement, it is important for students to be highly motivated in their learning, that they develop their interest and appreciation of the subject, and that they are confident in their learning. Seen in this light, the negative attitudes of Chinese students are worrying. Very often, Chinese teachers are so focused on student achievement, especially achievement in public examinations that they might overlook educational outcomes such as students' attitudes towards the learning of mathematics. But in this modern world of knowledge explosion and pervasiveness of information through the Internet, the capability of lifelong learning is much more important than a firm grasp of the knowledge learned in school, and it is obvious that positive attitudes are extremely important for lifelong learning, as they can propel students to continue their interest and study in mathematics even after they have left school.

Student attitude, together with academic achievement, is part of the attained curriculum (using the terminology of TIMSS). If students fail in this part of the attained curriculum, they fail to achieve what a comprehensive mathematics curriculum intends to achieve. No matter how well Chinese students are performing in their school days, if they graduate from school not liking mathematics and not having the confidence to handle it, it is unlikely that they will be engaged in lifelong learning. So eventually, even if they are able to beat their competitors during school days, they may lose out in the longer run when they enter the life of work. This is true even if they possess the so-called mathematics literacy as measured by PISA, which theoretically is the foundation for knowledge needed for work. Without positive attitudes towards mathematics and other subject matter, Chinese children will not be able to excel at work even if they have a solid foundation laid in their school days.

These results for the Chinese systems are similar to those of other high achieving East Asian countries, and Leung (2001) argued that the peculiar findings of high achievement and negative attitudes may be due to the common cultural values that they share. But the fact that this is due to culture should not imply that we should accept the low attitudes of students as they are.

In the secondary analysis of the TIMSS 2011 Hong Kong data, it was found that, "even in a culture where the attitudes towards learning are in general low, within the territory of Hong Kong, high mathematics achievement is still related to more positive attitudes towards learning, no matter for the younger Primary 4 students or the slightly older Secondary 2 students" (Leung & Wong, 2014, p. 216). One may argue that if the East Asian students who have low attitudes already perform very well in mathematics achievement, then they may score even higher if their attitudes are improved. The case of Singapore is telling here. Singapore, in general, did (slightly) better than the other East Asian countries (including China) in these international studies. At the same time, attitudes of Singaporean students were more positive than those of students in the other East Asian countries. Does the case of Singapore tell us that a combination of the cultural value that encourages success/hard work and a positive attitude towards mathematics will result in even higher mathematics achievement than that the Chinese students are performing at the moment?

If the answer is in the affirmative, how do we get there, i.e. how can we improve the attitudes of our students? From the analysis of the Hong Kong TIMSS 2011 results, it seems that there are some school characteristics that are conducive to a positive attitude towards mathematics and hence lead to higher mathematics achievement. As far as the Hong Kong TIMSS 2011 data are concerned, students, who find the school a safe place to study and stay in, who have a sense of belonging to the school and who like staying in the school, will develop a more positive attitude towards mathematics and will perform better in it (Leung & Wong, 2014). If these findings could be generalized to other Chinese communities, this may provide the key to improving the attitudes of students.

1.5.3 Classroom Teaching

How is the superior mathematics achievement of Chinese students, and perhaps their negative attitudes as well, related to the mathematics teaching in the Chinese classroom? Based on the international studies of mathematics classrooms, superficially mathematics teaching in the Chinese systems looks very traditional and backward, but a more fine-grained analysis shows that the quality of teaching is high. In particular, it seems that more advanced mathematics content is covered. Could this be the reason for the high achievement of Chinese students in mathematics on average? Could the teaching explain the relatively low problem-solving skills of students? Could the teaching have contributed to their negative attitudes? More in-depth studies than these large-scale international studies are needed to provide clearer answers to these questions. In any case, describing the Chinese mathematics classroom as traditional and backward may be too superficial a conclusion. We need in-depth studies to identify the strengths of Chinese mathematics teaching as well as to expose the weaknesses. It is through these in-depth studies that we will learn to keep the strengths, and improve on weak areas.

Based on the results of these international studies, the Chinese systems should keep up with their good tradition of teaching, including covering important mathematics concepts and ample well-designed practice problems with variations, and a high expectation for students to learn and achieve. High expectation, however, does not necessarily mean making mathematics learning a hardship for children. How should we maintain a high expectation to achieve without making mathematics too much of a hardship for students? Research tells us that making mathematics meaningful to students is the key to helping them enjoy mathematics. At the moment, in the Chinese systems, examinations are providing students with the major impetus to learn. As Leung (2001) suggested, in accordance with the Chinese cultural value, a fair amount of emphasis on examination is perhaps desirable. But to help students enjoy and feel confident about mathematics, the teacher must help students to derive meaning out the mathematics they learn, in addition to the incentive that comes from examination success.

How can we make mathematics more meaningful to our students? It seems from the TIMSS 2011 Hong Kong secondary data analysis that relating to students' everyday life experiences in teaching mathematics, an appropriate emphasis (and not over-emphasis) on examinations and high-quality homework will boost their attitudes towards mathematics and hence improve their mathematics achievement (Leung & Wong, 2014).

1.5.4 Implications

No education system is perfect, and there is inevitably a lot of shortcomings in mathematics education in the Chinese systems. But results of international studies discussed in this chapter show that Chinese communities are not in a state of crisis as far as mathematics education is concerned. We should not be complacent, but there is no need for immediate, drastic changes. In particular, we should not rush into importing or borrowing practices from (economically) "more developed" countries who do not even do as well as the Chinese communities in these international assessments.

As for areas of change, reasoning and problem-solving skills and positive attitudes towards mathematics are areas in which Chinese students are relatively weak. As argued above, in an era of lifelong learning, these higher-order skills and positive attitudes are extremely important. In this regard, these international studies ring a warning bell to mathematics educators in Chinese communities.

Academic achievement is, of course, only part of the benchmark for quality education. In nearly all countries in the world, the affective domain is part of the aim of education, and the curriculum objectives invariably include affective attributes such as liking mathematics and being confident in one's ability in it. So in up-keeping the mathematics achievement of students, we should at the same time aim at improving our students' attitudes towards this subject. Teachers should include the inculcation of positive attitudes towards mathematics as one of the important goals of their teaching and should proactively design teaching and learning activities to achieve this important goal. At the system level, teacher development activities should be organized to emphasize the importance of promoting positive attitudes towards learning, and to equip teachers with the know-how and skills in conducting activities in the classroom that will enhance students' attitudes towards learning. Workshops could be organized where teachers may share among themselves their successful experiences and strategies for enhancing the attitudes of students. Hopefully, with the joint efforts of policy makers, teacher educators and practising teachers, we may begin to see some positive changes in the instructional practices inside the Chinese classroom and start to reverse the pattern of negative attitudes in Chinese students towards mathematics and its learning.

1.6 Conclusion

In this chapter, we have reviewed results of international studies in terms of mathematics achievements, student attitudes and classroom teaching, and through this we have tried to gain a better understanding of mathematics education in the Chinese communities. How do we account for the curious findings of the high student achievement, low student attitudes towards mathematics and the seemingly traditional teaching in the Chinese communities, the so-called Chinese Learners' Paradox (Biggs, 1996)? Elsewhere, the author has argued that the underlying cultural values of these Chinese communities may be a plausible explanation (Leung, 2001). In particular, the "examination culture" is believed to be exerting influence on all the three areas. So understanding mathematics education in these systems also provides us with some appreciation of the cultural values of these Chinese communities.

We all realize there are limitations of these large-scale international studies because of the sheer size of the studies and the methodologies adopted (Leung, 2014). Notwithstanding these limitations, these studies are, in general, considered as being able to provide relative reliable and valid measures of student achievement in an international context, and the results provide an "'objective' measure of student achievement to answer the question of effectiveness of an education system" (Leung, 2011, p. 391). Indeed, international studies are often considered an "objective" yardstick in gauging the mathematics achievement, and thus the academic success, of a country. Using this yardstick, as discussed in the previous sections of this chapter, can we conclude that Chinese communities have a superior education system? According to some international agencies such as Pearson (2017), the education system in Hong Kong is indeed judged to be superior (ranking fourth in 2016, after Korea, Japan and Singapore).[2] As discussed above, the Chinese systems indeed performed very well in terms of students' mathematics achievement but, in terms of developing in students a positive attitude towards mathematics learning, the Chinese communities still have left a lot to be desired.

The high achievement of students should bring national pride to the Chinese communities, but there is certainly no scope for complacency. The most important lesson we should learn from these international studies is that the Chinese community should establish an identity of its own (Leung, 2001), and instead of relying on "Western" theories, we should strive to develop our own education theories that up-keep our strengths and hopefully improve on our weaknesses as well.

References

Beaton, A. E., Mullis, I. V. S., Martin, M. O., Gonzalez, E. J., Kelly, D. L., & Smith, T. A. (1996). *Mathematics achievement in the middle school years*. Chestnut Hill, MA: International Study Center, Boston College.

Biggs, J. B. (1996). Western misconceptions of the confucian-heritage learning culture. In D. A. Watkins & J. B. Biggs (Eds.), *The Chinese Learner* (pp. 45–67). Hong Kong: Comparative Education Research Centre.

Cai, J. (2000). Mathematics thinking involved in U.S. and chinese students' solving process-constrained and process-open problems. *Mathematical Thinking and Learning: An International Journal, 2*(4), 309–340.

Fan, L., & Zhu, Y. (2004). How have chinese students performed in mathematics? A perspective from large-scale international mathematics comparisons. In L. Fan, N. Y. Wong, J. Cai, & S. Li (Eds.), *How chinese learn mathematics* (pp. 3–25). New Jersey: World Scientific.

Gu, L., Huang, R., & Marton, F. (2004). Teaching with variation: A chinese way of promoting effective mathematics learning. In L. Fan, N. Y. Wong, J. Cai, & S. Li (Eds.), *How chinese learn mathematics* (pp. 309–347). New Jersey: World Scientific.

Hiebert, J., Gallimore, R., Garnier, H., Givvin, K. B., Hollingsworth, H., Jacobs, J., et al. (2003). *Teaching mathematics in seven countries. Results From the TIMSS 1999 video study*. Washingtonm DC: National Center for Education Statistics.

Huang, R., Mok, A. I. C., & Leung, F. K. S. (2006). Repetition or variation: Practising in the mathematics classroom in China. In D. Clarke, C. Keitel, & Y. Shimizu (Eds.), *Mathematics classrooms in twelve countries: The insider's perspective*. Rotterdam: Sense Publishers.

Lapointe, A. E., Mead, N. A., & Askew, J. M. (1992). *Learning mathematics*. New Jersey: Educational Testing Service.

Lee, S. Y. (1998). Mathematics learning and teaching in the school context: Reflections from cross-cultural comparisons. In S. G. Garis & H. W. Wellman (Eds.), *Global prospects for education: Development, culture, and schooling* (pp. 45–77). Washington DC: American Psychological Association.

Leung, F. K. S. (2001). In search of an East Asian identity in mathematics education. *Educational Studies in Mathematics, 47*, 35–51.

Leung, F. K. S. (2005). Some characteristics of East Asian mathematics classrooms based on data from the TIMSS 1999 video study. *Educational Studies in Mathematics, 60*(2), 199–215.

[2]Note that Pearson used results of international studies as one substantial indicator of the quality of an education system.

Leung, F. K. S. (2011). The significance of IEA studies for education in East Asia. In C. Papanastasiou, T. Plomp, & E. C. Papanastasiou (Eds.), *IEA 1958–2008—50 years of experiences and memories*. Nicosia: Research Center of the Kykkos Monastery.
Leung, F. K. S. (2014). What can and should we learn from international studies of mathematics achievement? *Mathematics Education Research Journal (MERJ), 26,* 579–605.
Leung, F. K. S., & Wong, A. S. L. (2014). *Factors that contribute to Hong Kong students' performance in trends in international mathematics and science study (TIMSS) 2011*. Report submitted to the Education Bureau of Hong Kong SAR on Further Analysis of the TIMSS 2011 Background Questionnaires. Hong Kong: EDB.
Mead, N. A. (1995). International assessment of educational progress. In International Comparative (Ed.), *Studies in education: Descriptions of selected large-scale assessments and case studies* (pp. 48–57). Washington DC: National Research Council.
Mok, I. A. C. (2006). Teacher-dominating lessons in Shanghai: The insider's story. In D. Clarke, C. Keitel, & Y. Shimizu (Eds.), *Mathematics classrooms in twelve countries: The insider's perspective*. Rotterdam: Sense Publishers.
Mullis, I. V. S., Martin, M. O., Beaton, A. E., Gonzalez, E. J., Kelly, D. L., & Smith, T. A. (1997). *Mathematics achievement in the primary school years*. Chestnut Hill, MA: International Study Center, Boston College.
Mullis, I. V. S., Martin, M. O., & Foy, P. (2008). *TIMSS 2007 international mathematics report*. Chestnut Hill, MA: TIMSS & PIRLS International Study Center, Boston College.
Mullis, I. V. S., Martin, M. O., Foy, P., & Arora, A. (2012). *TIMSS 2011 international results in mathematics*. Chestnut Hill, MA: TIMSS & PIRLS International Study Center, Boston College.
Mullis, I. V. S., Martin, M. O., Foy, P., & Hooper, M. (2016). *TIMSS 2015 international results in mathematics*. TIMSS & PIRLS International Study Center, Boston College. Retrieved from http://timssandpirls.bc.edu/timss2015/international-results/.
Mullis, I. V. S., Martin, M. O., Gonzalez, E. J., & Chrostowski, S. J. (2004). *TIMSS 2003 international mathematics report*. Chestnut Hill, MA: TIMSS & PIRLS International Study Center, Boston College.
Mullis, I. V. S., Martin, M. O., Gonzalez, E. J., Gregory, K. D., Garden, R. A., O'Connor, K. M., et al. (2000). *TIMSS 1999 international mathematics report*. Chestnut Hill, MA: International Study Center, Boston College.
Organisation for Economic Co-operation and Development. (2001). *Knowledge and skills for life: First results from PISA 2000*. Paris: OECD Publications.
Organisation for Economic Co-operation and Development. (2003). *Literacy skills for the world of tomorrow—Further results from PISA 2000*. Paris: OECD Publications.
Organisation for Economic Co-operation and Development. (2004). *Learning for tomorrow's world—First results from PISA 2003*. Paris: OECD Publications.
Organisation for Economic Co-operation and Development. (2007). *PISA 2006: Science competencies for tomorrow's world: Volume 1: Analysis*. Paris: OECD Publications.
Organisation for Economic Co-operation and Development. (2010). *PISA 2009 results: What students know and can do*. Paris: OECD Publications.
Organisation for Economic Co-operation and Development. (2014). *PISA 2012 results: What students know and can do (Volume I, Revised edition, February 2014): Student performance in mathematics, reading and science*. Paris: OECD Publications.
Organisation for Economic Co-operation and Development. (2016). *PISA 2015 results (Volume I): Excellence and equity in education*. Paris: OECD Publications.
Pearson. (2017, January 29). Retrieved from http://www.mbctimes.com/english/20-best-education-systems-world.
Scardino, M. (2008, December 11). The olympics of education: Fresh efforts to boost the UK in maths and science league tables will also help our economic health. *The Guardian.*
Stevenson, H. W., & Lee, S. Y. (1995). The East Asian version of whole-class teaching. *Education Policy, 9,* 152–168.

Stevenson, H. W., Lee, S. Y., Chen, C., Stigler, J., Fan, L., & Ge, F. (1990). Mathematics achievement of children in China and the United States. *Child Development, 61,* 1053–1066.

Stigler, J. W., Lee, S. Y., & Stevenson, H. W. (1987). Mathematics classrooms in Japan, Taiwan, and the United States. *Child Development, 58,* 1272–1285.

Stigler, J. W., & Perry, M. (1988). Cross cultural studies of mathematics teaching and learning: Recent findings and new directions. In D. A. Grouws & T. J. Cooney (Eds.), *Perspectives on research on effective mathematics teaching*. Erlbaum Associates: Hillsdale, NJ.

Chapter 2
Chinese Mathematics Education System and Mathematics Education Tradition

Dai Qin

2.1 Understanding Traditional Chinese Mathematics Education

Mathematics education in ancient China has a long history of over 3000 years, during which a lot of outstanding mathematicians have been cultivated; the first institute of higher education for mathematics in the world, named the *Ming Suan Ke* (Computation Department), was established in the Tang Dynasty, and splendid methods to promote mathematical thinking development. However, current mathematics education researchers and teachers in primary and secondary schools in China have not fully understood and applied the thoughts created in ancient China. Even though, in recent years, there has been some work linking mathematics education with mathematics culture and history, more emphasis has been placed on the history and culture of Western mathematics, which focuses on "concept games" instead of practical operability. The work which covers Chinese mathematics history and culture is limited to some typical cases—"Wu Bu Zhi Shu (Amount Calculation)," "Gou-Gu Ding Li (the Pythagorean theorem)," and some stories about well-known Chinese mathematicians such as Li Yan, Qian Baocong and Li Di. Admittedly, some works, like *Zhong Guo Gu Suan Jie Qu* (*Interesting Computation Cases in Ancient China*) by Yu Zuquan and *Shu Xue Liaozhai* (*Mathematical Stories*) by Wang Shuhe are very impressive, but they cannot be used in classroom teaching because of their limited function in rigorous exams. In a word, few educators can search the rich historical materials of Chinese mathematics for inspiration and apply it to mathematics education.

Benedetto Croce (1866–1952), a well-known Italian historian, said, "When there is a need with the development of life, the dead history will be resurrected and

D. Qin (✉)
Institute for the History of Science and Technology, Inner Mongolia Normal University, Huhehot, China
e-mail: daiqino@imnu.edu.cn

Y. Cao and F.K.S. Leung (eds.), *The 21st Century Mathematics Education in China*, New Frontiers of Educational Research,
https://doi.org/10.1007/978-3-662-55781-5_2

something from the past will become current.... Hence, most parts of the history viewed currently as chronicles and silent historical documents will be lit up in turn by new life and work well again" (Croce, 2005, p. IV). "From the perspective of chronology, no matter how long the history of a fact is, it is still a history related to current needs and forms, and those facts are continuously astonishing" (Croce, 2005, p. 6). The thought of "gaining new insights through reviewing old material" has particularly been advocated and practiced since ancient times in China, and importance has been attached to grasping new knowledge on the basis of reviewing old materials in individual learning. In the development of the nation, taking history as a mirror, Chinese people apply the knowledge, experience, and thinking methods gained in the past to current practice, which also serves the purpose of gaining new insights.

In recent years, through reviewing journal articles on mathematics education, masters' theses from normal universities, and other work in this field, I have found that a lot of young researchers present ideas that, directly or indirectly, deny traditional Chinese mathematics education. As Zhang Dianzhou said, "In recent years, the so-called traditional educational methods are almost the synonyms of 'backward' and 'obsolescent'" (Zhang & Zhao, 2012). I wonder how much these researchers know about traditional Chinese mathematics education. In my opinion, the tradition, the present, and the future are inseparable; "innovation" does not mean the inevitable separation from "tradition," but can be considered as a prerequisite to surpass and violate traditions, and there are some points between them where their values are compatible. As Ye Xiushan, a famous Chinese philosopher, said:

> History consists of the past, the present and the future. 'The past' results in 'the present', and 'the future' influences 'the present' as well. 'The past' and 'the future' are both included in 'the present'. 'The present' is not a geometrical 'point', but a 'side'. People live and work under the principles of 'the past' and the attraction of 'the future'. 'The past' has not gone, and 'the future' can be pursued. 'Values' and 'meanings' are not fragments, but extensions. (Ye, 2010, p. 149)

In the continuum of past, present, and future, "tradition" and "innovation" are two sides of the dialectical movement of culture development. We are expected to achieve a connection and balance between inheritance and development on the basis of traditional education so as to avoid the extremes. Without historical facts, any wise conclusions are not convincing. Therefore, it is necessary for those who intend to deny traditional Chinese mathematics education to make a careful study of it first.

Traditions are the sources of cultural memory of a nation, the driving force for innovation, and the stimulators for further growth. These traditions are the social factors passed from generation to generation; distinctive features from the past are integrated, directly or indirectly, with present features, and continue to play a role. No matter how sweeping the social changes are, the traditions of thoughts cannot be altered thoroughly. While the contents of school mathematics education in China are from the West, the education views and teaching and learning approaches follow the Chinese traditions of "honoring teachers and esteeming truth," "teaching benefitting teachers as well as students," and "teaching only the essential elements and ensuring plenty of practice," which display an inseparable connection between the past and the present.

Zhang Dianzhou said "Education in China has its own 'beauty,' and we need national confidence. Teachers in China are the watchmen of excellent educational traditions of China" (Zhang & Zhao, 2012). In my opinion, education in China has its own "truth" and "goodness" as well as "beauty," and its "beauty" is involved with its "truth" and "goodness"; furthermore, its "truth" and "goodness" are demonstrated in the form of "beauty." In short, abundant excellent content in the traditional education of China has vitality and integrates truth, goodness, and beauty, and this applies to mathematics education in China.

2.2 Brilliant Ancient Chinese Mathematics and Education Culture

China was not only a great civilization in ancient times, but also renowned for its mathematical genius. Ancient Chinese mathematicians made world-acclaimed achievements. Over its long history of development, a unique algorithm-centered and practical-oriented system has emerged with induction as the main method and problem collection as the primary model. With thousands of years of development behind it, Chinese mathematics education has accumulated rich experiences that provide a rich heritage for current education.

2.2.1 The Beginnings of Ancient Chinese Mathematics Education

Chinese mathematics education dates back more than 3000 years to the Shang Dynasty (1600–1048 BC). Over its long development, there has emerged a large number of outstanding mathematicians who dealt with the practical needs of everyday life, and who also contributed to the development of mathematics around the world. For example, traditional Chinese mathematics is the origin of Japanese mathematics: *Suanjing Shi Shu* (*Ten Mathematical Classics*) was once used as a textbook in Japan and Korea. In addition, ancient Chinese mathematics education created some world firsts. For example, the mathematics school built in the Tang Dynasty (AD 618–907) was the first advanced mathematics school in the world. The "Program for Learning Mathematics" in *Chengchu Tongbian Benmo* (*Multiplication and Division Developed from the Basics*), written in 1274 by Yang Hui, a mathematician and mathematics educator during the Southern Song Dynasty (AD 1127–1279), was the earliest mathematics syllabus in the world.

Mathematics education already existed in China during the Xia and Shang dynasties. The *Shuo Wen Jie Zi* (*Interpretation of Characters*), written by Xu Shen in AD 121, stated that "mathematics means calculation." In the Xi Zhou Dynasty (1046–256 BC), the "six arts" for national learning—rites, music, archery, riding, writing, and arithmetic—were formulated. "Rites" referred to political education, including moral norms, protocol, and the learning of religious and legal institutions in the slave societies. "Music" referred to arts education, including the learning of music, poetry, and dancing. "Archery" and "riding" were military training programs, and "writing" and "arithmetic" were part of the basic curriculum. "Six arts [techniques]" education aimed to combine general knowledge and military ability with a focus on rites, and was the highest education level attainable by China's slave society. Writing and arithmetic were minor arts, taught in primary schools, while rites, music, archery and riding, were major arts, taught in colleges. The idea of teaching mathematics as an art was a concept unique to ancient China. The inclusion of mathematics in the "six arts" education shows clearly that mathematics learning was an indispensable part of an official's education. Mathematics education was included as early as in the Xi Zhou Dynasty, which pointed the way for its future development. The mathematics education system was set up in China during the Qin Dynasty (221–206 BC).

Before the Sui and Tang dynasties, the mathematics education system was outlined by the following steps (Li, 1954, p. 253) which, despite their simplicity, laid a solid foundation for mathematics education in later ages.

- At six, a child learns numbers and the names of cardinal points. When he is ten, he goes to boarding school to get an education, learning writing and calculation. (*The Book of Rites*)
- A child is ready for education at eight when he has adult teeth, and he must go to school to learn writing and calculation. (*Bai Hu Tong*, by Ban Gu, AD 32–92)
- The sixth [of the six arts that must be taught to children] is the set of nine skills in mathematics. (*Zhou Li, Rites of Zhou*)
- At eight, a child goes to school to learn the Heavenly Stems and Earthly Branches, names of the cardinal points, as well as writing and calculation. (*Han Shu*, by Ban Gu, AD 32–92)
- In ancient times, a child went to school at eight to learn the Heavenly Stems and Earthly Branches, the names of the five directions, as well as writing and calculation. (*Ru Li Lun*, by Wang Can, AD 177–217)

2.2.2 *Ming Suan Ke*: The First Mathematics School in the World

Nine Chapters on Mathematics Arts (hereafter referred to as *Nine Chapters*) was used as a textbook for 500 years in private mathematics education in China. During this Period, there was no such thing as public mathematics education provided by the government. Despite its short-lived supremacy (AD 581–618), the Sui Dynasty had a profound influence upon the mathematics education of China. In AD 589, after the country was untied, the government of Sui enacted many laws and regulations. It restored national education and, for the first time in Chinese history, it started mathematics education. As well, it set up an education system for mathematics, recruited teachers, and enrolled students. “There were two mathematics masters, two assistants, and eighty students in the Imperial College” (Li, 1997). This marked the birth of state-run mathematics education in ancient China.

After the establishment of the Tang Dynasty, the Ming Suan Ke mathematics school was set up in the Imperial College in AD 656 after dozens of years of painstaking preparation. The curriculum, examination methods, and textbooks were all decided at the same time. In the school, there were usually about 30 students, 10 being the minimum. The students came from the families of eighth- or ninth-grade officials and common people’s families. The mathematics masters were ninth-grade officials, the lowest grade in the Tang Dynasty (Li, 1997, p. 305). Ming Suan Ke was run intermittently all the way through the Five Dynasties. It was the first mathematics school in the world, and Emperor Gaozong of the Tang Dynasty ordered that *Suanjing Shi Shu* (*Ten Mathematical Classics*) be the official textbook. People who learned mathematics and passed entrance examinations could get jobs in the government.

The culture of the Sui and Tang dynasties was extremely appealing to Japan and Korea at that time, especially to Japanese rulers and officials. They imitated China as extensively as they could, from its political system to its literature, arts, architecture, clothes, food, and written characters. Their mathematics education system was imitated in its entirety from that of the Tang Dynasty.

2.2.3 The Peak of Ancient Chinese Mathematics Culture

In the Song Dynasty (AD 960–1279) and the Yuan Dynasty (AD 1279–1368), the development of Chinese mathematics reached its peak and many world-class achievements were made. Thanks to the high level of mathematics education at that time, many famous mathematicians, such as Jia Xian, Qin Jiushao, Li Ye, Yang Hui, and Zhu Shijie, emerged.

In the 300-year history of the Northern and Southern Song dynasties, national mathematics education existed only for a short Period, without any significant developments. The government of the Northern Song Dynasty did not provide mathematics education until late in the dynasty, and mathematics education existed sporadically from 1083 to 1120. The achievements in this Period were not outstanding in any way, but two points are worth mentioning. First, in 1084, woodblock printing was adopted in printing some mathematics textbooks passed down from the Tang Dynasty. There is historical proof that *Zhou Bi Suanjing, Nine Chapters, Sun Zi Suanjing, Wu Cao Sunnjing, Zhang Qiujian Suanjing, Xiahou Yang Suanjing, Haidao Suanjing, and Ji Gu Suanjing* were printed using this method; however, it is not certain whether *Wu Jing Suanshu* and *Shu Shu Ji Yi* (*Memoir on Some Traditions of the Mathematical Art*) were printed (Li, 1955, p. 276). *Zhui Shu* was not printed, as it had been lost before then (Li, 1958). Second, the *Mathematics Education Guidelines* were formulated during the Yuanfeng Period (AD 1078–1085) and were amended and issued as an imperial order in 1107. These have been passed down from generation to generation until today.

During the 152 years of the Southern Song Dynasty from 1127 to 1279, the government made no effort to revive mathematics education. However, two facts deserve some attention. First, a scholar named Bao Huanzhi made every effort to collect the mathematics manuals printed during the Yuanfeng Period, and he found and copied *Shu Shu Ji Yi* in a temple in Hangzhou. He brought these copies to Changting County in Fujian Province when he went there to assume the duty of sheriff. He then reissued these faithful replicas in 1212–1213. Five-and-a-half of these copies still exist: *Zhou Bi Suanjing, Nine Chapters, Sun Zi Suanjing, Zhang Qiujian Suanjing, Wu Cao Suanjing*, and *Shu Shu Ji Yi*. Second, the "Program for Learning Mathematics" proposed by Yang Hui during the Southern Song Dynasty was a very precious work in Chinese mathematics education. It closely resembles the teaching syllabus that is currently being used.

The Jin Dynasty was established by Nüzhen (Jurchen), an ancient nationality in China. It challenged the Southern Song Dynasty in 1127 and was later destroyed by Mongols in 1234. There was no national mathematics education during the Jin Dynasty, but private mathematics education flourished all the way into the Yuan Dynasty.

From the twelfth to the thirteenth centuries, mathematics research and study were very popular in some areas of northern China, such as Shanxi, Hebei, and Shandong. This brought about an important academic achievement—Tianyuan shu, or the Tianyuan method, a method for the numerical solution of high-degree polynomial equations with one unknown. At that time, there was an intellectual group led by Liu Bingzhong in Wúan County, Hebei Province. Zhang Wenqian, Guo Shoujing, and Wang Xun were members of this group. They primarily studied science, including mathematics. Mathematician Li Ye set up the Fenglong Academy in Yuanshi County, Hebei Province, where he enrolled students and taught them mathematics as well as the humanities.

Zhu Shijie was a mathematician who lived during the thirteenth and fourteenth centuries. He made a living by teaching mathematics and traveled around the country for dozens of years to teach students. In 1299, Zhao Yuanzhen sponsored the publication of Zhu Shijie's *Suanxue Qimeng* (*Introduction to Mathematical Sciences*) in support of Zhu's teaching. Another of Zhu's masterpieces, *Si Yuan Yu Jian* (*Jade Mirror of Four Unknowns*), was published in 1303.

The Mongol emperors of the Yuan Dynasty, Kubilai Khan and Mogul Khan, both paid considerable attention to mathematics. Mogul Khan studied Euclid's *Elements*, and Kubilai Khan hired mathematician Wang Xun to teach the crown prince. Kubilai Khan also required officials' children to learn mathematics. During the Yuan Dynasty, lower-rank officials were also required to master mathematics so that they could meet the demands of their jobs.

2.2.4 *Transition from Traditional to Western Mathematics Education*

Hongwu, the first emperor of the Ming Dynasty, set mathematics as a subject for education in 1369, just after the establishment of the dynasty in 1368. In February 1392, the government reiterated the mathematics syllabus, and examination content: "*Nine Chapters* must be learned and mastered proficiently, and should be tested" (Li & Dai, 2000, p. 95).

Private mathematics education in the Ming Period was very active. Cheng Da Wei wrote *Suan Fa Tong Zong,* and this became an important textbook for private education, significantly influencing other countries, like Japan.

During the 300 years from the end of the Ming Dynasty to the end of the Qing Dynasty, Western mathematics was introduced into China, which brought about a transition from the traditional mathematics education system to the modern Western mathematics education system. This transition was completed by the end of the Qing Dynasty. This Period of mathematics education had its own features; therefore, it could be seen from the traditional mathematics education to the transition Period of Western mathematics education.

Although, in the Sui, Tang, and Yuan Periods, some mathematical knowledge was introduced from foreign countries, it had only a trace of impact. From the end of the Ming Dynasty, China began to translate a large number of foreign mathematics books. It was very different from the past.

Toward the end of the Ming Period, Western European mathematics entered into China. All six volumes of Euclid's *Elements* were translated and calculations with figures and mathematical tools were imported. They had a wide influence and shocked the intellectuals of the age greatly.

At first, scholars had differing attitudes toward the introduction of Western elementary mathematics. Some of them, such as Xu Guangqi (also known as Seu KwangKe) and Li Zhizhao, studied it eagerly. However, most scholars, such as Li Dupei, either ignored it or integrated it with Chinese mathematics. A few objected to the idea openly. Over time, the number of people who learned and researched Western mathematics grew steadily. Some scholars wrote work based on the new knowledge. One well-known example was Mei Wending and his family (Li & Guo, 1988). His basic approach was to combine Chinese and Western knowledge. He composed work on geometry, trigonometry, and algorithms, among others.

Kang Xi, an emperor in the Qing Period, had a great interest in mathematics and left some reports after learning Eastern European mathematics and surveying. He edited *Shu Li Jing Yun* (3 volumes, 1721), and this literature eventually became the mathematics textbook in the late Qing Period.

By the end of the Qing Dynasty, the new mathematics education system had taken on its final form after nearly 50 years of preparation. The introduction of Western scientific knowledge marked the beginning of contemporary Chinese mathematics education. After that, Western mathematics and mathematics education gradually took over and transformed traditional Chinese mathematics education ideas and methods thoroughly.

In 1857, Li Shanlan (1811–1887) and A. Wylie (1836–1887) translated the latter nine volumes of *Elements*, *Algebra* written by Augustus De Morgan (1806–1872) and *Daiweiji Shiji*. One of the textbooks for mathematics education in churches was *Shu Xue Qi Meng* written by A. Wylie. This textbook also played a positive role in China's accepting modern mathematics. In the 1870s, Hua Hengfang (1833–1902) and British missionary John Fryer (1839–1929) co-translated many mathematical documents about algebra, trigonometry, calculus, and probability.

In 1862, Tong Wen Guan was established in Beijing and Suan Xue Guan was founded. Li Shanlan became the principal. The education system of Suan Xue Guan was of eight-year duration and was in use for 30 years.

At the end of the nineteenth century, several mathematics magazines were published. *Suan Xue Bao* was published in 1897 by Huang Qingcheng. In 1899, Zhu Xianzhang published this magazine. In 1900, Du Yaquan issued *Zhong Wai Suan Bao* in Shanghai. The founding of these magazines also promoted the popularization of mathematical knowledge and the development of mathematics education.

In 1902, the Qing government promulgated the *Ren Yin education system*. This system enacted an mathematics education system that was similar to the present one. The *Gui Mao education system* was established in 1904 again. The *Ren Yin education system* and *Gui Mao education system* imitated Japan's *Implementation Regulations for the Secondary School Order* (Meiji 32). Almost all of the mathematics textbooks of this time were translations of Japanese mathematics textbooks.

2.2.5 *The Characteristic of Ancient Chinese Mathematics Culture: The Cultural Characteristics of* Nine Chapters

Nine Chapters is a classic of traditional Chinese mathematics. It dominated the development of the ancient traditional Chinese mathematics culture. It had a profound impact on the development of both ancient Chinese mathematics and world mathematics with its structure, form, and content. *Nine Chapters* was a comprehensive masterpiece of mathematics that dated back to before the Han Dynasty (206 BC–AD 220) and brought together most of the mathematics achievements at that time. It was an encyclopedia-like mathematics masterpiece. *Nine Chapters* showed that ancient Chinese mathematics had already reached a very high level during the early Christian era. It was innovation in many aspects and advanced even in world terms. For example, the place-value system notation described in *Nine Chapters* did not appear in India until the end of the sixth century. Calculations with fractions were relatively advanced in China during the *Nine Chapters* era, but were not seen in India until the seventh century. Square root algorithms, which were developed in

Western countries only at the end of the fourth century, and cube root algorithms, which were yet to be developed elsewhere, were discussed in *Nine Chapters*. The book also covered other topics such as positive and negative numbers, some calculation rules, and the systems of linear equations and quadratic equations, which did not appear in Western countries until much later.

It is still uncertain who wrote *Nine Chapters*. Experts believe that it was not written independently by a single author, but compiled by several mathematicians. Before Liu Hui annotated *Nine Chapters* during the Wei and Jin Period (in the mid-third century), it already had an unshakable authoritative position in traditional Chinese mathematics. Liu Hui's annotations helped people to achieve a better understanding of *Nine Chapters*.

Nine Chapters comprising this work were Surveying of the Land, Millet and Rice, Distribution by Progression, Diminishing Breadth, Consultation on Engineering Works, Imperial Taxation, Excess and Deficiency, Calculating by Tabulation, and Gou-Gu (right-angled triangles), with a collection of 246 mathematical questions. Most chapters started by posing a question, which was followed by the answer and the technique for solving the problem; some of these techniques were mathematics theorems or formulae. In total, 202 techniques were presented in the book, 69 of which were of general importance. These techniques were the foundation for traditional mathematical theories in China.

The fundamental characteristic of traditional Chinese mathematics, with *Nine Chapters* as representative, can be summarized as practical applicability and algorithm-centered calculability.

2.2.5.1 The Characteristic of Practical Applicability

The practical applicability of *Nine Chapters* determined the features of traditional Chinese mathematics, which was also a reflection of traditional Chinese philosophy. In other words, the practical applicability of traditional Chinese mathematics was rooted in traditional Chinese philosophy.

First, the practical needs of society at that time determined the emergence and development of *Nine Chapters*. *Nine Chapters* was a practical mathematics masterpiece, meticulously compiled based on the research and collation of ancient mathematical materials. Its content was related closely to practical calculation issues in people's daily lives, such as the calculation of land area, food exchange, goods distribution, taxation, penalties, attendance entry, or civil engineering. Most of the selected 246 questions related to the practical problems of the era. Some were everyday mathematical puzzles, and about 190 questions dealt with economic activities. These questions recorded some important historical materials of the Period's social economy. The chapters had parallel content, and the content and calculation methods in each chapter progressed from simple to complex.

Mathematical knowledge was regarded as valuable, was subsequently developed insofar as it met social needs, and was helpful in solving everyday problems; otherwise, little heed would have been paid to it, and it might even have been abandoned. For example, although *Nine Chapters* was an encyclopedia of the mathematics achievements since the Qin Dynasty, its authors did not include the geometrical knowledge of the Mohist School (a school of thought in the Spring and Autumn and Warring State Periods, 770–221 BC) into their own mathematics system, since the geometry of this school was abstract mathematical knowledge about the concepts of points, lines, and surfaces and their logical relationships, and this knowledge has no direct applicability to practical problems.

Second, to meet social needs and evaluation standards for applicability, the rationale and methods for compiling *Nine Chapters,* some of which are still in use today, were not defined or explained in the book; that is, the nature of these concepts was not disclosed. The logical relationships among these concepts were not so clear and, therefore, they seemed parallel to one another. People understood them intuitively when learning and using mathematics, based on their mathematical experience. Ancient Chinese mathematicians were more interested in the applicability of these concepts than in disclosing the relationships among them, nor were they aware of the importance of making clear these relationships to learners.

Third, there was no basic mathematical knowledge, such as the introduction of counting rods or the nine-by-nine multiplication table (in rhyme), in the book. This could be attributed to the mathematical level of either the learners at that time or the authors of the book. It is likely that the users of *Nine Chapters* had already understood the above-mentioned basic mathematical knowledge before they read the book or that they could grasp these concepts while learning and using the calculation methods. Moreover, such an arrangement was determined by the practical nature of the book; there was no need for a person to understand these terms, basic concepts, and common calculation methods when solving daily problems. The fact that the mathematical concepts and judgments in the book were presented with any logical reasoning shows that it was not a general textbook, but one for senior mathematics scholars or a practical mathematics manual for government officials. As mathematics historian Li Di said, "In the process of its composition, *Nine Chapters* was never independent of the government's economic department (it may have also been kept in the national library). It was a very practical book whose purpose was to serve the economy of the time" (Li, 1997, p. 109).

Fourth, the origins of the terms in *Nine Chapters* also reflected its practical characteristics. Most of the content and terms used in the book were related directly to social production. These were no abstract concepts, but a reflection of real aspects of existence. This was entirely different from *Elements* of Euclid, in which none of the issues discussed were practical; rather, they were all about the relationships among concepts. It is widely known that "geometry deals with points, lines, planes, angles, circles, triangles, and the like. For Euclid and for the Greeks whose work Euclid was presenting, those terms represented not physical objects themselves, but abstract concepts derived from physical objects. Actually, only a few properties of the physical objects are reflected in the mathematical abstractions to which they give rise.... To be precise about what his abstract terms included, Euclid began with some definitions" (Kline, 1995, p. 42). However, for *Nine Chapters*, there was no need to define self-evident ideas as doing so did not have any practical application, but reflected the historical background from which they originated as well. For instance, the shi and fa used in writing fractions were both related to real life. "In ancient China, the dividend was called shi and the divisor fa. The dividends in ancient mathematics were all real items, such as grain or silk, which is what shi means in Chinese. The divisor was actually a certain standard, which is what fa means in Chinese". That is why the divisor was called fa (Gu, 1963, p. 5). Almost all of the techniques in the book are linked to items used in daily life.

Fifth, the primary aim of ancient Chinese mathematics education was practicality. This suggests that mathematical knowledge was indispensable to real life and that people needed to learn it to complete real-life tasks successfully. "In a word, the aim of mathematics education was to train a certain skill and apply it in real life. It was not aimed at training the scientific spirit or methods to improve the quality of scholarship" (Li, 2005, p. 24). With such a goal, people did not need to carry out in-depth research on mathematics; rather, they only needed to grasp sufficient mathematical knowledge to meet their needs. All ancient Chinese mathematicians discussed, to a greater or lesser extent, the practicability of mathematics in their books. Cheng Dawei described the function of mathematics in *Suan Fa Tong Zong* (*Complete Collection of Algorithms*): "Mathematics is something that an intelligent child is able to understand while a foolish elder fails to comprehend. Mathematics tops all the skills in the human world. To be well educated but ignorant about mathematics is like staying in a dark room" (Mei & Li, 1990, p. 63).

Sixth, the social status and administrative influence of mathematicians also played a key role in making traditional mathematics practical. One important reason accounting for the difference between traditional Chinese mathematics and ancient Greek mathematics was that traditional Chinese mathematics texts were written by economic officials, while Greek mathematics researchers were scholars.

The scholars of ancient Greece who researched and compiled mathematics knowledge attempted to use it to describe the world and train the minds of certain talented people. They did not care about the application of mathematics in real life. Since scholars like Aristotle were also logicians, they applied logic methodology in mathematics research and produced deductive mathematical models, which were typically represented in Euclid's *Elements*. Some people in ancient China also saw the possibility of formulating deductive mathematical models, but the economic officials made it impossible. They were in charge of compiling mathematical knowledge and gathered mathematics problems into collections of questions such as *Suanshu Shu* (*Writings on Reckoning*). Their purpose of compiling was for everyday use. *Nine Chapters* is a typical representative of their work and well represents the oriental mathematical model (Li, 2000).

The influence of technology and administration held back the minds of mathematics researchers in the development of traditional Chinese mathematics for over 2000 years. Many world-class achievements were made during this process but, unfortunately, the model of *Nine Chapters* remained unchanged.

The applicability of *Nine Chapters* was also a fundamental characteristic of traditional Chinese mathematics. It was radically different from the deductive feature of ancient Greek mathematics. Generally speaking, ancient Chinese people always thought highly of realism and were good at discovering and abstracting questions from reality and, hence, analyzing and solving them. They set up ancient Chinese mathematics based on extensive real-life practice. This was fundamentally different from Greek geometry, which was divorced from reality, and which adopted formalism that followed the thought of pure logical deduction.

So far, the practicability of *Nine Chapters* has been discussed from many different perspectives; however, not everything in the book was practical. Some puzzles in the book had no practical use, which inspired the writing of *Ce Yuan Hai Jing* (Sea Mirror of the Circle Measurements). Some puzzles in ancient Chinese mathematics were of global significance. For example, the "problem of unknown quantities" in *Sun Zi Suanjing* is equivalent to the Chinese Remainder Theorem, or Sun Zi (Sun Tzu) Theorem, in modern mathematics.

2.2.5.2 The Fundamental Characteristics of Algorithm-Centered Calculability

Algorithm-centered calculability is another important feature of traditional Chinese mathematics. In general, in the history of China, science has always had a strong relationship to technique, and it was in fact difficult to differentiate the two. If we had to do so, the difference might be that technique, compared to science, was stronger in its combination of local customs and production. Therefore, it had a firmer and more real hold on people's minds. In China, science was not approached through theoretical exposition. Rather, people acquired knowledge through life experiences. As a result, the knowledge acquired in this way was not necessarily sufficiently logical, and the core of it is calculation. The main task of astronomy and mathematics was to develop calculation techniques.

Practicability generates the need for calculation. In other words, the impetus of the advanced calculation level in ancient China was in the daily needs of real life.

Counting rods played a crucial role in the development of ancient Chinese mathematics and mathematics education. Counting rods were a peculiar counting and calculation tool of traditional Chinese mathematics, used to denote numbers in mathematics and many other scientific areas, and they acquired this name because people used small bamboo or wood strips to perform calculations. This kind of calculation tool was unique to China. It appeared in China no later than the Spring and Autumn Period (770–476 BC), and it was recorded in *Lau Tzu* that "people good at calculating can calculate without using rods."

Denoting numbers with counting rods was straightforward and efficient. The use of counting rods boosted the development of the decimal system in China. "Ancient Chinese mathematics made great achievements in numerical calculation. This should be attributed to the system of numeration, which used counting rods with place value" (Qian, 1992, p. 9).

Decimal number denotation is the most outstanding and ingenious element of modern culture. It is as important as breathing, but the average person knows so little about it, just as he/she knows little about the chemical composition of air. It is a miracle that just ten digits can denote any number, no matter how large or small. Calculating with counting rods, or rod-arithmetic, was a distinctive way and system of calculating created by the Chinese people and was very different from calculating with a pen. It was convenient to add, subtract, multiply, and divide with rod-arithmetic, which was also simple to understand.

Counting rods exerted a great influence on the formation of the ancient Chinese mathematics model. As Li (1986) said, "Counting rods are not merely pure number operations, but a set of strip-style calculations. Chinese mathematicians not only used counting rods in different places to denote different values and invented the decimal notation, but also used strips in different positions to form specific mathematical patterns to describe certain types of practical problems."

The features of traditional Chinese mathematics and its great achievements in algebra were not accidental—they were actually related closely to the counting rods. When comparing Chinese and Indian mathematics, Japanese scholar Yoshio Mikami (1875–1950) said:

> The main contribution of Chinese mathematics is the development of algebra. Other than this, it is not distinctive in other areas. At three different points in history, Chinese algebra was top in the world. The first was the achievements of *Nine Chapters* in the Han Dynasty, *Ji Gu Suanjing (Continuation of Ancient Mathematics)* in the Tang Dynasty, and various mathematical masterpieces in the Song and Yuan Dynasty….The main reason was that the Chinese have used counting rods since ancient times, and counting rods have a great deal to do with the development of Chinese mathematics. Only when we consider the development of Chinese algebra together with the use of counting rods can we understand the relationship between them (Dai, 2003, pp. 74–207).

In ancient Chinese mathematics, the place-value system was not only used to denote the digits of a number, but also represented different numbers in calculating procedures, like the method of separation of variables in modern mathematics.

Despite their many advantages, counting rods also had some disadvantages that were hard to overcome. For instance, it was hard to avoid mistakes when denoting big numbers with them, and it was difficult to identify mistakes in long calculations. Therefore, they did not facilitate the abstraction of Chinese mathematics and the setting up of a strict, logical theoretical system.

2.3 The Soul of Traditional Mathematics Education Thoughts of China

As an inseparable part of Chinese traditional education, traditional mathematics education in China shared its guiding ideology. In other words, the thoughts of China's traditional education are the soul of its traditional mathematics education, which are embodied in the eternal themes of "honoring teachers and esteeming truth," "teaching benefiting teachers as well as students," and so on. In this respect, in Chapter One of my work *Mathematics Education in China: Tradition and*

Reality, mathematics education values of ancient China are expounded in great detail. In order to give a better understanding of the key thoughts of traditional mathematics education of China, some points are picked out here.

First, the traditional mathematics education of China advocates both the dominant role of students and the leading function of teachers in the teaching process. Since ancient China, teachers have been respectable and have held an ethically lofty social status. The following are some brilliant expositions of ancient Chinese philosophers in this aspect.

> Only if teachers are respected, truth and knowledge can be esteemed and acquired. (*Note of Learning in Book of Rites*)
>
> The country in which teachers are highly respected will thrive, and the country in which teachers are belittled will decline. (*A Conspectus to Xuncius*)
>
> A teacher in one day, is a father all the life. (*Taigong Family Education in Lost Ancient Books found in Mingsha Mountain*)
>
> For the academic students, nothing is more important than following the teacher's instruction. (*Revised Constitution of LiuYang Mathematics School*)

"Esteeming truth" means that knowledge is presented according to the reality of students, which can be realized only if the dominant role of students is guaranteed. However, some people misunderstand this as denying the principal status of students during the teaching process. Being strict with students cannot be confused with denying their dominant roles in education. Meanwhile, it should be reminded that an enjoyable classroom atmosphere is not equated with the realization of students' dominant roles.

Second, Confucius, a great educator and ideologist in ancient China, presented the idea of "teaching benefitting teachers as well as students" over 2000 years ago, which concisely indicated the teaching thought emphasizing both the dominant role of students and the leading function of teachers. In a word, the idea advocates that teaching and learning can promote each other; a mutual improvement can be achieved between teachers and students, and between teachers' teaching practice and their self-learning, which disapproves the dichotomy of "student-centered" or "teacher-centered" patterns and encourages the balance and harmony between teaching and learning. It has different expressions in various teaching settings and times. For instance, "Among any three people walking, I will find something to learn for sure" (*The Analects of Confucius*). "So pupils are not necessarily inferior to their teachers, nor teachers better than their pupils. Some learn the truth earlier than others, and some have special skills—that is all" (*On Teaching, Hanyu*). This concept is always in the practice and development of Chinese traditional mathematics education.

As for learning, Confucius valued the combination of learning and thinking: "Learning without thinking leads to confusion; thinking without learning ends in danger" (*The Analects of Confucius*). As an old Chinese saying says, "Have a master at home, train skills on your own." All of these encourage the learners' initiative of learners.

As for teaching, Confucius emphasized teachers' devotion. He advised teachers "to be insatiable in learning and to be tireless in teaching" (The Analects of Confucius). If a student cannot learn well, a teacher must take responsibility for this to some extent.

In mathematics teaching, a teacher's role is to lead the students through working out the teaching plan, fulfilling the teaching design, and implementing teaching access. In brief, mathematics education in China emphasizes both the leading function of teachers and the dominant role of students, striking a balance and harmony between teaching and learning, between teachers and students.

The idea of "teaching benefitting teachers as well as students" can be realized by the discussion and communication between teachers and students, which is similar to Socrates' elicitation teaching theory. Many typical cases of mathematics teaching in ancient China were in the form of conversation and discussion. The following case from *Zhou Bi Suan Jing* will illustrate this point vividly.

> In this case, Rong Fang asked Chen Zi about the solution to a mathematical problem, but Chen Zi instructed him to learn the methods of thinking instead of telling him the answer immediately. After Rong Fang's tirelessly repeated thinking, Chen Zi explained the causes of the problem and its solution.
>
> Once Rong Fang asked Chen Zi, "I have heard that you are so talented that you know the height and the size of the sun, the distance light can travel and a person can cover in a day; you even know the positions of all of the stars and how big the earth and the universe are, no matter whether they are close or far. Is that true?"
>
> Chen Zi said, "Yes."
>
> Rong Fang said, "Though I don't know about it, I hope you can tell me. Could you please teach me how to do it?"
>
> Chen Zi said, "Certainly. This is what mathematics can do. You are good enough at mathematics to know about it as long as you keep thinking."
>
> So Rong Fang went home and thought it over, but he couldn't get any answer. Then he interviewed Chen Zi again and asked, "I kept thinking about it but I failed, could you tell me now?"
>
> Chen Zi said, "You are thinking but not deeply pondering over it. This is also the skill to calculate the distance or the height, but you cannot get it because of your lack of analogy. Those who find it difficult to grasp the skill always learn but not extensively, or learn extensively but not intensively, or learn intensively but not analogically. It is the analogical skill that makes a difference between fools and wise men. In other words, only the person with analogical skills is qualified for intensive learning. As a result, there is no use explaining to you the abstruse problem now. Please go back again and ponder over it."
>
> Rong Fang went back again and pondered over it for many days without any solution. Then he visited Chen Zi and said, "I have thought it over but I am not intelligent enough to figure it out. Please tell me this time." Then Chen Zi instructed him how to calculate the distance between the sun and the earth with *Gou-Gu Ding Li* in the form of analogy.

Regarding this case, Zhao Shuang commented in his notes that a teacher would not explain unless a student was desperately anxious and determined to learn and, after being given an instance, the student was expected to infer other things from it. This viewpoint is quite similar to Liu Hui's ideas put forward in his notes of Su Mi

(Millet) Chapter and Equation Chapter in *Nine Chapters on Mathematical Procedures*, which emphasize the importance of analogy in the mathematical application.

Although the cases mentioned above are special ones, they reflect the philosophy of mathematics learning methods. The following can be inferred. In the mathematics teaching of ancient China, heuristic teaching methods were applied through the give-and-take of conversation between teachers and students, and this interaction was highly valued. This was similar to Socrates' "elicitation teaching theory" used in his geometry teaching mentioned in *Menon* by Plato of ancient Greek; Chen Zi advocated independent study, and Rong Fang clarified the importance of reflection in mathematics learning. Chen Zi did not pass on his knowledge to Rong Fang until the latter pondered on it over and over again for a long time.

In brief, it was emphasized in *Zhou Bi Suan Jing* that, in mathematics learning, inferences about other cases should be drawn from one instance; in addition, the drill and cultivation of thinking ability as well as some basic knowledge are essential. The heuristic education in *Zhou Bi Suan Jing* is comparable to Socrates' "elicitation teaching theory," and both of them have played a positive role in mathematics education history in the East and West, then and now.

2.4 Typical Cases in Chinese Traditional Mathematics Teaching Practice

That "three non-collinear points determine a plane" is a self-evident truth. Similarly, Chinese conventional thinking, educational ideology, and unique thinking methods of traditional mathematics determine Chinese traditional mathematics education as a perfect whole due to their dynamic integration produces. The mutual promotion and interdependence of these factors in the teaching practice lead to the internal positive cycle and vitality of the whole. Nevertheless, the whole is not a closed system but one advancing with the times, which adopts new ideas and methods constantly to realize its self-improvement and self-transcendence.

In Chinese traditional mathematics education, the cases of intelligent solutions to practical problems are too numerous to mention one-by-one. The following are some cases to illustrate this.

2.4.1 Methodology of Wholeness

Case 1 The proof of Gou-Gu Ding Li (Pythagorean theorem) by Zhao Shuang.

In *Zhou Bi Suan Jing*, Gou-Gu Ding Li is recorded as the square of Gou plus the square of Gu, and after being extracted by the square root, the chord length is gained (Vol. II, Zhou Bi Suan Jing) (Jiang and Xie 1996).

Gou-Gu Ding Li was proved by Zhao Shuang with his "Gou-Gu Yuan Fang Tu" in his annotations on *Zhou Bi Suan Jing*; this was created after rotations and symmetry transformation of some times according to the out-in complementary principle. "The square of Gou plus the square of Gu is the square of the chord, then it extracts a root, and the chord length is the result. Further illustrations are given that, according to the chordal graph, Guo multiplies Gu and then multiplies two, plus the square of the difference between Gou and Gu, and the result is the square of the chord as well."

【 朱:Zhu 股:Gu 勾:Gou 弦:Xian 黄:Huang】

In the chordal graph, the product of Gou multiplied by Gu is illustrated as a red rectangle named "Zhu Shi," and four rectangles if doubled. The square of the difference between Gou and Gu is presented as a small yellow square in the middle called "Zhong Huang Shi." Therefore, two "Zhu Shi" plus one "Zhong Huang Shi" is a "Xian Shi"—the square of the chord (Fig. 2.1).

The above proof can be illustrated by modern mathematical symbols.

If a, b, and c, respectively, represent Gou, Gu, and Xian in Gou-Gu graph (a right triangle), one "Zhu Shi" is $\frac{1}{2}ab$, four "Zhu Shi" is $2ab$, and "Huang Shi" is $(b-a)^2$. So $c^2 = 2ab + (b-a)^2 = a^2 + b^2$, i.e., $c^2 = a^2 + b^2$.

Case 2 The Proof of Guo Gu Ding Li by Liu Hui.

Liu Hui structured two different rectangles (as a whole) with the out-in complementary principle and dealt with the relationship between the base line, the height, and the size of a triangle from a global perspective, i.e., the size of a triangle is equal to the half size of the rectangle.

As shown in Fig. 2.2, if two right angle sides of a right-angled triangle are, respectively, a and b, and the hypotenuse is c, on each side of the right-angled triangle three squares can be structured counterclockwise with respective side lengths of a, b, and c, so the size of the square with the side length of c is c^2, i.e., $c^2 = \frac{1}{2}[(b-a)+b] \cdot b + \frac{1}{2}ab + a^2$, so $c^2 = a^2 + b^2$.

The proof methods of Gou-Gu Ding Li used by Zhao Shuang and Liu Hui are typical examples of geometrical proof methods in ancient China, which view the

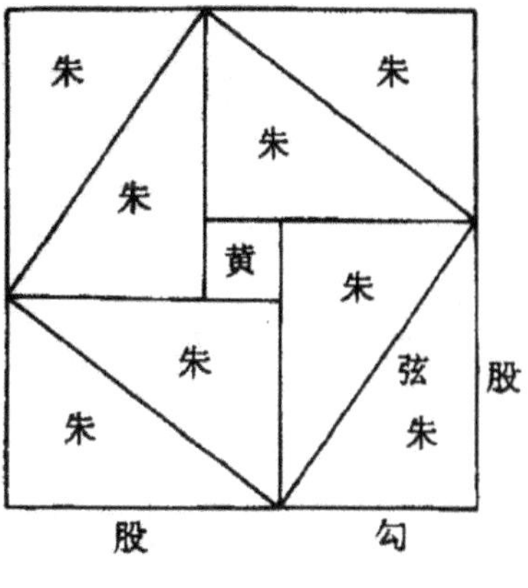

Fig. 2.1 Diagram proof by Zhao Shuang

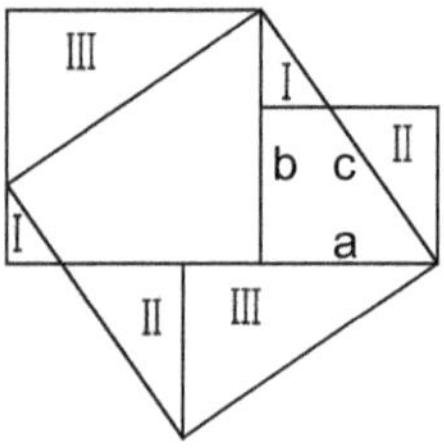

Fig. 2.2 Diagram proof by Liu Hui

internal relationships between conditions and conclusion and tackle it as a whole with intuition, so that the proof procedure can be explained briefly. The method called "the out-in complementary principle" implies the feature of wholeness in Chinese traditional thought, and geometrical proof involved attempts to structure orderly wholeness. The principle has two traits. First, the wholeness ideology is applied. A new holonomic shape is structured according to the size and shape of the known geometric form, and then, a conclusion is drawn in terms of its size and volume. In other words, the features of parts are handled from the global perspective. Besides, a conclusion is drawn according to the relationship between the wholeness and parts with a quantity standard, which means that geometric transformation is conducted through the congruence relationship between geometric forms (or size and volume) in order to structure the wholeness. However, only the size and volume are taken into consideration, but other factors such as angles and relations between the locations of lines are not. In other words, Liu Hui and other mathematicians in ancient China tackled the relationships of these factors by direct viewing in the geometric transformations, which resulted in some seemingly ambiguous proof methods. This research approach is quite different from that used in ancient Greek. For example, Zhao Shuang structured a wholeness, a square, after rotations and moving transforms of several times, then proved Gou-Gu Ding Li on this basis, which is quite different from the method used in *Elements of Geometry* by Euclid in ancient Greek to prove the Pythagorean theorem.

Case 3 Problems of "meetings of wild ducks and wild geese" in *Nine Chapters on the Mathematical Art.*

> It takes seven days for wild ducks to fly from the South Sea to the North Sea, and nine days for wild geese from the North Sea to the South Sea. If they depart simultaneously from the South Sea and the North Sea, how long does it take for them to meet?
>
> The problem was solved in this way:
>
> The sum of days is used as the divisor, and the product of days multiplied as the dividend.… In this way, if it takes 7 days for wild ducks to arrive and 9 days for wild geese, it takes 63 days for wild ducks to arrive 9 times, and for wild geese 7 times. The day on which all of them arrive is their meeting day. So 63 days divided by 9 plus 7 is the number of days it takes for them to meet. This can be illustrated in the following numerical table:

$$\begin{matrix} & \text{凫} & \text{雁} \\ \text{日} & 7 & 9 \\ \text{至} & 1 & 1 \end{matrix} \xrightarrow[\text{齐其至}]{\text{同其日}} \begin{bmatrix} 63 & 63 \\ 9 & 7 \end{bmatrix} \xrightarrow{\text{并齐至为共至}} \begin{bmatrix} 63 \\ 9+7 \end{bmatrix}$$

[日: travel time 至: arrival time 凫: wild ducks 雁: wild geese

同其日: the same set days

齐其至: respective travel time

并齐至为共至: the sum of their respective travel time and the same set days]

i.e., they meet 16 times within 63 days, so the number of days it takes for them to meet once is:

$$\frac{63}{7+9}=\frac{63}{16}=3\frac{15}{16}\text{ days.}$$

Another explanation given by Liu Hui is that the wild ducks cover 1/7 of the whole journey every day, and the wild geese 1/9. According to the requirement of "the set days and the respective travel time," they are written, respectively, as 9/63 and 7/63. Supposing it takes 63 days for them to travel from the south to the north, totally 16 times for both wild ducks and wild geese. So the number of days it takes to meet is $\frac{63}{16}=3\frac{15}{16}$ days.

This question displays the basic fraction methodology used by mathematicians in ancient China, which is called "Qi Tong Shu," i.e., finding common denominators and equivalent fractions in order to carry out addition and subtraction. The solution to the question reflects the methodology of wholeness in handling practical problems as well.

2.4.2 *Methodology of Symmetry*

2.4.2.1 Yang Hui's Methodology of Symmetry

In the research on mathematics education methodology in ancient China, Yang Hui, a well-known mathematician and mathematics educator of the Southern Song Dynasty (1127–1279), should be introduced above all. One of his contributions to mathematics education was his work *Mathematics Teaching Program,* a vital document in the mathematics education history of ancient China, which was first published in another work of his, *Yang Hui Suan Fa* (*Yang Hui's Arithmetic*). It was also the first document on mathematics teaching discovered in the world, falling into the same category of mathematics teaching plans found later. It recorded a very complete mathematics knowledge hierarchy, definite learning process and goals, specific hierarchical analytical processes of textbooks and reference books, teaching methods involving intensive explanations and more practice; and emphasis on mathematics principles as the root of all methods.

Yang Hui's thoughts about mathematics education embodied those of the sages of the past, and some teaching ideas such as "gaining new insights through reviewing old materials," "following in proper order and improving gradually," and "teaching only the essential and ensuring plenty of practice" were fully incorporated into his teaching practice. In addition, he attached great importance to the combination of mathematics thought and the fostering of students' creative thinking skills. In conclusion, Yang Hui left later generations a legacy of ideas which integrate mathematics thoughts and Chinese traditional education methods.

Symmetry Approach in Computation Teaching

The aesthetic ideology of mathematics can be appreciated in Yang Hui's works, especially his symmetry approach.

Case 4 "The first group of numbers includes one, three, five, seven and nine, and the second group includes two, four, six, eight and ten. If the result is 55, what are the calculation processes? The method is as follows. The sum of each pair is 11, and if multiplied by 10, the product is 110, half of which is 55" (*Xu Guo Zhai Qi Suan Fa I*). Its illustration in the form of modern mathematics is as follows:

$$S = 1+2+3+\cdots+9+10,\ S = 10+9+\cdots+3+2+1$$

The addition of each symmetrical pair of numbers in each group from left to right is:

$$2S = (1+10)+(2+9)+\cdots+(2+9)+(1+10) = 11\times 10 = 110$$

$$S = 110 \div 2 = 55$$

The arithmetic procedure was once recorded in detail in the biography of Gauss, the well-known mathematician: When he was in primary school, Gauss figured out this symmetrical calculation method without any hints from the teacher. Also, in the biography of Wu Zaiyuan, a famous mathematician in the Period of the Republic of China, it was recorded that he was very excited after he figured it out on his own as a child.

A new method was created by Yang Hui with this symmetrical principle (Dai, 2003, p. 204). It was clarified that the sum of each pair of numbers, which has the same distance to the first and the last number (Symmetry), is equal to the sum of the first and the last number (Uniformity). The discipline makes it convenient to carry out the calculation. As well, it is very inspiring for both mathematics teaching and research; for instance, the formula for the sum of the numbers in front of the arithmetic progression used in primary school education is deduced according to this discipline.

Symmetrical Methods of Area Calculation Teaching

Case 5 Symmetrical ideology is also applied to the deduction of the isosceles triangle area in *Nine Chapters on the Mathematical Art*, called the central symmetry principle. "Multiply half of the bottom side by the height. If half of the side is known, a rectangle can be created according to the out-in complementary principle. Also, it can be done with half of the height multiplied by the side" (Fig. 2.3).

Yang Hui gave more details about the deduction of the isosceles triangle area and its various supplementary conditions, which made it more comprehensive and flexible. "If the bottom side can be halved easily (i.e., a natural number), multiply half of it and the height. If the height can be halved easily, multiply half of it and the bottom side. If neither the bottom side nor the height can be halved easily, multiply the bottom line and the height and then reduce by half."

This is illustrated in the geometric figures, as shown in Figs. 2.3, 2.4 and 2.5. (Given ΔABC, D and E are, respectively, the midpoints of sides AB and AC.)

In the formula: $S = (\frac{1}{2}a)h$, $S = a(\frac{1}{2}h)$, and $S = \frac{1}{2}(ah)$ (S is the area of triangle, a is the bottom side, and h is the height).

In fact, while deducing the area formula for the geometric figure, the mathematicians of ancient China unconsciously adopted a simple center-transfer method of elementary geometry. Liu Hui and Yang Hui supplemented the incomplete figure to create a perfect integration—a rectangle, according to the out-in complementary principle. They based the area of a triangle on the concept of the rectangle.

Yang Hui's aesthetic ideology of symmetry is very useful in mathematics teaching. For instance, Fig. 2.3 is always used to teach the formula for the area of the triangle in primary schools in China. Some insightful teachers may use Fig. 2.3 first and then inspire the student with Figs. 2.4 and 2.5, which is more helpful for arousing students' interest and fostering their thinking ability.

Fig. 2.3 Diagram derivation of the formula for area of triangle (1)

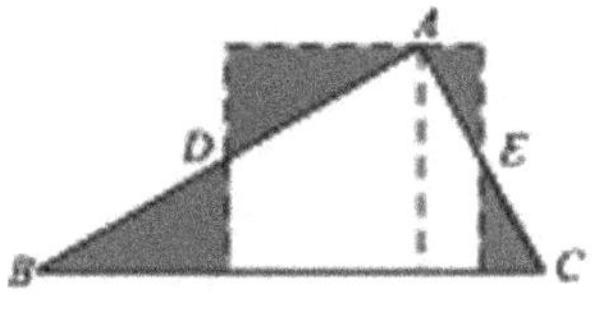

Fig. 2.4 Diagram derivation of the formula for area of triangle (2)

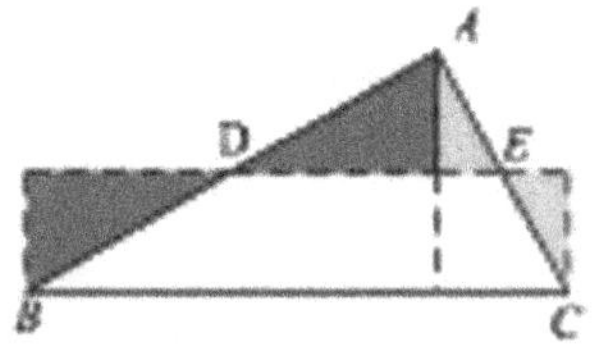

Fig. 2.5 Diagram derivation of the formula for area of triangle (3)

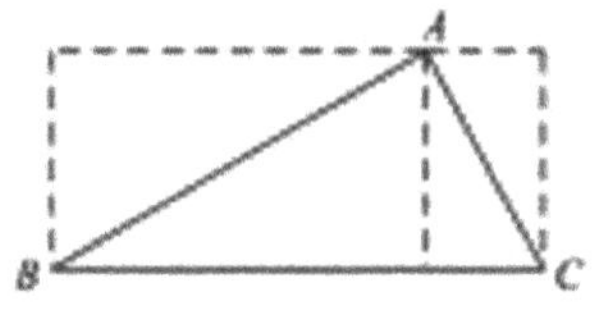

2.4.2.2 Zhen Luan's Problem-Solving Teaching

Case 6 Zhen Luan, a mathematician in the Northern and Southern dynasties, recorded some typical cases of problem-solving teaching using the symmetrical method in his notes of *Shu Shu Ji Yi*. "How can we know the width of a river with marks instead of any calculation?"

The solution is as follows: "Supposing someone sets three parallel lines on the north side of the river, and the lines in the north and south have a distance of one Zhang (a length unit in ancient China). Standing in the north of the middle line he takes a straight look at the north bank (M) and marks the point (A) on the southern line, then takes a straight look at the south bank (N) and marks the point (B). The length between the two points is transferred to the counterparts (A′ and B′) on the northern line. Afterward, standing in the south of the middle line, he takes a straight look at the north side through the two points (A′ and B′), respectively, and gets the other two points (M′ and N′). So the distance between these two points is the width of the river."

In this case, an axial symmetry principle of geometry is used (Li, 1997, p. 263). The corresponding geometric figure is as follows (in Fig. 2.6A, A′, B, B′ are the marked points):

[北表:the north line 中表:the middle line 南表:the south line 记号:marked points 北岸:the north bank 南岸:the south bank 河宽:the river width].

2.4.3 *Formulae in Verse in Mathematics Teaching*

It was believed that a perfect memory is helpful for understanding. There is a variety of ways of developing a strong memory and the formula in verse is one of the favorites of the Chinese. In ancient China, some mathematics problems were adapted to formulae in verse for the convenience of mastery and instruction. Rhythm formulae orally display the connection between quantity and space form as well as the rules. The earliest rhythm formulae appeared in the preface of *Zhou Bi Suan Jing*, and many were found in the works of Yang Hui and Zhu Shijie during the Song and Yuan dynasties. In *Suan Fa Tong Zong* by Cheng Dawei of the Ming

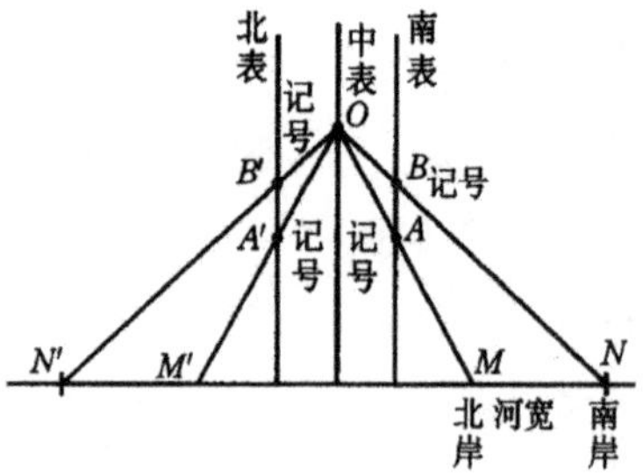

Fig. 2.6 Diagram of the measurement of river width by Zhen Luan

Dynasty, the abacus operation was accompanied by rhythm formulae as well as some mathematics questions.

These rhythm formulae are the organic integration of arts and mathematics, embodying the romance of ancient mathematics. Additionally, they played a role in mathematics learning and research, and business as well. Furthermore, they promoted the popularity of mathematics knowledge, although limited to some common sense.

Case 7 "Put a Ju (a measuring tool) flat to measure the horizontal and vertical level; set the Ju upright to measure the height; put it upside down to measure the depth; put it lying flat on the ground to measure the horizontal distance; put it encircled to get a circle; combine two Ju to get a rectangle" (Fig. 2.7).

The first four verses tell about the measuring methods with Ju, and the last two illustrate the formation of a circle and a rectangle. This theory is the same principle of similar triangles in plane geometry in secondary schools.

The measuring method of height is as follows: Put the side AC of the square flat and overlap the horizontal line of AE, keeping the side BC vertical. Look at the top of F, and the sight line of AF and BC intersect at D. So $\Delta ACD \backsim \Delta AEF$ $\therefore \frac{AC}{CD} = \frac{AE}{EF}$, height $EF = \frac{CD \times AE}{AC}$.

2.4.4 *Learning Mathematics Through Games*

In ancient China, people created plenty of mathematics games according to the characteristics of children's ages. Mathematics games help arouse children's interests and cultivate their thinking abilities. Generally, there are two kinds of games: number games and jigsaw (or take-apart) puzzles. "Qi Qiao Ban" (seven-piece puzzle) and "Yi Zhi Tu" (Puzzle Chart) are two popular jigsaw puzzles, which build graphs of the same size. The function of "Qi Qiao Ban" was recorded in the preface of *Qi Qian Ban He Bi* in 1803. "Nobody knows who created the seven-piece puzzle. The talented play it for entertainment."

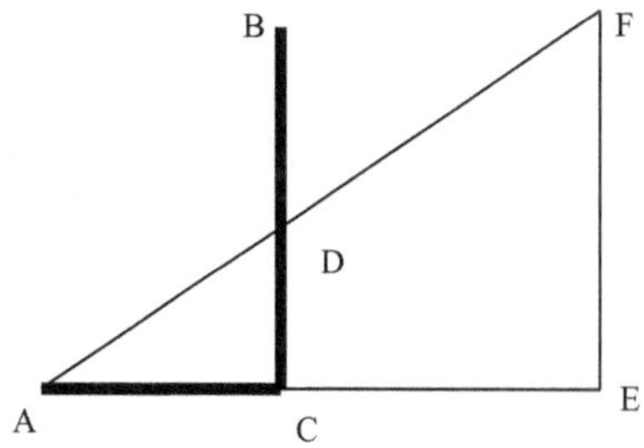

Fig. 2.7 Diagram of Ju

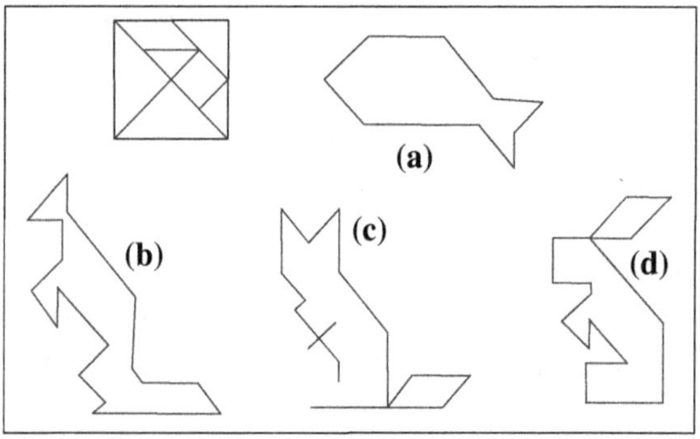

Fig. 2.8 Tangram shapes (1)

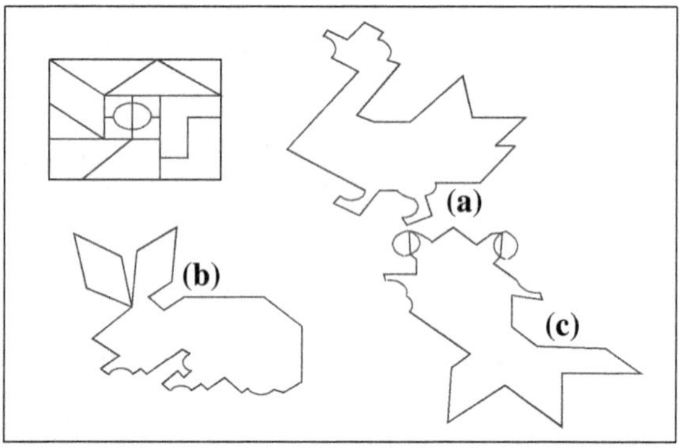

Fig. 2.9 Tangram shapes (2)

Case 8 Try to divide the figures (a, b, c, and d) as shown in Fig. 2.8 into the same parts in the square which consists of 5 kinds of figures, big-, small-, medium-sized, square, and inclined.

Case 9 Try to divide the three figures (a, b, and c) as shown in Fig. 2.9 into the same parts in the square.

2.5 Conclusion

In 1676, Newton said "If I have seen farther than others, it is because I was standing on the shoulders of giants." "The shoulders of giants" are the scientific traditions and achievements in Europe. Traditions are sources and the basis of new sources. Based on its own traditions, the development of mathematics education in China can still gain from the advanced experiences from other countries and positive elements of new ideas and thoughts. It is not perfect yet and its flaws should not be neglected.

From the perspectives of the past, the present, and the future, traditions and reformations should be integrated. The irrational factors in traditions must be sublated and new ideas and methods must be introduced gradually to avoid the waste of educational resources. It is helpful for the development of Chinese mathematics education to grasp traditional mathematics knowledge and understand traditional mathematics education deeply. As for traditional mathematics education, instead of denying it blindly or exaggerating it excessively, a mathematics educator is expected to have abilities of selection and judgment, like an excellent photographer who can take a good picture even with an old camera in contrast to an unskilled shutterbug who cannot do this even with an advanced one.

It is believed that a wonderful future for mathematics education can be achieved with a comprehensive and deep understanding of traditions, constant practice, and creativity.

References

Croce, B. (2005). 作为思想和行动的历史 [The history of thoughts and action]. Beijing: China Social Science Press.

Dai, Q. (2003). 儒家思想与中国传统数学 [The confucian school and chinese traditional mathematics]. Beijing: The Commercial Press.

Gu, S. S. (1963). 中国古代教育家语录类编 [Classified quotations from ancient chinese educators]. Shanghai: Shanghai Education Press.

Guo, S. C., & Liu, D. (1998). 算经十书*(二)* [Ten mathematical classics(II)]. Shenyang: Liaoning Education Press.

Jiang, X. Y., & Xie, Y. (1996). 周髀算经 [(Critical edition of) Zhou Bi Suanjing]. Shenyang: Liaoning Education Press.

Kline, M. (1995). 西方文化中的数学 [Mathematics in western culture] (Z. G. Zhang, Trans.). Taipei: Jiuzhang Press.

Li, D. (1958). 缀术的失传时代问题 [An investigation to the time when Zhui Shu was lost]. *Communications in Mathematics, 11*, 33–34.

Li, D. (1997). 中国数学通史*(上古到五代卷)* [General history of chinese mathematics] (Volume on early times to five dynasties). Nanjing: Jiangsu Education Press.

Li, D. (2000). 古代ギリシアと古代中国における異なる数学モデルの形成された背景 [Background to the formation of different mathematical models in Ancient Greece and Ancient China]. *Osaka Journal of Mathematics Education, 30*, 35–39.

Li, D., & Dai, Q. (2000). 中国数学教育史纲 [Historical survey of chinese mathematics education]. In Y. Kiyoshi, S. J. Zhong, & D. Li (Eds.), 中日近现代数学教育史*(第四卷)* [History of chinese and japanese modern and contemporary mathematics education] (Vol. 4, pp. 92–103). Osaka: Hankai Publishing House.

Li, D., & Guo, S. R. (1988). 清代著名天文数学家梅文鼎 [Mei Wending-a famous astromathematician in the Qing Dynasty]. Shanghai: Shanghai Scientific and Technical Publishers.

Li, J. M. (1986). 试论中国传统数学的特点 [On the characteristics of traditional chinese mathematics]. In W. J. Wu (Ed.), 中国数学史论文集*(二)* [Collected articles on chinese mathematical history(II)] (pp. 9–18). Jinan: Shandong Education Press.

Li, Y. (1954). 中算史论丛*(四上)* [Series of articles on chinese mathematical history] (Vol. 4, Part A). Shanghai: The Chinese Scholarship Society & The Commercial Press.

Li, Y. (1955). 中算史论丛(第四集) [Series of articles on chinese mathematical history] (Vol. 4). Beijing: Science Press.

Li, Z. H. (2005). 中国近代数学教育史稿 [A history of chinese modern mathematics education]. Jinan: Shandong Education Press.

Mei, R. Z., & Li, Z. H. (1990). 算法统宗校释 [Collation and translation of Suan Fa Tong Zong]. Hefei: Anhui Education Press.

Qian, B. C. (1992). 中国数学史 [Chinese mathematical history]. Beijing: Science Press.

Ye, X. S. (2010). 美的哲学 [The philosophy of beauty (revised edition)]. Beijing: World Publishing Corporation.

Zhang, D. Z., & Zhao, X. P. (2012). 中国教育是不是有"美"的一面? [Is there a charming part of chinese education?]. *Mathematics Teaching, 3,* 50.

Chapter 3
Chinese Mathematics Curriculum Reform in the Twenty-First Century

Lidong Wang, Qimeng Liu, Xiaofeng Du and Jian Liu

Abstract Curriculum reform is a fundamental factor in pushing forward educational development and reform. In this chapter, we examine the development and implementation of Chinese mathematics curriculum standards, with a focus on the development mechanism and characteristics of curriculum policy and its impact on public schools as well as the educational systems in China during the early twenty-first century. Our goal is to present to the world the current situation of mathematics curriculum reform and the development in mainland China (i.e., China excluding Hong Kong, Macao, and Taiwan) since 2000.

Curriculum reform is a fundamental factor in pushing forward educational development and reform. In this chapter, we examine the development and implementation of Chinese mathematics curriculum standards, with a focus on the development mechanism and characteristics of curriculum policy and its impact on public schools as well as the educational systems in China during the early twenty-first century. Our goal is to present to the world the current situation of mathematics curriculum reform and the development in mainland China (i.e., China excluding Hong Kong, Macao, and Taiwan) since 2000.

3.1 The Background of New Century Chinese Mathematics Curriculum Reform

From an international perspective, we live in an age witnessing a rapid development of science and extraordinary changes in people's lifestyles. New knowledge, innovative technology, socialization, and globalization have related modern

L. Wang (✉)
The High School Affiliated to Renmin University, Beijing, China
e-mail: wldyaoyuan@163.com

Q. Liu · X. Du · J. Liu
Beijing Normal University, Beijing, China

Y. Cao and F.K.S. Leung (eds.), *The 21st Century Mathematics Education in China*, New Frontiers of Educational Research,
https://doi.org/10.1007/978-3-662-55781-5_3

mathematics closely to all areas of human existence (The Research Group of Mathematics Curriculum Standard, 1999). Since the 1980s, many countries around the world have hoped to improve the mathematics literacy of their own citizens through various efforts, including reflecting on the history of education and developing New Curriculum Standards (Dong, 2006). Many of the world's major countries and regions have implemented new rounds of mathematics education reform, including the Principles and Standards for School Mathematics in the USA (National Council of Teachers of Mathematics [NCTM], 2000) and the National Curriculum in Great Britain (Cockcroft, 1994).

Social and economic development in China [especially the development of information technology, digital technology, lifelong learning, and democratization (The Research Group of Mathematics Curriculum Standard, 2002)] has raised the bar for mathematics literacy. New demands for modern citizens have required corresponding changes in public schools, especially in mathematics curriculum and instruction (Ma, 2001). From June 1996 to 1997, the division of basic education in the Ministry of Education organized a survey to investigate the status of the implementation of compulsory education in all subjects, including mathematics, across the nation. The data and facts collected from this survey demonstrated that the curriculum used at that time achieved certain goals (e.g., basic knowledge and basic skills training); however, many problems were identified. For example, the old curriculum was characterized as "complex, difficult, partial, and old." Students suffered from rote memorization and drill practice. At the same time, teachers struggled with "draining students in a sea of problems" (Liu, 2009). There was too much emphasis on using test scores to screen students. The old curriculum was highly centralized, with little flexibility for local adaption, and it did not meet the different social and economic requirements of a diverse student body (Liu, 2009). At the same time, the community also began to reflect on education in schools; for instance, a series of articles titled "My Concerns about Chinese Language Education" appeared in the 11th issue of the "Century Watch" column of the Beijing Literature magazine in 1997. These articles had a great impact on the community. The trends in international and national education that were mentioned above demanded curriculum reform. Similar to the previous education reforms, the current one adopted a top-down approach: However, we cannot negate the fact that it also reflected certain concerns raised from the community.

3.2 Mathematics Curriculum for Compulsory Education (Grades 1–9)

The Twenty-First-Century Revival Action Plan in Education (Ministry of Education of the People's Republic of China, 1999) called for the "implementation of a cross-century project of quality education, promotion of education to improve the national education quality and students' innovation capability" and noted the need to "spend about 10 years experimenting with a nationwide curriculum for basic education in the twenty-first century," as well as the need to "reform the curriculum system and the evaluation system based on the experiences we gain." In March 1999, the Department of Education first commissioned an existing research group to explore a national mathematics curriculum standard ahead of the other disciplines (hoping to provide first-hand experiences in the curriculum development process for other disciplines).[1]

3.2.1 The Development of a New Standard for Compulsory Education

The Mathematics Curriculum Standards for Full-time Compulsory Education (draft) was completed and put forth for extensive comments from the community in March 2000. The mathematics standards research group mentioned above consisted of mathematics and mathematics education scholars, researchers, and staff members from local provinces (cities) and school teachers. About 70% of the research team members worked in higher education institutes and about 30% of them worked in public schools. This research group developed new mathematics standards by studying the research results and best practices from both the Chinese and the international mathematics education communities. The research team members also solicited comments from scholars and experts in various fields including mathematics, psychology, mathematics education, and school teachers. The comments received by the team ranged from discussions of the nature of mathematics and educational goals to issues about methods for handling the definition of multiplication (Zhang & Liu, 1999). The development process adopted a procedure of open discussion so that the resulting curriculum policy could benefit from the wisdom of different parties, with a careful consideration of diverse values (Song & Xu, 2010, p. 121).

The development of the mathematics curriculum played an important role in this round of curriculum reform in fundamental education, which provides the idea of basic value, the mechanism of implementation, and the way to develop the standard for other subjects in fundamental education. The Ministry of Education formally promulgated and implemented Mathematics Curriculum Standards for Full-time Compulsory Education (Trial Version) (MCSFCE) in June 2001.

3.2.2 The Features of Standards for Compulsory Education

In addition to focusing on additions and deletions of some content topics, the MCSFCE differed from the products of previous curriculum reform in several fundamental aspects, such as the basic curriculum ideas, curriculum objectives,

[1]The "Prospects in the Twenty First Century Mathematics Education Project" research group.

curriculum implementation (including guidance on textbook development), teaching suggestions, evaluation recommendations, and even curriculum management. It provided detailed descriptions in some dimensions. For example, the traditional syllabus only provided a brief description of teaching content and objectives. Most of the descriptions of teaching objectives were included in the textbook developed by the state. MCSFCE changed both the scope and depth of the role that the state plays in the curriculum by providing descriptions of learning content, learning processes (special attention), and teaching recommendations (including several cases for some content). This provided a standard for the transformation from one single national textbook policy to a policy of diversity; a national committee certificated and authorized the different versions of textbooks, according to the curriculum standards.

To examine some of the differences between the old syllabus and MCSFCE in more detail, consider the following descriptions of how students and teachers should approach the Pythagorean Theorem.

The old syllabus included some dimensions such as

Master the Pythagorean Theorem. (Students) know how to use the Pythagorean Theorem to solve for the third side given the measurement of the other two sides. (Students) know how to use the converse of the Pythagorean Theorem to determine if a triangle is a right triangle. Conduct patriotic education by introducing the research on the Pythagorean Theorem done by ancient Chinese mathematicians.

The MCSFCE included some dimensions not covered in the previous syllabus, such as suggestions for evaluations and recommendations for textbook development:

Explore the proof process of the Pythagorean Theorem. (Students) know how to use the Pythagorean Theorem to solve simple problems. (Students) know how to use the converse of the Pythagorean Theorem to determine if a triangle is a right triangle. The recommendations for textbook development suggest introducing several well-known proofs (such as the Euclidean proof and Zhao Shuang[2] proof) and some well-known problems so that students are aware that mathematical proof can be flexible, beautiful, and sophisticated. Students should also be aware of the Pythagorean Theorem's rich cultural connotations. At the same time, some teaching suggestions include guidance on the teaching activities and teaching process of the Pythagorean Theorem.

As mentioned above, the MCSFCE proposed a basic reform idea: "Mathematics for All." In other words, "Everyone can learn valuable mathematics; everyone can learn the necessary mathematics; different people benefit from different mathematical development" (Ministry of Education of the People's Republic of China, 2001). This concept was totally different from the underlying idea of the old syllabus (Zhang & Song, 2004). The MCSFCE suggested following the psychology of learning mathematics and using real-life experience to motivate student development. Students were to experience the process of mathematical modeling, which would allow for the interpretation and application of the problem-solving process.

[2]A very famous Chinese mathematician in the early 200s.

Thus, as was the hope of mathematics education reformers elsewhere in the world, students would be enabled to grow in mathematics understanding, mathematics thinking ability, attitudes toward mathematics, and appreciation of mathematics (NCTM, 1989, 2000).

Even though, in terms of curriculum objectives, MCSFCE inherited qualities from traditional Chinese mathematics education which emphasized training in basic knowledge and basic skills ("The Two Basics") (Zhang, Li & Tang, 2005), the MCSFCE also emphasized learning goals for the growth of mathematical thinking ability, problem-solving skills, attitudes toward mathematics, and the appreciation of mathematics.

The MCSFCE highlighted the nature of mathematics and the "non-formalized aspect" of mathematics content, including applications of intuitive geometry and a spiral curriculum (Zhang & Song, 2004). At the same time, emphasis was placed on the cultural value of both pure and applied mathematics, real-world applications of mathematics, the importance of human development, the technical attributes of mathematics, and the connections between mathematics and calculators (and computers). MCSFCE defined mathematics as a language to describe the real world. It was considered a process of theory abstraction from nature using qualitative/quantitative methods that also involved the application of theories to solve real-world problems.

In terms of specific curriculum content, the MCSFCE was arranged in several sections, including "Number and Algebra," "Space and Figure," "Statistics and Probability," and "Practice and Synthetic Application." The focus was on the development of students' number sense, symbol sense, space concepts, statistical concepts, and the application of awareness and reasoning abilities. In the number and algebra section, the MCSFCE added the concept of negative numbers and applications of calculators and strengthened the role of estimation. The emphasis on the use of the abacus, complicated operations, and simple numbers was decreased (Kong & Hu, 2002). In terms of geometry (space and shape section), the topics of translation, rotation, symmetry, and other geometric transformations were increased to a certain extent to replace the traditional Euclidean geometry system. The coverage of topics in orientation, measurement, space, and shapes was also increased, as was emphasis on the real-world application of measurement and estimation, and the application of mathematics topics in everyday life. For example, from the first stage (grades 1–3), students began to encounter a variety of geometric shapes (cars), observed from different perspectives (front, left/right side, and above) (Kong, Liu & Sun, 2001). The MCSFCE especially increased attention to probability and statistics, reflecting the basic mathematical literacy requirements for citizens in modern society.

The MCSFCE proposed the use of critical-thinking skills, inquiry, and cooperation in mathematics teaching and learning (Zhang, 2008), pointing out that the mathematics learning process is full of observation, experiment, simulation, inference, and other exploratory and challenging activities. One emphasis of the MCSFCE was that textbooks should make connections with other disciplines by incorporating science, social studies, and other relative subjects to teach mathematics. The textbooks should also provide space for student investigations and communication. Accordingly, teachers were urged to use concrete examples and demonstrations to guide students in the learning process and encourage them to communicate ideas via discussions. According to the MCSFCE, teachers should encourage students to think critically and independently. Also, they must recognize individual differences generated by the culture, learning environment, family background, and different thinking styles.

The MCSFCE additionally put forward clear guiding principles for development and evaluation by focusing on the process and different assessment methods, notably recommending that assessment should be used to inform teaching (Kong & Sun, 2001). It also provided recommendations for evaluation according to grade bands. For example, the evaluation schema for grades 1–3 emphasized the assessment of students' mathematics learning processes, mastery of basic knowledge and basic skills, and their ability to identify and solve problems. In particular, it was felt that multiple evaluation methods should be used.

In general, the intensity of this curriculum reform was greater than in all previous reforms. Some innovative features served as an important impetus for the curriculum reform, but this also brought greater challenges and difficulties in its implementation. The central characteristic of the new curriculum was the development of students, which provided the inner motivation for the pursuit of fairness and prompting quality in the new century educational curriculum.

3.2.3 *The Implementation of Standards for Compulsory Education*

Before the release of the MCSFCE, a set of textbooks based on the idea of the new curriculum had been designed by a research group for experimental use (the majority of the members were to part in the development work of the MCSFCE later). Since 1994, this group had conducted two rounds of experiments; more than 60,000 students from more than 10 provinces (including both well-developed school districts to undeveloped school districts) participated, which provided abundant empirical experience for the later implementation of the MCSFCE.

The Ministry of Education started a national curriculum reform conference to convene the implementation of the new curriculum in July 2001. Several decisions were made at the conference. First, the overall objectives and strategies for the implementation of the new curriculum in public schools were determined. Second, the strategies to spread the curriculum reform to all Chinese public schools were developed. Third, professional development and teacher training programs were set up. The positioning of the Trial Version of the curriculum standards necessitated a multi-stage process for spreading the new curriculum. The first stage was to set up the goals, then to conduct preliminary experiments before the nationwide implementation, and finally to broaden the experiment gradually.

In the initial round of experimental implementation of the curriculum, school participants were recruited on a county basis in 2001. First, applications to be volunteer schools were submitted by counties and were examined before being approved by the Ministry of Education. Forty-two regions (3300 elementary schools, 400 secondary schools) participated in the first round of the national curriculum reform with about 270,000 first graders (1% of the population of first graders nationwide) and about 110,000 seventh-grade participants (0.5% of seventh graders) in 2001. Starting in 2002, each province developed a curriculum reform plan at the province level and determined their experimental regions. There were a total of 570 experimental regions with 20% of Chinese first graders and 18% of the seventh graders participating in the new curriculum. Subsequently, more schools from an additional 1072 counties became experimental regions at the province level, bringing in about 40–50% of the student population of each grade. Including the earlier participants in 2001 and 2002, there were 1642 experimental regions with about 35,000,000 students participating in the new curriculum in 2003. Based on the results of these pilot tests, the new curriculum entered the phase of nationwide promotion. By 2004, 90% of the school districts in China were using the new curriculum. As of 2005, except for a few places, the new curriculum had been implemented all over mainland China (Ma, 2009).

3.2.4 The Revision of Mathematics Curriculum Standards for Full-Time Compulsory Education (Trial Version)

Since the implementation of the MCSFCE (Trial Version), the work of developing it has never been interrupted. After the first round (three years) of mathematics curriculum reform, the revision process began. Based on the experience, account was taken of the problems arising from the implementation of the standards, as well as comments from society (including severe criticism from some mathematicians). In May 2005, the Ministry of Education organized the revision group for Mathematics Curriculum Standards for Compulsory Education and officially began the revision process.

There were 14 members in the revision group, from different backgrounds including universities, coaching offices, and primary and secondary schools. About half of them had worked on the design of MCSFCE (Trial Version). Through the process of surveys, situation analysis, and discussions of special issues, the Mathematics Curriculum Standards for Compulsory Education (2011 Version) (MCSCE2011) were finished in 2010 and approved in May 2011. The standards were published officially in December 2011 (Ministry of Education of the People's Republic of China, 2012, p. 34).

MCSCE2011 was developed from the Trial Version; several revisions were made (Zhu, 2012), such as the basic curriculum ideas, curriculum objectives, content standards, and suggestions for curriculum implementation.

The following several paragraphs summarize several aspects of the important revisions, such as the structure, the expression, the concrete content, and suggestions for curriculum implementation (Ministry of Education of the People's Republic of China, 2012, p. 34).

1. For the value of mathematics and the function of mathematics education, MCSCE2011 discussed the research object of mathematics and the relationship between mathematics and human society and then gave the fundamental characteristic of mathematics, which were different from the statement of the Trial Version.
2. MCSCE2011 expanded the 6 core concepts (Sense of Number, Sense of Symbol, Idea of Space, Idea of Statistics, View of Application, and Ability of Inference) into 10 core concepts (add Perceptual Intuition of Geometry, Idea of Modeling, Operations Ability, and changing the Idea of Statistics into View of Data Analysis).
   The new concepts were very important concepts in mathematics education research. For example, the Perceptual Intuition of Geometry was considered as one of the important factors which impacted on the mathematical development of primary and secondary students (Kong & Shi, 2012).
3. For the curriculum objective, MCSCE2011 gave the "Four-Basics": Fundamental Knowledge, Fundamental Skill, Fundamental Idea, and Fundamental Activity Experience from the "Two Basics," Fundamental Knowledge and Fundamental Skill. The Fundamental Idea generally included the Idea of Mathematical Abstraction, the Idea of Mathematical Inference, and the Idea of Mathematical Modeling. Fundamental Activity Experience refers to the individual experience the students gain by experiencing mathematical activities personally.
   The Fundamental Activity Experience was one of the characteristics of MCSCE2011. This issue was considered by Chinese scholars since the 1980s, but did not receive due attention. After the introduction of MCSCE2011, many scholars began to explore this issue (Guo & Shi, 2012).
4. Revisions were made to the concrete contents and the requirements, across all the domains (Shi, Ma & Liu, 2012). The content domains of "Space and Figure" and "Practice Synthetic Application" were revised into "Space and Geometry" and "Synthetic and Practice." The word "Geometry" emphasized the abstraction of concrete figures and space and also explained the general laws behind figures and space. The word "Synthetic" emphasized that an important stage of learning was knowing the relationship between the knowledge and concepts that students learned, and "Practice" was a higher requirement. Some concrete content was omitted, such as the requirements of the trapezoid and position relationship between circles.

With the base established by the implementation of the MCSFCE (Trial Version), MCSCE2011 was implemented at one time. Since the autumn semester, all beginning grades (for primary and middle schools) began to implement the New Curriculum Standards (not only mathematics).

A survey conducted in 2013 indicated that middle school teachers generally approved the MCSCE2011. They thought the teaching objectives were clear, and the amount and difficulty of the content were satisfactory.

Some changes appeared in the high-risk examinations. For example, the entrance examination to high school in Beijing adapted the concrete content and new rubrics were introduced focusing on the Mathematical View, Mathematical Activity Experience, and Mathematical Ability.

Some scholars thought that the issues of assessment, hardware, and the teachers' views were still obstacles to the implementation of the new curriculum (Zhu, 2012). A survey revealed that mathematics teachers generally did not learn to use the MCSCE2011 well.

MCSCE2011 discussed the relationship between plausible and deductive reasoning and the relationship between the real-life world and systems of knowledge. Its objectives highlighted the development of students' creative and application abilities and added the ability to discover and raise problems (Ministry of Education of the People's Republic of China, 2012, p. 84).

The two versions of standards consolidated and perpetuated the achievements of the new century mathematics curriculum reform and played an important role in giving impetus to the healthy and continuous development of mathematics education in China.

3.3 Mathematics Curriculum for High School Education (Grades 10–12)

The mathematics curriculum reform for secondary education was inspired by the same educational beliefs that motivated the compulsory education reform. The core secondary mathematics curriculum committee was formed in June 2000. The committee included 13 members from different backgrounds. They were mathematicians and mathematics educators from universities, mathematics education scholars from research institutes, mathematics teachers, staff members from national testing centers, and delegates from publishing houses (Song & Xu, 2010).

3.3.1 The Development of Standards for High School Education

In the process of standards development, the research team studied mathematics curricula in several developed and developing countries around the world. The study included the following topics: trends of current research in mathematics, current demands on public education, learning in secondary school, international comparison studies, and current teaching and learning in China (Song & Xu, 2010).

The team conducted surveys, interviews, and classroom observations in several provinces. The participants in these initial studies were teachers, students, principals, and guidance officers in secondary schools.

The results indicated that China had accomplished a great deal in mathematics education through cultivating talented students. However, several challenges in secondary mathematics education also confronted the mathematics education community, such as too much emphasis on basic knowledge and basic skills. The learning objectives were focused on three skills (mathematics, logical reasoning, and spatial imagination). Due to the impact of the old mathematics syllabus regulations on learning objectives, teaching materials, examinations, and practical concerns of teachers, the high school mathematics curriculum was characterized as complex, difficult, deep, and narrow. It raised public concerns about limited real-world connections. The curriculum was outdated, with limited electives, and there was no room for modern mathematics teaching strategies. The primary evaluation methods came from test scores and final examinations, which had a significant negative impact on students (Lv & Zhong, 2001).

The research team formalized the reform theory, curriculum objectives, and corresponding high school mathematics curriculum standards based on the research results of previous studies. The development process for the secondary school curriculum standards was similar to that for compulsory education. The research team solicited suggestions from all parties including mathematicians, mathematics education experts, scholars from research institutes, secondary school teachers, and experts from related disciplines such as educational psychology (National High School Mathematics Curriculum Standard Group, 2002). At the same time, the research team conducted several studies, in more than 30 high schools, of some newly added content (such as algorithms) and mathematical investigations (including curriculum design and pilot teaching). These research results provided both evidence and experiences for the later development and revisions of the standards (Song & Xu, 2010, p. 123).

After 30 revisions, the draft version of Mathematics Curriculum Standards for Secondary Education came out at the end of 2002. The final version, MCSSE (Trial Version), was formally published and promulgated in April 2003, after the Ministry of Education completed the document review.

3.3.2 Features of Standards for High School Education

MCSSE was fundamentally different from the curriculum guidelines developed in previous reforms. It shared similar characteristics to the MCSFCE, including the outline of structural changes. It deepened and specified some dimensions (e.g., curriculum content descriptions). MCSSE also included teaching suggestions, teaching materials, suggestions, and recommendations (Ministry of Education of the People's Republic of China, 2003).

MCSSE proposed "student-centered" curriculum ideas, such as cultivating mathematics literacy, increasing active learning, mastering basic knowledge and basic skills, integrating mathematics and information technology, developing critical-thinking skills, developing application and mathematical modeling skills, and the significance and values of a mathematics culture. MCSSE advocated that the high school mathematics curriculum should include a mathematics culture, through which mathematics literacy could be achieved.

MCSSE also advocated a modular structure (36 classes per module), with each module mutually independent, but also with logical connections. The new curriculum offered a variety of selections to meet the needs of individual students. The old curriculum only provided two elective courses at the high school level—mathematics for liberal arts majors and mathematics for science majors. The new curriculum provided more choices. Students needed to take five required modules before the elective courses. There were four elective series, where Series 1 (targeting students majoring in humanities and social science) and Series 2 (targeting students majoring in science, engineering, and economics) were basic elective courses. Students could continue to choose Series 3 or Series 4 after finishing courses in Series 1 or Series 2. Series 3 and Series 4 had a number of topics, with each topic requiring 18 classes. They were designed for students who were interested in mathematics and hoped to learn more. They involved several topics aimed at some important mathematical ideas, scientific value, application of mathematics, and the understanding of a mathematics culture, which reflected some important mathematical ideas, hoping to provide a mathematical base for students' lifelong development. Selective topics in Series 3 included the history of mathematics, information security and passwords, and spherical geometry (six topics). Selective topics in Series 4 included geometric proofs, matrices, and transformations (10 topics).

The intention was to expand these elective topics gradually, with careful monitoring of the quality of these courses. MCSSE encouraged schools to set up certain topics in Series 3 and 4. Schools also had opportunities to enrich and improve various additional elective courses based on the school-based curriculum and faculty resources (18 classes for each credit).

In addition to the new electives (which mostly appeared in Series 3 and 4), the MCSSE also contained several new topics, including orthographic views, spatial coordinates, algorithms, block diagrams, random numbers, and statistics. It also presented this new content using new ways of representation. For example, in three-dimensional geometry, the new textbook took the whole-part approach, rather than the traditional logical approach of point, line, plane, and solid. In terms of geometry objectives, the new textbooks followed a cognitive order from overall perception to the details of point, line, and plane. The new curriculum also presented probability and statistics in the order of statistics, probability and counting techniques, rather than the traditional order of counting techniques, probability and statistics (Cao, 2008, p. 34).

In addition to the curriculum based on mathematical knowledge, the MCSSE designed the series of mathematical exploration, mathematical modeling, and mathematical culture, which was required to be integrated into the regular curricula.

3.3.3 The Implementation of Standards for High School Education

With the promulgation of the MCSSE, the high school curriculum reform entered an experimental deployment stage. The high school curriculum policy was promoted under a step-by-step experimental expansion model. Different from the compulsory education case, the experimental deployment of the high school reform began in large regions such as provinces, self-regulated regions, and municipalities. In fall 2004, four provinces, self-regulated regions, and municipalities became the first experimental zones of the high school curriculum. The curriculum reform received strong criticism and even opposition in 2005, which slowed down the deployment process (Cao, 2005; Wang, 2005; Zhang, 2000). By fall 2012, the high school curriculum had been adopted at entry-grade level in all high schools.

A survey was used to summarize the implementation of MCSSE 10 years after it was published. The sample size was 13 provinces, 446 mathematics teachers, and 5685 students (Lv et al., 2015). The results showed that the implementation of the multi-objective was good. The students' problem-solving and creative-thinking abilities and the ability to collect, clean, and analyze information had increased gradually. As well, some teachers thought that the skills of operation, logical reasoning, and spatial imagining had decreased in varying degrees. The teaching method had changed in a positive direction, but the space left for the students' self-learning was still not enough. The learning method tended to be diverse, but the loading of learning was still heavy. The limitations of the examination system were still obvious, especially for the selective Series 3 and 4. For example, for Series 3, since it was not included in the college entrance examination, 70.6% of teachers reported that their schools had not set this series, and for Series 4, only 6.8% teachers thought their students could select curricula freely. Furthermore, the examination system limited the development of a multi-assessment system.

Some scholars (Li and Ni, 2010) argued that in the implementation of the new curriculum, the content and difficulty, the order of teaching of different modules and the class hours all had some kinds of obstacles which were necessary to overcome.

3.3.4 The Revision of Mathematics Curriculum Standards for High School Education (Trial Version)

With the publication of MCSCE2011, the revision work of the MCSSE was started in November 2014, 10 years after it was first published.

The revision raised a new central concept of "key competencies," which was one of the trends of the international curriculum reform. The model of key competencies was applied to promote the curriculum reform (Xin, Jiang & Liu, 2013). Mathematical key competencies formed the most fundamental component, which decided the main line of the curriculum. The key competencies at high school level included mathematical abstraction, operation, deductive reasoning, mathematical modeling, intuitive imagination, and data analysis.

Based on the existing published literature (Hong et al., 2015), the "Curriculum Plan of High School (Revision)" was based on subjects and did not distinguish the students according to science or the social liberal arts. The requirement for graduation credit was 144, and 88 for the essential curriculum, with no less than 42 for selective Series 1 and no less than 14 for selective Series 2.

It was intended that the new curriculum would include essential series, selective Series 1 and selective Series 2. The essential series consisted of "Preparing Knowledge" (set, logic language, equivalence and inequality, etc.), "Function and Sequence" (the concept of function and the principles, fundamental functions, sequences, and the application of functions), "Vector and Geometry" (solid geometry, two-dimensional vectors, and the application of vectors: solving for triangles), "Statistics and Probability" (random sampling, error modeling, estimation, classical probability, and geometric probability, which emphasized the fundamentals and modernization of the content.

Series 1 included "Function and Derivative" (derivatives and applications, optimizing, inequality), "Vectors and Geometry" (solid vectors and solid geometry, analytical geometry, conics, etc.), and "Statistics and Probability" (counting principles, conditional probability, discrete random variables, Bernoulli model, and linear regression), which emphasizing the fundamentals of the content.

Series 2 was divided into five categories, A, B, C, D, and E. Category A was for students who chose a science direction, including calculus with one variable, three-dimensional geometry, three-dimensional linear algebra, and models of statistics and probability. Category B included calculus, linear algebra, and statistics and probability, which had less content than A, and emphasized application and mathematical modeling. Category C (social science) included logic, social surveying, and mathematical modeling, which emphasized application. Category D was "Beauty and Mathematics," which included mathematics in sport, in music, and in art. Category E was the school-based curriculum, an adaptation of the Advanced Placement Curriculum, including calculus with one variable, integration with one variable, linear algebra, and statistics and probability.

Based on the existing published literature, the revised high school standard was changed a lot from the MCSSE, including the organization of the curriculum, the division between science and liberal arts, and the introduction of AP. The teaching method and other directions were changed according to the change of curriculum.

It needs to be noted that the revision was still ongoing, so systematic revisions still needed to be made after the publication of the final version.

3.4 Contentions with the Mathematics Curriculum Reform

The curriculum reforms of the early twenty-first century led to deep changes in ambitions, curriculum content, teaching methods, textbooks, and assessment methods. These changes had prompted the development of mathematics education in China. As well, both the preparation and deployment processes of the curriculum reform had caused various theoretical and practical contentions in the mathematics education community. All aspects of the curriculum reform were subject to some contention (see, e.g., Cao, 2005; Wang, 2005; Zhang, 2000), especially the requirements for the objectives and content of the curriculum, such as the issues of "Calculation," "Mathematical Systems," "Geometry," and "Uniformity and Diversity."

3.4.1 Contentious Issues with Calculation

During the process of the mathematics curriculum reform, most of the discussion about the calculation area included the Two Basics (basic knowledge and basic skills). Both of these had always been the priorities of the mathematics curriculum in China (Cao, 2005). Chinese students were very good at calculation, routine problem solving, and reasoning but lacked skills in solving authentic and open problems, application, and innovation. That was why the curriculum reform was imperative. Song and his Xu (2010) pointed out that, with the popularity of the calculator and computer, the requirement for mental arithmetic and rough estimation should be weakened. While taking advantage of the modern technology to learn mathematics would become a part of the Two Basics, Cai and Xin suggested that another important slogan for the New Curriculum Reform should be "everyone should learn useful mathematics," which stressed problem-solving ability and the close relationship between mathematics and daily life. As a result, mathematical modeling, and random mathematics captured attention in the New Curriculum Reform and complicated calculations and convoluted processes of reasoning were given less attention. This change conformed with the new trend in education. On the other hand, some voices from the front line of education reflected that, in the calculation teaching progress, teachers were determined to implement the concept of curriculum standards, advocating the diversification of the algorithm. After a phase of implementation, the teachers, to some extent, found that there had been positive developments with respect to emotion, attitudes, and values. But the gaps among students in terms of calculating became larger and larger. This means that the students who were originally active thinkers would perform better, while others who were not quick-minded would feel overwhelmed and lose their own control and judgment when encountering a wide variety of computing methods (Zheng, 2003). It was good for the development of the student's mathematical abilities to reduce the amount of tedious calculations, restrict the level of arithmetic, emphasize number sense, use digital expression and communication, and utilize functions and equations as important models to depict the real world. However, what were the fundamental knowledge and skills that every child should master? What was the actual status of the practice of mathematics education in primary and secondary schools as a result of the basic education innovation? What kind of initiatives could keep the balance between the basic orientation and innovation orientation? These questions required more research and evidence.

3.4.2 Contentious Issues with Mathematical Systems

The discussion about how to construct a mathematical system was not a special case in the New Curriculum Reform. Tracing back to the period of the Cultural Revolution, there were excessive links between mathematics education and production activities. During the next 20 years, there was another kind of trend that laid particular stress on the mathematical knowledge system, regardless of the combination of knowledge and practice. The "Practice and Synthesis Application" domain referred to in the New Curriculum Standards could be explained from four aspects: (a) guiding the students to focus on practice and application; (b) making the abstract mathematical concepts deeply rooted in children's real lives; (c) coming to the internal integration of mathematics; and (d) achieving the cross-disciplinary synthesis between mathematics and other subjects. Research indicated that it was imperative to set up the "Practice and Synthesis Application" domain. This was effective in promoting interest in learning, realizing the value, accumulating the experience, and developing conscious and innovative applications. It was also essential to develop skills of problem solving and posing (Li, 2009). Nevertheless, the traditional mathematics curriculum had been laid out according to a mathematical logic system which overemphasized the academic aspects of mathematics. Although the formalization was the basic feature of mathematics, going too far was as bad as not going far enough (Cao, 2005). Some scholars had questioned the curriculum system erected by the New Curriculum Standards. Jiang (2005) considered that the direction of the New Curriculum Standards deviated from the norm. The current curriculum system was not consistent or systematic, and many of the grassroots teachers were at loose ends. As the most typical basic subject with a strong academic background, the mathematics curriculum setting needed to be "bold hypothesis, careful to confirm" for researchers in dealing with the relationship between the theory and application or the discipline and children. This referred to a series of questions such as "how to look upon the nature of mathematics and mathematics education," "what it meant for mathematics to be systematic," "how to understand mathematics as an education task," and "how to understand the world of mathematics as the vehicle for children's learning," all of which deserved further discussion.

3.4.3 Contentious Issue with Geometry

Geometry was always at the core of the mathematics reforms in primary and secondary schools. Affected by the modernization movement of mathematics education initiated by Bailey and Klein in the twentieth century, significant changes had been made in the New Curriculum Standards to break the Euclidean geometry system, and great importance was attached to the visual geometric (Qi, Sun & Zhang, 2012). In the past, most of the content in the curriculum system was confined to "number, formulae and operations," and "proof of plane geometry." Students could hardly see the whole picture of mathematics, let alone experience the overall process of it. There was a great breakthrough in the new curriculum. The content in transformation geometry, measuring geometry, and experimental geometry had been strengthened; the term "Plane Geometry" had been replaced by "Space and Geometry." Zhong et al. (2003) said that mathematics was not only made up of a heap of boring, tasteless, and profound symbols and formulae. The extreme stringency would ultimately drown the lively mathematical thinking of students in the sea of logic. At the same time, some critics were very cautious for this modification. They believed that it was a kind of "supplanting," many people did not know how to deal with this change. "The plane geometry in the middle school curriculum had been scattered into too many pieces, it was painful," said Jiang (2005). Other daunting challenges still need the efforts of several generations: What kind of educational values occur in plane geometry? Is it possible to achieve the goal of strictly logical proof and rational thinking in plane geometry? How can we develop the rational spirit in teenagers?

3.4.4 Contentious Issues with Unity and Diversity

Unity and diversity have been referred to as a "twice-told story" in the field of philosophy. Under the background of the curriculum reform, there were still many different opinions in clarifying the relationship between unity and diversity in mathematics education. Some views expressed that the teaching philosophy and methods of mathematics teachers had changed to some degree (Huang, 2011). Multiple teaching and learning methods gradually evolved. In many schools, the mathematics teachers used group learning or panel discussions to create active classroom atmospheres and improve student's learning initiatives with satisfactory effects. Huang (Huang 2002) also mentioned that the mathematics curriculum reform should be adapted to meet the demands of the mathematical form and address areas of diversity. The associations with reality were multi-dimensional, and this should be presented and reflected in the same style by means of the mathematics curriculum. Teachers used this kind of association creatively, to form fruitful teaching mechanisms and strategies that would encourage the development of every student. But some scholars held different points of view. In their thoughts, the traditional distinctions between geometric algebra and trigonometry had been replaced by the so-called Integrated Mathematics, which organized the relative course learning content encompassing real life. In spite of being comprehensive and a better reflection of the practical significance of mathematics, "Integrated Mathematics" did not help students master the knowledge they needed appropriately or easily (Zheng, 2007). A further criticism was that there should not be a unified curriculum in our country, because this could result in targeting the middle levels, and people with the real talent might be left out. All of the issues mentioned above required a strategic understanding about the nature of compulsory education, the connotations of curriculum standards, the relationships between mathematics for examinations or mathematics for all, and especially for helping to achieve the basic requirements of education for the majority of students while exploring, protecting, and cultivating students with great talent.

As the high school curriculum reform rolled across the nation, it triggered some controversies in concepts and practices (Chen, 2006; Liu, 2010; Lv & Guo, 2008; Lv, et al., 2015). For example, in one province, only 2.9% of the teachers believed that the concepts and targets of the new curriculum could be realized completely. It was believed that excessive use of the parsing (vector) method in tackling geometry problems would deprive the cultivation of student's spatial concepts (especially in the test-oriented context, where mechanized and efficient problem-solving methods were prevalent). The mathematics culture of the new curriculum lacked sufficient theoretical research foundations. The optional courses were not ideal either; in particular, some content referred to higher mathematics such as matrices and transformations, Euler's formula and the classification of closed surfaces. Furthermore, the assessment reform, higher school administration system, and the curriculum reform were not synchronized.

3.5 Epilogue

All new initiatives are subject to questioning and some rejection in the process of development. In the past 10 years of curriculum reforms, including the Mathematics Curriculum Standards for Full-time Compulsory Education (draft) or Mathematics Curriculum Standards for Compulsory Education (2011 Version), the fundamental research was far from enough. In fact, the existing output of research in primary and secondary school education in the Chinese context was too little to allow for the shaping of a persuasive, rational, and substantial data-based curriculum standard. But this lack of sufficient research was not a reason to delay; certainly, no countries would wait until all questions had been resolved fully before they moved forward (Liu, 2010). It was an exploratory process which needed to be refined and improved continuously. In the very beginning, there was a need for emphatic discussions about the problems to be addressed in this reform and to explore possible solutions even if they were not fully mature. The curriculum was expected to have different functions. As the curriculum promoter, the government needed to participate in the academic arguments. Normative procedures were needed to make sure that the communication was unobstructed and the problems could be solved. Regardless of the decision making or the execution, it needed to be a democratic process (Yu, 2005).

The path of reform was an exploratory process. It was necessary to synthesize theory and practice from mathematics, education, psychology, and many other disciplines, pooling resources from all areas and levels, from the most academically high-achieving to the rural schools. The success of the curriculum reform demands rigorous academic attitudes, national responsibility, and steady work.

References

2005年中国数学会教育工作委员会扩大会议实录 [Transcript of 2005 Mathematics Education Committee of Chinese Mathematics Society Enlarged meeting]. (2005). *Mathematics Bulletin, 4*, 1–12.

Cao, Y. (2005). 义务教育数学课程改革及其争鸣 [The curriculum reform of compulsory education and argument] (pp. 28–30). Special Issue: Mathematics Bulletin.

Cao, Y., Huang, Q., & Ma, B. (2008). 数学教学论 [Theory of mathematical instruction]. Beijing: Higher Education Press.

Chen, W. (2006). 给过分强调用向量方法解立体几何题唱唱反调 [Comment on the phenomena that too much emphasis on vector-method in solid geometry]. *Mathematics Bulletin, 45*(9), 30–32.

Cockcroft, W. H. (Ed.). (1994). 数学算数: 英国学校数学教育调查委员会报告 [Mathematics counts: Report of the committee of inquiry into the teaching of mathematics in schools under the chairmanship] (L. Fan, Trans.) Beijing: People Education Press.

Dong, Y. (2006). 高中数学课程改革的现状及对策 [The present situation of high school mathematics curriculum reform and the countermeasures of high school mathematics teachers]. *Journal of Tianjin Normal University (Elementary Education Edition), 7*(3), 60–64.

Guo, Y., & Shi, N. (2012). 数学基本活动经验:提出、理解与实践 [Basic Mathematics Activities Experiences: Proposing, Understanding and Practicing]. *Journal of the Chinese Society of Education, 4*, 42–45.

Hong, Y., Zhou, J., Wang, S., & Bao, J. (2015). 《普通高中数学课程标准(修订稿)》的意见征询—访谈张奠宙先生 [Advice on ordinary high school mathematics curriculum standards (Revised Draft)—Interview records of Zhang Dianzhou]. *Journal of Mathematics Education, 24*(3), 35–39.

Huang, X. (2002). 数学课程改革的国际视角及启示 [Implication from mathematics curriculum reform in an international perspective]. 课程.教材.教法*[Curriculum, Teaching Material and Method], 6*, 73–75.

Huang, T. (2011). 数学课程改革向何处去——关于基础教育数学课程与教学改革的调查报告 [Whither mathematics curriculum reform—A survey of mathematics teaching and curriculum reform of basic education]. *Journal of Mathematics Education, 20*(3), 12–16.

Jiang, B. (2005). 新课标让数学课失去了什么 [What mathematics loses with new curriculum standard]. *Guangming Daily, 5*.

Kong, Q., & Hu, S. (2002). 数与代数领域加强与削弱的内容介绍——《课程标准》与《大纲》内容比较之—[The characteristics in number and algebra field in new math standards]. *Subject Education, 9*, 1–5.

Kong, F., Liu, X., & Sun, X. (2001). 《义务教育阶段国家数学课程标准》"空间与图形"的特点 [Characteristics of national standards of mathematical courses in the phase of compulsory education about space and figure]. *Journal of Mathematics Education, 10*(3), 58–62.

Kong, F., & Shi, N. (2012). 关于几何直观的含义与表现形式 [Meaning and Form of Geometric Intuition]. *Curriculum Teaching Material and Method, 32*(7), 92–97.

Kong, F., & Sun, X. (2001). 我国义务教育阶段数学课程与教学的评价发展趋势———兼谈《义务教育阶段国家数学课程标准》评价的主要特点 [The trend of mathematics curriculum and instructional assessment of Chinese compulsory education]. *Shandong Education Research, 10,* 17–18.

Li, Q. (2009). 初中数学"实践与综合应用"领域课程研究 [Curriculum Research of the Topic 'Practice and Comprehensive Applications' in Middle School Mathematics] Doctoral dissertation. Chang Chun: East North Normal University.

Li, Q., & Ni, Y. (2010). *Impact of curriculum reform: evidence of change in classroom practice in the Mainland China*, Paper Presented at the Annual Meeting of [21] American Educational Research Association, Denver, USA, April 30–May 4, 2010.

Liu, J. (2009). 课程发展的中国案例—我国新一轮基础教育课程变革历程回顾 [The Chinese case of curriculum development]. *Basic Education Curriculum, 1*(2), 67–73.

Liu, J. (2010). 浅析高中数学课程中的数学文化 [The mathematics culture in the high school mathematics curriculum]. *Mathematics Bulletin, 9,* 10–15.

Lv, S., Cao, C., Jin, X., & Wang, S. (2015). 普通高中数学新课程实施现状研究 [Implementation Status of the New Mathematics Curriculum for Senior High School]. *Journal of Mathematics Education, 24*(3), 6–12.

Lv, S., & Guo, X. (2008). 甘肃省义务教育阶段7 ~ 9 年级数学新课程实施现状调查——从教师视角的研究 [Survey on the new mathematics curriculum implementation of 7–9 grade in Gansu Province]. *Journal of Mathematics Education, 17*(5), 31–35.

Lv, S., & Zhong, Z. (2001). 高中数学课程实施现状研究 [Research on the present situation of mathematical curriculum implementation in senior high school]. *Journal of Mathematics Education, 10*(4), 88–92.

Ma, Y. (2001). 义务教育阶段数学课程改革背景、理念与目标 [The background, ideas and ambition of mathematics curriculum reform of compulsory education]. *Modern Education of Middle and Primary School, 11,* 11–14.

Ma, Y. (2009). 基础教育课程改革:实施进程、特征分析与推进策略 [Compulsory Education Curriculum Reform: The Implementation Characteristic Analysis and Promoting Strategies]. *Curriculum, Teaching Material and Method, 4*, 110–113.

Ministry of Education of the People's Republic of China. (1999). 面向*21*世纪教育振兴行动计划学习参考资料 [The learning material of the 21st century revival action plan in education]. Beijing: Beijing Normal University Press.

Ministry of Education of the People's Republic of China. (2001). 全日制义务教育数学课程标准*(*试验稿*)* [Mathematics curriculum standards for full-time compulsory education (trial version)]. Beijing: Beijing Normal University Press.

Ministry of Education of the People's Republic of China. (2003). 普通高中课程方案*(*实验*)* [The high school curriculum plan (experimental version)]. Beijing: People Education Press.

Ministry of Education of the People's Republic of China. (2012). 义务教育数学课程标准 *(2011* 年版*)*解读 [Explaining the Mathematics Curriculum Standard for Compulsory Education (2011)]. Beijing: Beijing Normal University Press.

National Council of Teachers of Mathematics. (1989). *Curriculum and evaluation standards for school mathematics*. Reston, VA: National Council of Teachers of Mathematics.

National Council of Teachers of Mathematics. (2000). *Principles and standards for school mathematics*. Reston, VA: National Council of Teachers of Mathematics.

National High School Mathematics Curriculum Standard Group. (2002). 《高中数学课程标准》的框架设想 [The envisaged framework of high school mathematics curriculum standard]. *Curriculum, Teaching Material and Method, 4,* 1–7.

Qi, C., Sun, X., & Zhang, D. (Eds.). (2012). 数学课程论与数学课程教材改革 [Mathematics Curriculum Theory and Mathematics Curriculum Textbooks Reform]. Beijing: Beijing Normal University Press.

Research Group of National Mathematics Curriculum Standards, Zhang, C. & Liu, J. (1999). 集思广益话改革——关于"研制国家数学课程标准"来稿汇总与分析 [Talks about the reform–analysis of collected manuscripts with regard to Developing National Mathematics Curriculum Standard]. *Journal of Educational Studies, 12,* 1–5.

Shi, N., Ma, Y., & Liu, X. (2012). 义务教育数学课程标准修订过程与主要内容 [The revision process and main contents of mathematics curriculum standards for compulsory education]. *Curriculum, Teaching Material and Method, 3,* 50–56.

Song, N., & Xu, B. (Eds.). (2010). 数学课程导论 [Theory of mathematics curriculum]. Beijing: Beijing Normal University Press.

The Research Group of Mathematics Curriculum Standard. (1999). 关于我国数学课程标准研制的初步设想 [The initial ideas of mathematics curriculum standards]. *Curriculum, Teaching Material and Method, 5,* 17–21.

The Research Group of Mathematics Curriculum Standard. (2002). 全日制义务教育数学课程标准(实验稿)解读 [Analysis on mathematics curriculum standards for full-time compulsory education (trial version)]. Beijing: Beijing Normal University Press.

Wang, K. (2005). 甘肃天水地区部分中学教师座谈会记录 [Transcripts on symposium with some middle school teachers]. *Mathematics Bulletin, 3,* 4.

Xin, T., Jiang, Y., & Liu, X. (2013). 我国义务教育阶段学生核心素养模型的构建 [Model construction of key competencies of students in compulsory education period]. *Academic Journal of Beijing Normal University, 1*(1), 5–11.

Yu, H. (2005). 课程改革的路必须走下去 [The curriculum reform must go on]. *People Education, 24,* 21–24.

Zhang, D. (2000). 中国数学会中小学数学教育改革研讨会记录 [Transcripts of mathematics education reform seminar of Chinese mathematical society]. *Mathematics Bulletin, 44*(Special Issue), 1–11.

Zhang, D. (2008). 中国数学教育在改革与反思中前进 [Chinese mathematics education is developing with the background of reform and reflection]. *Mathematics Bulletin, 12,* 22–26.

Zhang, D., Li, S., & Tang, R. (2005). 中国大陆的"双基"数学教学 [The "Two basics:" Mathematics teaching and learning in Mainland China]. In L. Fan, N. Wong, J. Cai, & S. Li (Eds.), *How Chinese learn mathematics (Chinese version)* (pp. 189–207). Nanjing: Jiangsu Educational Press.

Zhang, D., & Song, N. (2004). 数学教育概论 *[The introduction of mathematics education].* Beijing: Higher Education Press.

Zheng, Y. (2003). 简论数学课程改革的活动化、个性化、生活化取向 [Activities, Personality and Life: Orientation of Mathematics Curriculum Reform]. 教育研究 [Educational Research]. 281(6), 90–94.

Zheng, Y. (2007). 从课程改革看数学教育理论研究. [Reflections on the Research in Mathematics Education on the basis of Mathematics Education Reform]. *Journal of Mathematics Educatio, 16*(1), 40–43.

Zhong, Z., Sun, M., & Liu, K. (2003). 转变教育观念 正确理解新数学课程标准 [Partially pursue to determinacy is wrong trend to understand new mathematics curriculum criterion]. *Journal of Mathematics Education, 12*(3), 95–99.

Zhu, L. (2012). 《义务教育数学课程标准 (2011年版)》修订了什么 [What has the "mathematics curriculum standard for compulsory education (2011)" Revised]. *Journal of Mathematics Education, 21*(3), 7–10.

Chapter 4
Education Equity in China: An Analysis of Local and Migrant Students' Mathematics Learning in Shanghai

Yan Zhu

Abstract As a big "melting pot" in China, Shanghai has become one of the most popular cities for migrants. Shanghai government has tried several measures to help migrant students to be well integrated into the local education system. Through a secondary analysis of the PISA 2012 Shanghai-China data, this study compared the mathematics attainments in both cognitive and noncognitive aspects of students with different migrant backgrounds so as to provide some evaluation of the effect of the local government's measures on its integration progress from a perspective of educational equality. While second-generation students performed similarly well to the local students, the first-generation students performed significantly poor. The variations related to migrant backgrounds were larger at the junior secondary level than the senior level. The differences in students' perceptions of themselves as mathematics learners related to migrant backgrounds were much smaller than those in academic performance. Discussions and interpretations of the results are also included.

4.1 Equity in Education

Equity in education is an important concern for most countries, regardless of whether they are developed or in the process of developing. It is believed that unequal education implies that human potential is being wasted. According to Grubb, Field, Jahr and Neumüller (2005), the equity issue is both a social and an individual problem. They suggested that, from a social standpoint, a large number of under-educated individuals fails to contribute to national prosperity, which can generate social costs, either directly through welfare costs or through their indirect impact on social problems; from an individual standpoint, a lack of adequate schooling and school-based competencies usually leads to lower earnings, higher levels of unemployment, and the many correlates of poor economic conditions. Moreover, researchers have argued that educational inequality and its many con-

---

Y. Zhu (✉)
East China Normal University, Shanghai, China
e-mail: yzhu@kcx.ecnu.edu.cn

Y. Cao and F.K.S. Leung (eds.), *The 21st Century Mathematics Education in China*, New Frontiers of Educational Research,
https://doi.org/10.1007/978-3-662-55781-5_4

sequences are almost never completely random. It usually affects some groups more than others. In this sense, group inequality can then be more serious than an inequality that can be attributed to random elements or to individual attributes which are thought to be distributed randomly within a population.

In China, the elimination of educational inequality is an elusive goal. One important issue of inequity that has often caught public and media attention is the education of the children of migrant workers. These children either accompany their parents to their new urban homes or stay with other family members in their rural areas (the "Left Behind Children Syndrome"). In the former case, there is a struggle for the children to find places in urban public schools, where they are often denied due to their official rural household status; in the latter one, the children face potential neglect along with a whole range of developmental and sociological problems (Li & Wang, 2014). In China, there are around 30 million children of school age belonging to migrant families, which is 20% of the entire student population at the basic education level; that is to say, one in every five school children come from migrant families (OECD, 2011).

The massive movement of people from rural-to-urban areas has caused great pressure on the urban education systems, due to the need to integrate rural children into the urban mainstream (Postiglione, 2015). Werwath (2011) remarked on the massive rural-to-urban migration phenomenon in China—the greatest that the world has ever seen. As one consequence, in 2004, the term "education equity" appeared for the first time in the text of a central government document. Moreover, educational and social problems associated with these children have also become a major issue that China's government pledges to tackle in its 2020 Education Plan.

4.1.1 Migrant Education in Shanghai: **Achieved Progress**

As a metropolis, Shanghai had 9.90 million migrant residents and 14.25 million local residents by the end of 2013 (Shanghai Statistics Bureau, 2014). While the size of the floating population in Shanghai did not change significantly during this period, the number of school-age children increased greatly. According to Yin (2013), more than 70% of the migrant population were migrant workers who had brought more than 400,000 children with them. It is well known that integrating migrant students into the receiving education system is a matter of central concern to many places worldwide. Shanghai recognizes the importance of making sure that this large minority of students gets a quality education (NCEE, no date).

Chen and Feng (2014) commented that, of all the major migrant-receiving cities, Shanghai is probably one of the most accommodating in terms of meeting migrant children's education needs. It is believed that this could be related to the fact that Shanghai's spectacular economic growth can be very much attributed to the contributions of migrant workers, so their children should be treated well (OECD, 2011). Moreover, Shanghai has also established the notion that migrant children are "our children" and work constructively to include them into its educational development.

Followed by the national "Two Mainly" policy (i.e. the education of migrant children is mainly the responsibility of the recipient region and migrant children should be offered compulsory education mainly in public schools) (Tucker, 2014), Shanghai's local government launched a "three-year action plan for the education of migrant children" in 2008, which aims to open up public schools further to migrant children and to subsidize migrant schools. By 2011, all migrant schools in Shanghai's central districts had been shut down and migrant students in these districts were transferred to public schools. Between 2006 and 2010, the municipal government investigated nearly 10 billion yuan (US$1.64 billion) in school infrastructure in the suburbs to meet the increasing demands of children from migrant and relocated families. In February 2010, Shanghai further declared that it would become the first city in China to provide nine years of free compulsory education to migrant children. According to government statistics, in 2010 97.3% of Shanghai's 400,000 children of migrant workers were receiving free education, either in regular public schools, public schools dedicated solely to migrants, or special private ("minban") schools for migrants (Steele, 2010).

4.1.2 *Migrant Education in Shanghai:* Continuing Challenges

Types of schools attended by migrant children. Overall, about 70% of migrant children are enrolled in public schools in Shanghai. However, the participation rate of these children in elite primary schools is relatively low. According to Li, the corresponding percentages in eastern, middle and western areas were 6.3, 6.9, and 12.9%, while the percentages for private schools attended by migrant children were 23.5, 29.5, and 2.6%, respectively (OECD, 2013a). In fact, the entry to such schools is usually dominated by parents in professional occupations and officials (Wu, 2009). These schools, formerly known as "key schools", often receive extra funding and have better teachers. Even though Shanghai has abolished the system of key primary schools, some high-quality primary schools still set entrance examinations as a normal practice (Liang, 2012).

Accessibility beyond compulsory education. While almost all migrant children in Shanghai can go to primary school (either a public or a migrant one), not all of them can go to local middle schools. This is related to the fact that migrant middle schools are not allowed to operate, and there are not enough public middle schools. Consequently, migrant youths may only choose to attend vocational secondary schools. The Shanghai Education Committee justified local secondary schools' refusal to admit migrant children on the grounds that "If we were to open the door to them, it would be difficult to shut it in the future; local education resources should not be freely allocated to immigrant children" (Ren, 2012).

Since 2013, Shanghai has implemented a new residence permit management approach. It allows the qualified migrant students[1] without Shanghai *hukou* (which is a local household registration) to go to senior secondary schools, follow the Shanghai curriculum, and take Shanghai university entrance examinations. Meanwhile, the majority who are unqualified for the permit application will still be forced to return to their home regions to continue their senior secondary education and then take the university entrance examinations in their home regions.

Receiving different curricula. Shanghai has its own curriculum and examinations, and they are different from the national ones which are employed by many other regions in the country. Correspondingly, regardless of the types of schools in which they are studying, migrant students usually do not receive the same standard of education as their Shanghainese peers. In other words, they are often considered to be outside the system and grouped in different classrooms due to the differing education standards. Moreover, most migrant families are unable to afford extra tuition required to keep up with their Shanghainese peers.

Conditions of private schools for migrants. Private schools for migrant students still have funding less than half of that received by schools for Shanghai local students. Class sizes are typically much larger than schools for locals, and teachers have heavier teaching schedules. Ding's (2004) survey of 59 migrant schools in Shanghai reported that 78.3% of the teachers in such schools had monthly incomes of 700 yuan or less, which was far below the average monthly income for local office workers (2815 yuan in 2004, City of Shanghai, 2005). It was found that many of these teachers used migrant schools as a stepping stone; that is, once they found better jobs, they would quit teaching. Correspondingly, the teachers who chose to stay may not have had the necessary teaching experience or qualifications. A survey conducted by Chen and Feng (2012) showed that teachers in state schools had much better conditions than those in migrant schools, in terms of teaching experience (years), teachers' education, and their monthly salaries.

4.2 Research Questions and Purposes

As one comprehensive study of students' schooling experiences, including both cognitive and non-cognitive aspects, the OECD has conducted surveys with one domain as a primary focus every three years since 2000; this is known as PISA. In 2012, the fifth round of this survey was conducted with mathematical literacy as the major domain. In the study, mathematics literacy was defined as "an individual's capacity to formulate, employ, and interpret mathematics in a variety of contexts" (OECD, 2013b, p. 25). In this sense, PISA offers an invaluable opportunity to map

[1]These students' parents usually have a sufficiently high education status and income with a stable work unit so as to be able to apply for Shanghai residence permit. The permit allows their child to remain in Shanghai for their senior secondary school education.

the skills and competencies of young people as they enter the job market for the first time (Borgonovi & Jakubowski, 2011).

This PISA data also offer a great opportunity to investigate differences among students with various migration backgrounds in Shanghai. The purpose of the current study was to conduct a closer investigation of the role of migration backgrounds on Shanghai students' mathematics attainment from both cognitive and non-cognitive perspectives. In particular, this study aimed to answer two sets of research questions via a secondary analysis of the PISA 2012 Shanghai-China data:

(1) Do Shanghai 15-year-old local students perform differently from their migrant peers in mathematics? Do the two groups of students assess themselves differently as mathematics learners?
(2) Do Shanghai 15-year-old second-generation migrant students perform differently from their first-generation migrant peers in mathematics? Do the two migrant group students assess themselves as differently as mathematics learners?

By answering the two sets of research questions, the intention of this investigation was to paint a clear picture about education equity in Shanghai–China from a perspective of mathematics education in relation to students' migration backgrounds. As indicated by the research questions, this study was not only looked into students' cognitive attainments, but also their non-cognitive attainments from their school learning experiences. Furthermore, the comparison between locals and migrants aimed to reveal how well migrant children adjust to a new culture and social situation, while the comparison between the two migrant groups was made to provide insights into the extent to which different school systems succeed in supporting migrant students' learning.

4.3 Research Methods

4.3.1 Data Source

The data for the present analysis were taken from the Shanghai-China sample of PISA 2012. The PISA samples were obtained through a two-stage stratified random sampling approach, with schools as the primary sample units. Once schools were selected at random, students (at the age of 15) were then drawn randomly from each sample school (Adams & Wu, 2002; Ma, Ma, & Bradley, 2008). The Shanghai-China sample, retrieved from the PISA official website, contained 5177 students from 154 schools. Both the final sampling weights and replicate weights for students were used in this study to make the sample reflective of the population.

4.3.2 Measures

The PISA study used student and school questionnaires to collect background data and to construct indicators of social, cultural, economic, and education determinants of student achievement (UNESCO, 2003). Three sets of questions from the student questionnaires were used to identify the students' migration backgrounds in the present analysis, including their birthplaces (COBN_S) as well as those of their parents (COBN_F and COBN_M). The types of schools attended (STRATUM) and grade levels located (ST01Q01) were also included in the present analysis.

To assess 15-year-olds' literacy in reading, mathematics, and science, PISA 2012 assigned each student randomly with one of 22 rotated booklets containing questions about one or more of the three testing domains; all the booklets included mathematics materials. In this sense, each student only attended to a part of the assessment item pool so that the raw test scores could not be compared across students. To resolve the problem, a multiple imputation approach was used to estimate the unobservable latent achievement for all students. As a result, five plausible values (PV1MATH to PV5MATH) were produced for each student for their overall mathematics performances, and all of these values were used in the present study to calculate parameter estimates.

To measure students' perceptions about themselves as mathematics learners, three sets of items reflecting their mathematics self-beliefs were included in this analysis: mathematics self-efficacy (ST37Q01 to ST37Q08), mathematics anxiety (ST42Q01, ST42Q03, ST42Q05, ST42Q08, ST42Q10), and mathematics self-concept (ST42Q02, ST42Q04, ST42Q06, ST42Q07, ST42Q09). PISA 2012 defined students' mathematics self-efficacy as the extent to which they believe in their own ability to solve specific mathematics tasks; for instance, they were asked how confident they were when solving an equation like $3x + 5 = 17$ (ST37Q05), ranging from "1: very confident" to "4: not at all confident". Anxiety towards mathematics was about students' feelings of helplessness and stress when dealing with mathematics, such as "I get very nervous when doing mathematics problems" (ST47Q05). Mathematics self-concept referred to their beliefs in their own mathematics abilities (e.g. ST42Q07: "I have always believed that mathematics is one of my best subjects"). For the latter two sets of items, the students were asked to report their agreement, on a scale ranging from "strongly agree" (1) to "strongly disagree" (4).

Based on students' responses to these three sets of items, PISA further constructed three indices pertaining to the three non-cognitive notions related to mathematics learning (i.e. MATHEFF: math self-efficacy, ANXMAT: math anxiety, and SCMAT: math self-concept). All of these indices were standardized to have a mean of 0 and a standard deviation of 1 across the participating education systems. Moreover, the standardization procedure applied to the indices meant that positive values on the index indicated students who had more positive perceptions than an average student, while negative values indicated less positive perceptions. In the study, both the individual items and constructed indices were used for the comparisons of students' non-cognitive attainments with different migration backgrounds.

4.3.3 Data Process and Analysis

Following PISA's definition of students' immigrant status, this study classified Shanghai students' migrant status into three categories: (1) local students (those with at least one parent born in Shanghai regardless of their own birthplaces), (2) second-generation migrant students (those born in Shanghai but whose parents were not born there), and (3) first-generation migrant students (those born outside Shanghai and whose parents were also born outside Shanghai).

There were five types of schools in Shanghai for 15-year-old students defined in the PISA student questionnaire. In this analysis, they were first regrouped into two categories (i.e. academic schools vs. vocational schools). Within upper secondary schools (academic type), two sub-categories were further created (i.e. ordinary schools vs. model schools). In addition, students were asked to identify their grade levels, ranging from 7th grade to 12th grade, which were used to regroup them into either lower secondary or upper secondary groups so as to differentiate these in the compulsory education stage.

To explore the possible differences in students' cognitive and non-cognitive attainments related to mathematics learning and to their migrant backgrounds, a series of t-tests (on locals and migrants) and ANOVA analyses (of locals, first-generation, and second-generation) was carried out. When significant differences were revealed in the ANOVA analyses, post hoc analyses were then conducted. When variances in group distributions occurred, chi-square tests were used. For all the tests, effect sizes[2] were reported whenever the differences were found at a statistically significant level.

4.4 Results and Findings

According to the PISA 2012 Shanghai-China data, about 26.9% of 15-year-olds had some migrant backgrounds, including 17.8% first-generation and 9.1% second-generation (see Table 4.1). This overall figure is very close to 27.7% from the national population census representing the proportion of residents without Shanghai *hukou*. In addition, it can be known that the majority of the migrant students (66.2%) belonged to the first-generation, which means that both the students and their parents were not born in Shanghai.

From Table 4.1, one can see that, when moving from junior secondary to senior secondary, the percentages of students with migrant backgrounds dropped by 10%.

[2]Effect size for a *t*-test and post hoc test in ANOVA is expressed by Cohen's d (small: $d = 0.20$, medium: $d = 0.50$, large: $d = 0.80$), when a significant t or F is detected, whereas effect size for an ANOVA is express by ω^2 (small: $\omega^2 = 0.01$, medium: $\omega^2 = 0.06$, large: $\omega^2 = 0.14$) with a threshold ω^2 being 0.03. Effect size for a Chi-square test is expressed by Cramer's V (small: $V = 0.10$, medium: $V = 0.30$, large: $V = 0.50$).

Table 4.1 Percentages of students with different migrant backgrounds by school levels and streams

	Junior secondary (%)	Senior secondary			Overall (%)
		Academic (%)	Vocational (%)	Both (%)	
Native	67.7	79.0	75.0	77.5	73.1
First-generation	23.4	11.2	16.8	13.3	17.8
Second-generation	8.9	9.7	8.2	9.2	9.1

At the senior secondary level, more first-generation migrant students opted to study in the vocational stream. In fact, about 47.3% of the first-generation migrant senior secondary students studied in vocational schools and the corresponding proportions of the other two types of students were 36.3% (local) and 33.5% (first-generation).

4.4.1 Students' Mathematics Achievement with Different Migration Backgrounds

It was found that native Shanghai students ($M_{\text{native}} = 621.33$, $\text{SD}_{\text{native}} = 93.71$) performed significantly better than new ones ($M_{\text{migrant}} = 592.64$, $\text{SD}_{\text{migrant}} = 113.30$), t (83569) = 37.001, $p < 0.001$, $d = 0.29$. Further investigation found that the difference actually existed between first-generation migrant students and the other two groups (see Table 4.2). In particular, the native students scored 44.13 points higher than their first-generation peers with the magnitude of the difference approaching medium ($d = 0.45$). Similarly, the second-generation migrant students scored 45.87 points higher than their first-generation peers ($d = 0.41$). Although the second-generation migrant students scored 1.74 points higher than the local students, the former group's score distribution was more dispersed, particularly given that the size of the local group was about 8 times that of the second-generation. Moreover, only a small proportion of the local Shanghai students was born outside Shanghai (3.5%), and an additional comparison shows that they had similar performances as their locally-born peers, with a 3.33 points difference ($t[60941] = -1.617, p = 0.11$).

It is known that PISA used an age-based (i.e. 15-year-olds) rather than a grade-based definition for its target population; that is, a definition that is not tied to the institutional structures of national education systems. The data showed that there were 39.6% of Shanghai 15-year-olds studying at Ninth Grade and 54.2% at Tenth Grade.

As argued earlier, migrant students may be exposed to more discrimination when they move beyond compulsory education. That is, many of them cannot stay on in Shanghai after completing compulsory education, and those remaining in senior secondary school either studied at vocational schools or had good family backgrounds (e.g. parents with higher education and stable incomes). As expected, greater variance was observed at the junior secondary level (F[2,

Table 4.2 Shanghai students' mathematics achievement in PISA 2012 by their migrant backgrounds

	Junior secondary	Senior secondary			Overall
		Academic	Vocational	Both	
Native	605.10 (89.17)	680.69 (68.69)	548.77 (74.61)	632.84 (95.13)	621.33 (93.71)
First-generation	553.82 (106.63)	696.14 (64.75)	515.31 (77.67)	610.58 (114.96)	577.20 (113.63)
Second-generation	596.18 (98.58)	698.75 (69.86)	536.43 (84.57)	644.40 (107.29)	623.07 (106.27)

Note Students' mathematics achievement was reported with mean and standard deviation (which in the parentheses)

37475] = 964.640, $p < 0.001$, $\omega^2 = 0.05$) and the between-group difference at the senior secondary level was trivial ($F[2, 46090] = 174.874$, $p < 0.001$, $\omega^2 = 0.01$). Furthermore, in all the three groups, senior secondary students demonstrated significantly better performances than their junior fellows and the corresponding magnitudes were much larger within the two migrant groups ($d_{\text{first}} = 0.47$ and $d_{\text{second}} = 0.52$) than the local one ($d_{\text{native}} = 0.30$).

Further, separate comparisons were conducted for the academic and vocational streams at the senior secondary level. The difference in the vocational stream was approaching threshold ($F[2, 17275] = 234.325$, $\omega^2 = 0.026$), whereas the difference in the academic stream was negligible ($F[2, 28812] = 142.870$, $\omega^2 = 0.01$). An interestingly opposite pattern was observed between the two different streams; that is, in the academic stream, the two migrant groups performed better than the local group, while a reverse relationship was found in the vocational stream (see Table 4.2). Furthermore, in the vocational stream, the magnitude of the difference between local students and first-generation migrants was approaching medium ($d = 0.45$), while the difference between local students and the second-generation migrants was quite trivial ($d = 0.16$).

In Shanghai, at the senior secondary level, there are model schools and ordinary schools for the academic stream. It is generally believed that the quality of model schools is better than that of ordinary schools. As senior secondary education is beyond compulsory education, students usually need to take a locally administered Senior Secondary School Entrance Examination (commonly known as *Zhongkao*), and their performances in the examination will have an important impact on which types of schools (e.g. academic vs. vocational, model vs. ordinary) they are eligible to enter. Within the academic stream, the model schools usually require a higher entrance score. A chi-square test on the distribution of the students with different migrant backgrounds in the two types of schools showed a negligible difference ($\chi^2[2, N = 28815] = 145.556$, $p < 0.001$, Cramer's $V = 0.07$). It can be seen that nearly 40% of native senior secondary students (academic stream) were studying in model schools, whereas the corresponding percentages for the two migrant groups were slightly higher (first-generation: 46%; second-generation: 49%).

It is interesting to observe that, for the students studying in model schools, second-generation migrant students performed the best ($M = 738.74$, SD = 62.27) followed by first-generation migrant students ($M = 719.64$, SD = 58.84) and then native students ($M = 715.12$, SD = 63.28). An ANOVA test showed that the between-group difference was negligible ($F[2, 11821] = 81.059$, $p < 0.001$, $\omega^2 = 0.01$). Similarly, for the students studying in ordinary senior secondary schools, migrant students (first-generation: $M = 673.55$, SD = 62.03; second-generation: $M = 664.39$, SD = 56.38) again performed better than the locals ($M = 658.44$, SD = 62.58), although with trivial differences ($F[2, 16988] = 47.125$, $p < 0.001$, $\omega^2 < 0.01$). One important fact should be taken into account when interpreting these comparison results; that is, the population size was much larger for the local group than for the two migrant groups. In addition, the standard deviations showed that the dispersions of students' performances by different migration backgrounds in the two types of schools were similar. It was expected that great differences would exist between students from model schools and those from ordinary schools, and this was supported by the data and also applied to all three student groups. In addition, the largest difference was observed within the second-generation group ($d = 1.26$).

4.4.2 Mathematics Self-efficacy of Students from Different Migrant Backgrounds

The comparisons showed that, regardless of the problem contexts, the local students were more confident than their migrant peers, even though the magnitudes of the differences were all trivial or small. The largest difference was observed on the item about "solving an equation like $2(x + 3) = (x + 3)(x - 3)$" (ST37Q07: $d = 0.32$), and the smallest was on the one about "understanding graphs presented in newspapers" (ST37Q04: $d = 0.12$). Due to the migrant group including both first-generation and second-generation students, the dispersal of these students' mathematics self-efficacy was consistently greater than that of the local students' as expected, although the combined size of the two migrant groups was smaller than that of the local group. The local–migrant difference on the constructed index of math self-efficacy (MATHEFF) was small, $t(23447) = 21.316$, $p < 0.001$, $d = 0.21$.

An analysis with a separation of the two migrant groups showed that the second-generation group was at a similar confidence level as for their local peers. On a four-point Likert scale, the absolute differences between the two groups ranged from 0.01 to 0.05. Out of the eight items, the second-generation group showed slightly more confidence on five items. In other words, the observed differences between local and migrant students came mainly from the first-generation group.

A further investigation of students at different school levels (see Table 4.3) revealed that the differences in mathematics self-efficacy among the student groups

Table 4.3 Shanghai students' mathematics self-efficacy in PISA 2012 by their migrant backgrounds

	Junior secondary	Senior secondary			Overall
		Academic	Vocational	Both	
Native	1.03 (1.04)	1.32 (0.99)	0.38 (0.97)	0.98 (1.08)	1.00 (1.07)
First-generation	0.59 (1.11)	1.48 (0.88)	−0.13 (1.03)	0.71 (1.25)	0.64 (1.17)
Second-generation	0.93 (1.15)	1.48 (0.80)	0.27 (0.94)	1.07 (1.02)	1.01 (1.10)

Note Students' mathematics self-efficacy (MATHEFF) was reported with mean and standard deviation (which in the parentheses)

were greater at the junior level, with a magnitude approaching threshold ($F[2, 25054] = 349.603$, $p < 0.001$, $\omega^2 = 0.027$), while the differences at the higher school level were negligible ($F[2, 30501] = 115.802$, $p < 0.001$, $\omega^2 < 0.01$). At the junior secondary level, the analysis showed that local students consistently had the highest confidence levels on all of the relevant items and the magnitudes of the differences among three migrant groups on three items (i.e. ST37Q05: "solving an equation like $3x + 5=17$", ST37Q06: "understanding the actual distance between two places on a map with a 1:10 000 scale", ST37Q07) exceeded thresholds. At the senior secondary level, a different pattern was observed with second-generation students who reported a higher level of mathematics self-efficacy than the other two groups on the majority of the items. However, at this school level, the magnitudes of the differences were all small.

A comparison of senior secondary students' mathematics self-efficacy in the academic and vocational streams indicated that the students on the academic track were significantly more confident than those on the vocational track in solving mathematics problems. In particular, on the constructed index (i.e. MATHEFF), the academic–vocational difference was large, $t(22445) = 70.275$, $p < 0.001$, $d = 0.83$. When looking into the differences related to students' migrant backgrounds at each stream type, the analysis revealed that students on the academic track were at a similar mathematics self-efficacy level and the two migrant groups were slightly more confident than the local group in solving all various mathematics problems (i.e. all the eight items). However, on the vocational track, the first-generation group was significantly less confident, particularly on three items (ST37Q05: $\omega^2 = 0.05$; ST37Q06: $\omega^2 = 0.03$; ST37Q07: $\omega^2 = 0.09$).

This was consistent with the expectation that the students studying in model schools showed a higher level of math self-efficacy than those studying in ordinary schools; the difference on the constructed index (i.e. MATHEFF) was approaching the threshold ($\omega^2 = 0.028$). Within each type of school, the analysis only found negligible differences related to students' migrant backgrounds.

4.4.3 *Mathematics Self-concepts of Students from Different Migrant Backgrounds*

Compared to mathematics self-efficacy, which pertains to specific mathematical problem contexts, mathematics self-concept is a more general perception that is about students' self-evaluation of their own abilities. While students with different migrant backgrounds showed some differences in their mathematics self-efficacy, these students, whether with or without migrant backgrounds, gave themselves similar evaluations of their mathematics ability ($M_{\text{native}} = -0.06$ vs. $M_{\text{migrant}} = -0.02$). Compared to the second-generation group, the local students' mathematics self-concept was closer to that of the first-generation students. In fact, on all the items and the constructed index, the second-generation group gave the highest rating (see Table 4.4).

Although the differences related to school levels were negligible [$t(55153) = 16.633$, $p < 0.001$, $d = 0.14$], it appears that junior secondary students evaluated their own mathematics ability higher than did their senior peers. Within each school level, the second-generation group consistently reported a higher level of mathematics self-concept than the other two groups, while the overall differences across the three groups were trivial.

The differences between students on different tracks were also small [$t(25908) = 17.187, p < 0.001$. $d = 0.20$], while those in the academic stream valued their own abilities slightly higher than those in the vocational stream ($M_{\text{academic}} = -0.04$ vs. $M_{\text{vocational}} = -0.21$). Further analysis within each stream showed that students with different migrant backgrounds did not have great differences in their perceptions of their ability. In this aspect, students with migrant backgrounds in the academic stream had a closer view than the local ones, while for the students on the vocational track, local students had a closer view with their second-generation peers in most cases.

Students from both model schools and ordinary schools also had similar views about their own mathematics abilities [$t(17740) = 11.621$, $p < 0.001$, $d = 0.17$] with the evaluation levels of students from model schools being slightly higher than

Table 4.4 Shanghai students' mathematics self-concept in PISA 2012 by their migrant backgrounds

	Junior secondary	Senior secondary			Overall
		Academic	Vocational	Both	
Native	0.02 (0.82)	−0.08 (0.84)	−0.17 (0.78)	−0.11 (0.82)	−0.06 (0.82)
First-generation	0.02 (0.87)	0.08 (0.98)	−0.37 (0.90)	−0.12 (0.97)	−0.04 (0.92)
Second-generation	0.08 (0.87)	0.13 (0.99)	−0.24 (0.84)	0.01 (0.96)	0.04 (0.92)

Note. Students' mathematics self-concept (SCMAT) was reported with mean and standard deviation (which in the parentheses)

those from ordinary schools ($M_{model} = 0.05$ vs. $M_{ordinary} = -0.10$). While students with different migrant backgrounds gave themselves similar levels of evaluation of their mathematics abilities in both types of schools, the analysis showed that the differences were more observable in the model schools. Moreover, on the majority of the items, the two migrant groups of students had a closer evaluation levels than the local one, though all the differences were negligible or small.

4.4.4 *Mathematics Anxiety for Students from Different Migrant Backgrounds*

The PISA data revealed that the local and the migrant students' overall mathematics anxiety (i.e. ANXMAT) were nearly at the same level, $t(26017) = 4.6$, $p < 0.001$, $d = 0.04$. In particular, it appears that the local students might have perceived themselves to get slightly more nervous about doing mathematics problems ($M_{native} = 2.83$ vs. $M_{migrant} = 2.86$) and also more worried about getting poorer grades in mathematics than their migrant peers ($M_{native} = 2.09$ vs. $M_{migrant} = 2.15$).

Looking further into the two migrant groups, the study found that the second-generation was the most anxious of the three groups when dealing with mathematics. However, when concerned with getting poor mathematics grades (ST42Q10), the local group ($M_{native} = 2.09$) became more worried than the two migrant groups ($M_{first} = 2.14$ and $M_{second} = 2.19$). Compared to the second-generation group, the other two groups had closer anxiety levels (see Table 4.5). In particular, on the constructed index (i.e. ANXMAT), the second-generation group's rating was about 0.13 standard deviation lower than the other two groups' ratings (see Table 4.5).

Consistent with the expectation, the junior students were less anxious than their senior peers about mathematics learning and performance. When taking students' school levels into account, some different patterns related to students' migrant backgrounds were observed. At the junior secondary level, first-generation students reported the highest anxiety level. However, the local students were more anxious about mathematics learning than migrant students at the senior secondary level.

A comparison between students in different steams found that those in the vocational stream were more anxious about their mathematics learning than those in the academic stream, while the latter group was more worried about getting poor grades. Regarding migrant backgrounds, in the academic stream the local students' mathematics anxiety levels were higher than their migrant peers. However, for the vocational stream students, the first-generation group was the most anxious one with regard to both learning and grades.

It was found that students from the ordinary schools had higher levels of mathematics anxiety than those from the model schools ($M_{ordinary} = 0.09$ and $M_{model} = -0.08$), $t(17760) = 12.417$, $p < 0.001$, $d = 0.18$. In fact, this relationship applied to all the relevant survey items. When the students' migrant backgrounds

Table 4.5 Shanghai students' mathematics anxiety in PISA 2012 by their Migrant Backgrounds

	Junior secondary	Senior secondary			Overall
		Academic	Vocational	Both	
Native	−0.08 (0.91)	0.09 (0.93)	0.18 (0.87)	0.12 (0.91)	0.04 (0.92)
First-generation	0.02 (0.97)	−0.19 (1.09)	0.40 (0.93)	0.07 (1.07)	0.04 (1.01)
Second-generation	−0.12 (0.92)	−0.20 (1.00)	0.22 (0.89)	−0.07 (0.99)	−0.09 (0.96)

Note Students' math anxiety (ANXMAT) was reported with mean and standard deviation (which in the parentheses)

were taken into account, the analyses consistently showed that, in both ordinary and model schools, the local students were always the most anxious group. Moreover, the two migrant groups were also more similar to each other than to the local group. Even though all the between-group differences were minor, the magnitudes were relatively larger in the model schools than in the ordinary schools.

4.5 Conclusion and Discussion

As a big "melting pot" in China, Shanghai has become one of the most popular cities for migrants since 1990. In 2010, about 39% of the total population were citizens from other provinces and did not have household registration in Shanghai (Huang, 2012). With the large amount of migrants, integration is a key for a receiving system in order to develop the social harmony and prosperity of society as a whole (International Organization for Migration, 2006). In this integration process, education plays a crucial role. Yet, as Borgna (2014) remarked, "education is a double-edged sword". It is not only a vital precondition for upward social mobility, but also legitimizes the reproduction of social inequalities related to students' different family backgrounds.

With the recognition of the importance of providing this large student minority with a quality education, the Shanghai government has tried several measures to help migrant students to be well integrated into the local education system. One aim of the present study was to provide some evaluation of the effect of the local government's measures on its integration progress from a perspective of educational equality. Through a secondary analysis of the PISA 2012 Shanghai-China data, this study intended to examine the similarities and differences between students with different migrant backgrounds in terms of their mathematics attainments, in both cognitive and non-cognitive aspects. In particular, for students' non-cognitive attainments, this study focused on students' own perceptions about themselves as mathematics learners.

The analysis revealed that local students did perform significantly better than their migrant peers. However, a follow-up analysis found that significant differences actually came from the first-generation group, while second-generation students' performances were similar well to those of the local ones. Given the fact that an age-based sampling method was used for PISA, 15-year-olds may study at the junior secondary or senior secondary level. Related to the migrant education policy in Shanghai, all migrant students are eligible to attend local schools to study for their compulsory education, while there are some restrictions when they move beyond the compulsory years. The latter is illustrated by the shrinkage of the size of the first-generation group as well as a relatively larger percentage of them taking a vocational track at the senior secondary level.

The results showed that the variations in the mathematics scores of students from different migrant backgrounds were much greater at the junior secondary level. In contrast, students from the three migrant groups at the senior secondary level had more similarities in assessment performance. One reason for this inconsistency could be related to the wide policy of inclusion for newcomers into local schools at the compulsory education level. In some sense, this implies that newcomers are eligible to join local schools regardless of their prior academic performances. Therefore, the existence of potential differences between and within newcomer groups is to be expected. In other words, the differences could be related to the newcomers' prior schooling experiences. Moreover, newcomers also need to make efforts to adjust themselves to the new learning environment, including different curricula, different teaching styles, and different peer relationships, as well as the new living environment.

It is known that senior secondary education is not compulsory in China. For all students, including both locals and migrants, it is necessary to sit for the locally administered Senior Secondary School Entrance Examinations and to attain certain entrance minimum marks in order to get into either the academic or vocational streams of senior secondary schools. It is worth to be mentioned that the marks required for the entry is much more higher for the academic stream of senior secondary school for all eligible students. In the case of migrant students, some further restrictions related to migrant policy are applied. Those who meet both of these requirements can remain in Shanghai to continue their senior secondary studies. In this sense, it is understandable that these students were found in this study to have had similar performance results to those of the local students. In fact, this study actually found that both types of migrant students studying in the academic stream performed better in the PISA math assessment than their local peers, for both model and ordinary schools. The fact that the migrant students (particularly first-generation) who can opt for local senior secondary schools (academic) are those who have also qualified for new Shanghai residence permits may help to shed some light on this finding. In contrast, migrant students who have not qualified for residence permits but still want to remain in Shanghai for their senior secondary education may choose the vocational track. To a certain extent, these students' academic qualification received less control; in other words, the entry requirement for the vocational schools is relatively looser. In this sense, these students' weaker performance than local students was somehow expected.

The differences between local and migrant students in relation to their perceptions of themselves as mathematics learners were much smaller than the differences in academic performance. The study looked into three specific aspects of students' perceptions. One of these was their confidence in solving mathematics problems with given contexts (mathematics self-efficacy), another was their self-evaluation of their overall mathematics ability (mathematics self-concept), and the third was their anxiety levels regarding their mathematics learning and results (mathematics anxiety).

Due to the fact that the survey items on the notion of mathematics self-efficacy were problem-context based, the students' corresponding confidence levels could be very close to their actual academic performances. As a result, the local students did show higher confidence levels than their migrant peers; this was observed at both junior secondary and senior secondary levels (vocational stream). At the senior secondary level (academic stream), both migrant groups, in general, showed more confidence than the local students. Consistent with the results related to academic performance, the confidence levels of students with different migrant backgrounds were more similar in the academic stream than the vocational. In particular, it appears that first-generation students studying in the vocational stream were less confident in solving some kinds of mathematics problems.

As mentioned, students' mathematics self-concept is a more general notion. Different from the findings on mathematics self-efficacy, local and first-generation students were closer on this dimension, while second-generation students gave themselves the highest ratings on all the individual items as well as the constructed index. In addition, when students moved from the junior secondary level to the senior, the differences among the three groups became larger. However, such a relationship was not observed in the students studying in the vocational stream. In fact, in this stream, the local students appeared to be more confident, which was also found with their mathematics self-efficacy.

Compared to students' self-evaluations of their mathematics abilities, either general or specific, those with different migrant backgrounds showed greater similarities in terms of their mathematics anxiety. Further analysis showed that, while second-generation students appeared to be slightly more anxious about their mathematics learning than the other two types of students, the local ones were concerned more about their mathematics results. School-level related differences were also found for mathematics anxiety; in particular, the first-generation group was the most anxious about mathematics learning experiences at the junior secondary level, while the local students studying in the academic stream reported a higher anxiety level than their migrant counterparts at the senior secondary level. Of the vocational stream students, the first-generation group was again the most anxious.

These inconsistencies could be related to the different subject requirements as well as learning targets. In particular, at the junior secondary level, first-generation students may need more time and effort to adapt themselves to Shanghai's local mathematics curriculum, which could be very different from what they had learned in their own regions. For those aiming to stay in Shanghai to continue their senior secondary education, the high level of anxiety was understandable to a certain extent. When the migrant students, particularly the first-generation ones, were able to find places in local senior secondary schools (academic type) through taking the Senior Secondary School Examination, they were at a more equivalent level with their local peers, which was also found with cognitive attainment. As mentioned, attending vocational schools was a more popular option for the first-generation students, as this is less competitive. Therefore, a similar pattern as the one at the junior secondary level was revealed with vocational stream students.

In sum, the present study has shown that first-generation migrant students were generally in a disadvantaged position in terms of their mathematics performances as well as their perceptions about themselves as mathematics learners. However, compared to the differences in cognitive attainment, the non-cognitive differences were less observable. This is true for students at the junior secondary school and also those studying in vocational schools. In other words, for students in the two types of schools, local students and second-generation migrant students showed more similarities. This result is slightly different from the PISA 2003 findings on migrant students' performance, which found that both first-generation and second-generation students often reported similar or even higher levels of interest and motivation in mathematics and more positive attitudes towards schooling (OECD, 2006).

Furthermore, when the analysis focused on students studying in the academic stream, the first-generation group was no longer at a disadvantage and, in some aspects, their performance was even better than the local students. This, to a certain extent, indicates that a higher level of equality related to students' migrant backgrounds was reached at the senior secondary level (academic stream). However, some cautions should be borne in mind as the first-generation students admitted to senior secondary level had met the high requirements of both academic examinations and residence permit applications, which may mean that they were a particularly high calibre of students.

References

Adams, R., & Wu, M. (Eds.). (2002). *Programme for international student assessment (PISA): PISA 2000 technical report*. Paris: OECD.

Borgna, C. (2014, May 12). *Immigrant integration: School failures and successes*. Retrieved from http://www.eutopiamagazinearchive.eu/en/camilla-borgna/issue/immigrant-integration-school-failures-and-successes.html.

Borgonovi, F., & Jakubowski, M. (2011). *What can we learn about the gender gap from PISA*. Paris: OECD. Retrieved from http://doc.iiep.unesco.org/wwwisis/repdoc/SEM313/SEM313_1_eng.pdf.

Chen, Y., & Feng, S. (2012). *Access to public schools and the education of migrant children in China*. IZD Discussion Paper No. 6853. Retrieved from http://ftp.iza.org/dp6853.pdf.

Chen, Y., & Feng, S. (2014). *Quality of migrant schools in China: Evidence from a longitudinal study*. Shanghai University of Finance and Economics School of Economics Working Paper, No. E2014001.

City of Shanghai. (2005, May 16). *Office workers earn pay raises*. Retrieved from http://www.shanghai.gov.cn/shanghai/node17256/node18151/userobject22ai16924.html.

Ding, C. (2004). Farmland preservation in China. *Land Lines, 16*(3), 9–11. Retrieved from http://www.lincolninst.edu/pubs/913_Farmland-Preservation-in-China.

Grubb, N. S., Field, H., Jahr, M., & Neumüller, J. (2005). *Equity in education thematic review: Finland country note*. Paris: OECD.

Huang, J. (2012). *Attitudes toward rural migrant workers—In-depth interviews with 24 educated Shanghai residents* (Unpublished master thesis). Sweden: Uppsala Universitet. Retrieved from http://www.pcr.uu.se/digitalAssets/67/67531_1jingjing-huang_master-thesis_2012.pdf.

International Organization for Migration. (2006). *IOM and migrant integration*. Retrieved from https://www.iom.int/files/live/sites/iom/files/What-We-Do/docs/IOM-DMM-Factsheet-LHD-Migrant-Integration.pdf.

Li, L. C., & Wang, W. (2014). Pursuing equity in education: Conflicting views and shifting strategies. *Journal of Contemporary Asia, 44*(2), 279–297.

Liang, Y. (2012, May 9). Fierce competition to get into top primary schools. *Shanghai Daily*. http://english.peopledaily.com.cn/203691/7812468.html.

Ma, X., Ma, L., & Bradley, K. D. (2008). Using multilevel modeling to investigate school effects. In A. A. O'Connell & D. B. McCoach (Eds.), *Multilevel modeling of educational data* (pp. 59–110). NC: Information Age Publishing Inc.

National Center on Education and the Economy (NCEE). (no date). Shanghai-China: Education for all. Retrieved from http://www.ncee.org/programs-affiliates/center-on-international-education-benchmarking/top-performing-countries/shanghai-china/shanghai-china-education-for-all/.

Organization for Economic Co-operation and Development. (OECD). (2006). *Where immigrant students succeed—A comparative review of performance and engagement in PISA 2003*. Paris: OECD.

Organization for Economic Co-operation and Development. (OECD). (2011). *Lessons from PISA for the United States, strong performers and successful reformers in education*. Paris: OECD. doi:10.1787/9789264096660-en.

Organization for Economic Co-operation and Development. (OECD). (2013a). *OECD economic surveys: China 2013*. Paris: OECD.

Organization for Economic Co-operation and Development. (OECD). (2013b). *PISA 2012 assessment framework—Key competencies in reading, mathematics and science*. Paris: OECD.

Postiglione, G. A. (Ed.). (2015). *Education and social change in China: Inequality in a market economy*. New York: Taylor & Francis.

Ren, D. (2012, August 25). Government mulls opening city schools to migrant children. *South China Morning Post*.

Shanghai Statistical Bureau. (2014). 上海统计年鉴*2013 [Shanghai statistics 2013]*. Shanghai: Shanghai Statistic Press.

Steele, M. (2010). *Update on Shanghai reforms for migrant education: Stepping stones*. Retrieved from http://steppingstoneschina.net/wp-content/uploads/2010/07/Update-on-migrant-schools-June-10-MS.pdf.

Tucker, M. S. (Ed.). (2014). *Chinese lessons: Shanghai's rise to the top of the PISA league tables*. Washington, DC: National Center on Education and the Economy.

UNESCO. (2003). *Measuring and monitoring the information and knowledge societies: A statistical challenge*. Montreal: UNESCO. Retrieved on November 20, 2011, from unesdoc.unesco.org/images/0013/001355/135516e.pdf.

Werwath, T. (2011, May). *Separate but equal? Children of migrant workers in Shanghai's public education system*. Paper presented at the Annual Meeting of the 55th Annual Conference of the Comparative and International Education Society, Fairmont Le Reine Elizabeth, Montreal, Quebec, Canada.

Wu, X. (2009, March). *The power of market mechanism in the school choice in China: An empirical study*. Presented to the conference "Penser les marches scolaires", Université de Genève. Retrieved from http://www.unige.ch/fapse/ggape/files/2214/1570/7325/Wu.pdf.

Yin, L. (2013, December 24). *Education for migrant children sparks concern*. Retrieved from http://www.womenofchina.cn/womenofchina/html1/features/education/16/8394-1.htm.

Chapter 5
Shadow Education of Mathematics in China

Lidong Wang and Kan Guo

Abstract Generally speaking, students can learn mathematics both in their formal schooling and outside school. Therefore, when we discuss students' mathematics learning and achievement, the role of out-of-school learning cannot be ignored. This chapter aims at highlighting the importance of shadow education in mathematics to the overall picture of Chinese mathematics education, especially the notable achievements of Chinese students in international comparative research, such as PISA.

Generally speaking, students can learn mathematics both in their formal schooling and outside school. Therefore, when we discuss students' mathematics learning and achievement, the role of out-of-school learning cannot be ignored.

Private supplementary tutoring is widely known as shadow education, since it mimics the mainstream; this has a long history in Asia and, in recent decades, it has expanded greatly, even when the quality of education is high (Bray & Lykins, 2012).

The purpose of this chapter is to highlight the importance of shadow education in mathematics to the overall picture of Chinese mathematics education, especially the notable achievements of Chinese students in international comparative research, such as PISA.

Three aspects of shadow education in mathematics in China will be discussed here.

1. Cultural setting: Selected aspects of traditional Chinese culture will be reviewed to help readers to understand why certain phenomena exist in China, and even elsewhere in Asia.
2. Current status: By review existing research, the scale, and categories will be illustrated.

L. Wang (✉)
The High School Affiliated to Renmin University, Beijing, China
e-mail: wldyaoyuan@163.com

K. Guo
Beijing Normal University, Beijing, China

Y. Cao and F.K.S. Leung (eds.), *The 21st Century Mathematics Education in China*, New Frontiers of Educational Research,
https://doi.org/10.1007/978-3-662-55781-5_5

3. Influence on student achievement: Based on the data from a large-scale survey, the effectiveness of shadow education on student achievement will be reported.

5.1 Cultural Setting of Chinese Students' Learning of Mathematics in Shadow Education

Some of the dilemmas identified in previous research have been linked to different cultural settings (Clarke, Wang, Xu, Aizikovitsh-Udi, & Cao, 2012) so, when describing Chinese shadow education in mathematics, it is essential first to understand something about the cultural background.

China is a country with an ancient history and continuous civilization (from about 2000 BC), and is very different from most other cultures in the world, especially from western cultures.

Most social values in China can be derived from Confucianism, Taoism, and Buddhism (Wong, 2004). Confucianism valued the element of Ren (Humaneness), Yi (Righteousness or Justice), Li (Propriety or Etiquette), Zhi (Knowledge), and Xin (Honest integrity), as well as Zhong (Loyalty), Xiao (Filial piety), and Jie (Moral integrity) (Runes, 1960). Confucianism also influenced other East Asian countries, such as Japan and Korea. Moreover, it was the official philosophy for about 2000 years throughout most of Imperial China's history. At this time, a complete mastery of Confucian texts (the four books, the five classics, etc.) was required to succeed in the Imperial Examinations (Keju). The national examination system, dating back to the seventh century in Imperial China, was designed to select the best potential candidates (especially important for the populace) to serve as administrative officials in the Imperial bureaucracy (Liu, 2007).

Traditionally, Chinese education has been influenced by Confucian values, which contribute to a specific belief system (Li, 2004), hold education, and learning in high regard, and value parental involvement in and commitment to the education of children (Wong, Wong, & Wong, 2012).

As a result, in the traditional career system education was valued as an appropriate preparation for promising jobs and happy lives, which also worked as an external motivation for education (Leung, 2001; Wong et al., 2012). The notion of "the examination culture" (Wong et al., 2012) reflects the important status of high-stakes examinations in China, such as the role of the Keju policy in the change of social status, which implies a practical and even utilitarian purpose of education.

Chinese parents and students tend to attribute high achievement to hard work, rather than talent; they typically think that study is a serious endeavor and expect children to put hard work and perseverance into their studies (Leung, 2001). Chinese students are likely to agree with their parents' high expectations. From a Confucian perspective, the major reason for a learner not achieving a desired learning outcome is his/her lack of effort (Wong, 2004). Thus, for example, in the US. context, American children might view mathematics as a relatively easy subject and believe that they have already met their parents' expectations, while Chinese–American parents might respond by providing more formal learning support (Wang & Lin, 2005).

Furthermore, Crystal and Stevenson (1991) found that Chinese parents tend to be more critical of and dissatisfied with their children's mathematics performances than American parents.

The research undertaken by Wang and Zhu (2009) indicated that Chinese urban families spend a high proportion of family income on their children's education. However, in rural areas this has become a social problem as Chinese parents try their best to support their children in pursuing school study. For example, some poor families spend the majority of their income, or even get into debt, to make sure their children receive good education. Many families, especially those from low socio-economic backgrounds, believe that "school education" is the only way to improve their current living conditions. They believe that if their children have education which leads to qualifications or professional status, they will have better chances of getting good jobs and will have a high social status. Even middle- and upper-class families share the same belief (Wang, Li, & Li, 2014).

In terms of mathematics teaching and learning in China, there are differences between the past and the present. In ancient China, mathematics (and even the sciences) was not considered important officially, and mathematicians (scientists) did not have a high social status. Chinese learners were engaged in Imperial Examinations (Keju) to become officials, which greatly influenced the East Asian world (Liu, 2007). Even though mathematics was not included in this examination system, there were numerous mathematical products contributed by folk mathematicians, such as *The Nine Chapters on the Mathematical Art.*

Mathematics today is regarded differently than it was in ancient China (Leung, 2001), with the modernization of society impacted by Western culture. It is now one of the most important school subjects (a compulsory subject for grades 1–12), which is typical in Western mathematics education systems. This situation can be traced back to the importation of the former Soviet Union educational system when the People's Republic of China was founded in the 1950s. Mathematics teachers often have a high status compared with their colleagues.[1] Nowadays, the academic selection function of mathematics (in school recruitment) plays a very important role in high-stakes examinations, not only in recruitment for universities but also in other levels, even compulsory levels. Under this pressure, Chinese families spend large human and economic resources on children's mathematics learning, such as buying learning materials, hiring private tutors, and supporting outside-school learning (Wang et al., 2014). It needs to be mentioned here that the Mathematics Olympics may play some kind of role, beyond that of ordinary school mathematics, in school recruitment, especially in gifted education.

However, the significant change between the situation of mathematics in ancient China and today may not have arisen from the views of Chinese parents and

---

[1]Teachers teach a single subject in mainland China.

teachers about mathematics, but rather from the role of the Keju policy in school recruitment (in which mathematics is an essential recruitment requirement for nearly all the different types of schools), which follows the "practical and even utilitarian purpose of education" mentioned above (Wang et al., 2014).

Socially and economically, China is the world's fastest growing economy, but the economic development is unbalanced. Both rich and poor families are eager to put their resources into educating their children to be successful in future competition. Some scholars have argued that shadow education is involved in the equity of education (Bray & Lykins, 2012).

The above situation can explain parents' motivation to make use of shadow education to help their children gain high mathematics achievement, and thus benefit them in social competition. It helps readers to understand why the current situation of shadow education of mathematics exists in China.

5.2 The Current Situation of Students' Learning Mathematics Through Shadow Education

This section describes the current situation of shadow education in mathematics in China, based on a review of several existing studies.

It should be noted that shadow mathematics education is a very widespread social phenomenon in mainland China, as in other Asian countries.

When the report of PISA2012 was published in December 2013, Shanghai (the biggest city of China) was announced as being in the first place in mathematics again. But the report also showed that the extra tutoring time spent by students in Shanghai was 2.08 h a week on average, which ranked ninth of the 65 countries and territories; the time spent doing homework was 13.85 h a week on average, which ranked first (about 3 times as much as the OECD's average time) and was even 4 h more than the second-ranked Russia.

A large scale survey reported that more than 70% of grade 4 and grade 8 students from Beijing (the capital of China) received supplementary tutoring (tutors and social schools), this being a higher rate compared to other school subjects (similar to English), and that 22.9% of grade 4 students and 15.1% of grade 8 students had after-school tutoring more than twice each week (Zeng & Zhou, 2012). A survey of 6474 students in Jinan (a big city in China) found that 28.8% of lower secondary students were receiving tutoring in mathematics, (Zhang, 2011). Another large-scale survey, based on data from three big city districts in China, showed that the average time spent on shadow education of mathematics by grade 7 students was 1.72 h each week (Wang, 2012). Data from Wuhan (a big city in China) indicated 39.6% of students (from grade 4 to grade 9) having shadow mathematics education (Peng, 2008), which is still a large percentage.

It has been reported that there are some primary school students who participate in as many as five different out-of-school classes (all for the Mathematics

Olympics) at one time! Recently, since the role of the Mathematics Olympics in school recruitment has become more important in some areas of China, the training has attracted a great deal of social attention, and sometimes disputes (Zhou, 2010). A considerable part of the training is operated by shadow education. A survey from Lanzhou (a city in the western part of China) indicated that taking part in informal schooling is the main way for students to access the Mathematics Olympics. Mathematics Olympics training has become an "education industry", and there are competitions, held by shadow schools, which have attracted much attention. Many scholars, including some internationally famous mathematicians, have argued that the scale has become aberrantly large, and might have negative effects on the development of mathematics education.

As a middle school mathematics teacher of gifted education, the first author has found that shadow education in mathematics is very prevalent. Sometimes, the majority of the students in the class have already learned the content from informal schooling before it is taught in the formal school.

It is not difficult to understand the above phenomenon based on the high value of education in traditional Chinese culture. As well, the huge unbalanced distribution of educational resources existing not only between the rural and the urban areas, but within big cities, also contributes partly to this phenomenon.

The characteristics of shadow education vary in China; it can take the form of one-to-one tutoring or be conducted in small classes or even large classes. Some services are provided by officially authenticated, private shadow schools, which hire tutors to conduct one-to-one tutoring at the clients' homes or in places provided by the schools. These shadow schools also hire tutors to teach small or large classes. Some services are provided by independent tutors, who may offer one-to-one tutoring in the home of either the tutor or the client; some parents may join together to hire a tutor to teach small classes and some individual tutors may organize small classes, all of which are not officially authenticated. We should notice that nearly all of these services require payment, so it has become a big industry.

Another issue is where the teachers of shadow education came from. In China, in-service teachers of formal schools are forbidden to teach for-profit shadow education, for example by working as tutors. However, we cannot claim that this policy is always obeyed. The other sources are university students and full-time shadow school teachers. The full-time shadow school teachers can be selected from graduates as well as former school teachers. Nowadays, shadow schools attract more and more university graduates choosing to do this as their initial jobs, while the competition in applying for job positions is more and more fierce and the income level is becoming more attractive.

Generally, shadow education can be classified into two purposes; one is to enhance the formal school learning according to the school curriculum, for example previewing and reviewing the school curriculum, which can be categorized as enrichment (helping high achievers reach new levels) and remediation (helping slow learners to keep up with their peers) (Bray & Lykins, 2012; Zhang, 2011). The other purpose is to expand the formal school curriculum, and the main content is the

Mathematics Olympics, which tend to benefit students in the recruitment process, from middle schools to universities.

Zhang (2011) summarized the effects of shadow education on formal schooling, suggesting that private tutoring may help school teachers to satisfy students' individual needs for both remedial and expanded learning. On the other hand, if teachers take the students who receive tutoring as the norm in their classes and do not cover the pre-taught materials in class, all parents are placed under pressure to invest in private tutoring for their children.

We argue that Zhang's summary may not have covered all of the negative effects on formal schooling. We have also had direct experience of students losing interest due to repeated learning. As well, some carefully designed investigation tasks might become ineffective when the students already know the results, and some misunderstandings of mathematical content from a shadow school may upset the mathematics teaching in the formal school. The quality of shadow education may not be ensured, since there are no systematic quality assurance policies or standards.

5.3 The Influence of Shadow Education on Students' Mathematics Learning

The following analysis illustrates the relationship between shadow education and students' mathematics achievement. The data were drawn from a project called "Middle-school Mathematics and the Institutional Setting of Teaching in China" (MIST-China). Three school districts from Beijing, Shenyang, and Chongqing were selected as representative of northern, northeastern, and southwestern China. Stratified random sampling was used to select schools, with key schools and ordinary ones sampled separately. Students in each participating class were divided randomly into two groups, one who participated in an algebra test and one a geometry test. Table 5.1 presents the participant demographics.

The students' mathematics achievements were measured by the algebra or geometry tests designed by Wang (2012) and based on Chinese mathematics curriculum standards.

We designed a framework to measure student achievement at different cognitive levels of mastering algebra and geometry separately. The Chinese national mathematics curriculum standard was designed according to a taxonomy of hierarchical cognitive levels, consistent with the ideas of many important studies (e.g., Stein,

Table 5.1 Participant demographics

	Number	Gender		District		
		Boys	Girls	Beijing	Shenyang	Chongqing
Algebra	1633	810	823	401	751	481
Geometry	1712	854	858	370	749	593

Grover, & Henningsen, 1996). We adopted a four-level hierarchical cognitive coding scheme based on the framework of the QUASAR Project (Stein et al., 1996), and the Chinese National Mathematics Curriculum Standard (Ministry of Education of PRC, 2001).

The tests were administered to grade 7 students in the sampled schools in May 2011. The students were given 40 min to complete 12 rubrics to indicate their mathematical abilities in algebra or geometry. The test documentation suggested strong content validity resulting from the extensive involvement of expert educators in the development and selection of the test items. Furthermore, all school districts use the same curriculum standard in China, the content of which is divided into three parts: algebra, geometry, probability and statistics. According to our previous studies, probability and statistics accounts for only 7.3% of the curriculum standard and 9.9% of the textbooks (Kang & Cao, 2013; Wu & Cao, 2013), and the percentage is less in grade 7, hence we did not assess students in this. The results indicated that the internal consistency reliability (Cronbach's alpha) of the tests were 0.701 for algebra and 0.705 for geometry, with inter-scorer agreements of more than 99%. PARSCALE was used in the IRT scoring and the scores resembled a normal distribution.

The students' questionnaire in MIST-China included a question that measured the time they spent per week in shadow education (personal tutoring or remedial class), with options from "almost none" to "4 h or more". The results of the survey showed that about a quarter of the students hardly participated in mathematics tutoring and that those who did participate in extracurricular classes spent about 2 h per week on average (see Fig. 5.1), which was similar to the results of PISA 2012 in Shanghai. Pearson's correlation analysis showed a significant but weak relationship between students' shadow education time and their algebra or geometry scores ($r = 0.101$ and 0.132, respectively; $ps < 0.01$) and the shadow education explained less than 2% of students' variance differences in mathematics achievement.

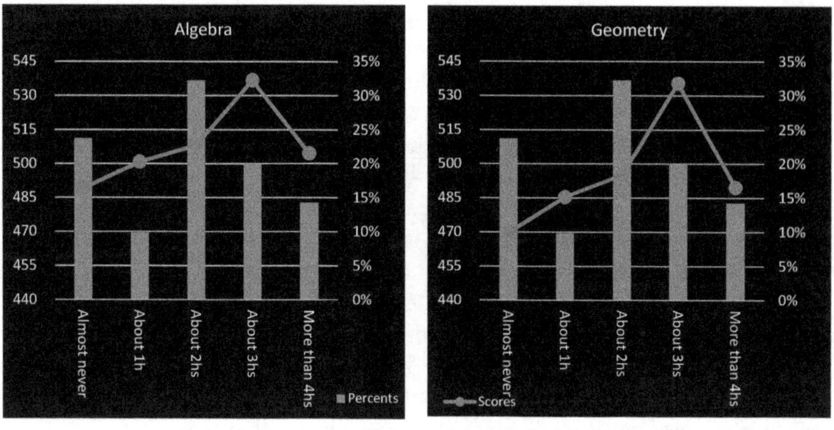

Fig. 5.1 Students' shadow education time and mathematics achievement (MIST-China)

Therefore, it would seem that shadow education is not of great relevance to students' mathematics achievement.

It is important to note, however, that the correlation analysis just explains the effect of a linear relationship between shadow education and students' achievement. The analysis of variance indicated that there may be a nonlinear relationship. Figure 5.1 shows the "inverted U-shaped" trend between students' achievement and shadow education time. The follow-up tests consisted of all pairwise comparisons among the five options. The pairwise comparisons between some shadow education times ("about 3 h", "almost none") were all statistically significant $p < 0.001$, but the pairwise comparisons of 1, 2 or 4 h were non-significant.

The best-fitting line between the Shanghai students' shadow education time (including homework, tutoring, and extracurricular tutoring classes) and their mathematics achievements from the PISA2012 database is a quadratic curve, which is much better than liner fitting ($R^2 = 0.271$ and 0.189, respectively). The quadratic curve is also in accordance with the "inverted U" trend. Thus, the relationship between students' shadow education times and achievement does not reflect "more will pay off". It is very difficult to explain the phenomena of education or society clearly because the relationship between students' shadow education time and their mathematics achievement is not simply cause and effect: on one hand, learning for too long may cause much pressure and have a bad influence on students' physical and mental health, resulting in poor outcomes; on the other hand, students' poor grades and low learning efficiency may lead them to spend too much time in shadow education. Cooper, Robinson, and Patall (2006) summarized the research conducted in the USA on the effects of homework from 1987 to 2003, and then suggested that moderate homework can improve the students' understanding of knowledge and help improve grades, but that too much homework time can make them bored and lose interest in learning. The "inverted U" trend is consistent with another traditional philosophy in China, "Overdone is worse than undone" (Fig. 5.2).

It is not difficult to find from these two cases that the shadow education—mathematics achievement relationship is represented by an "inverted U" trend, which is really "no pain, no gain" but not "more will pay off". That is, the investment in shadow education time cannot promote mathematics achievement to grow continuously. The PISA 2006 report pointed out that students' shadow education time is related negatively to their academic performance, and suggested that the possible reason is that those who participate in shadow education are mainly there for the purpose of enhancement rather than increased performance. This means that they spend more time in shadow education because of their low grades, not because the long time leads to poor performance. To verify this conjecture, the different types of shadow education were investigated using the students in the MIST-China database who spent more than 4 h in this type of education. Six options were provided, "*Mathematics contest training*", "*Advanced learning of course content*" "*Recitation*", "*Interest class (interesting mathematics, mathematical stories, research learning,* etc.*)*", "*After-school care class (tutor completes homework, answers some questions)*" and "*Others*". As shown in Fig. 5.3, most of

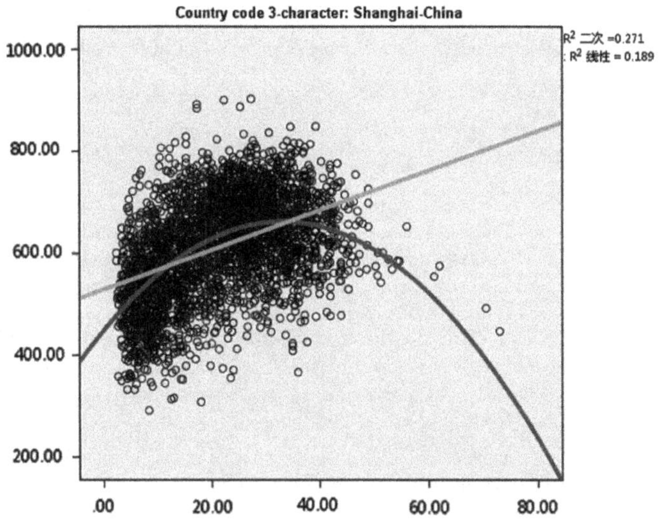

Fig. 5.2 Mathematics achievement and shadow education time (PISA2012Shanghai)

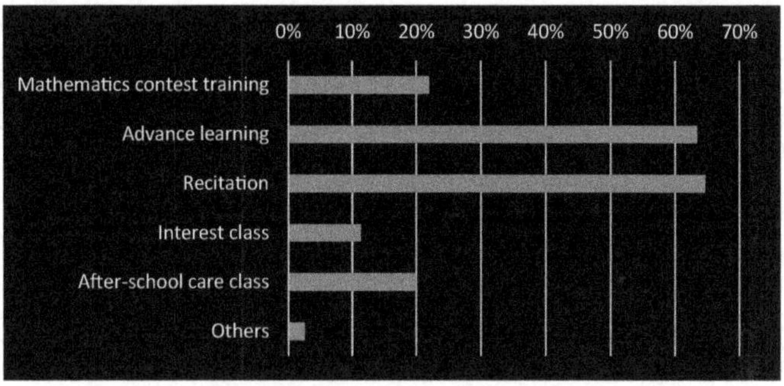

Fig. 5.3 Time and types of shadow education of the students who spent more than 4 h

the students who spend much time in shadow education tended to attend advance learning or recitation in the school course. Fewer students participated in the more optimal courses, like mathematics contests or interest classes.

In conclusion, the "inverted U" relationship was found between shadow education and mathematics achievement in Chinese students. While we should be cautious to conclude that a student will certainly fall in grades when shadow education time is extended, this "inverted U" relationship was found as a population characteristic. Our results at least suggest that the excellent students may not spend more time in learning than others. Previous research has supported the idea that excellent students usually have healthy lifelong habits, such as taking exercise and reading books for one hour per day. Teachers and parents should encourage students to improve their learning interests, learning habits, and learning methods. The traditional Chinese thought "no pain, no gain" should be rethought.

References

Bray, M., & Lykins, C. (2012). *Shadow education: Private supplementary tutoring and its implications for policy makers in Asia.* Hong Kong: Asian Development Bank and Comparative Education Research Centre (CERC), Faculty of Education, The University of Hong Kong.

Clarke, D. J., Wang, L., Xu, L., Aizikovitsh-Udi, E., & Cao, Y. (2012). International comparisons of mathematics classrooms and curricula: The validity-comparability compromise. In T. Y. Tso (Ed.), *Proceedings of the 36th Conference of the International Group for the Psychology of Mathematics Education (PME36)* (Vol. 2, pp. 171–178). Taipei, Taiwan.

Cooper, H., Robinson, J. C., & Patall, E. A. (2006). Does homework improve academic achievement? A synthesis of research, 1987–2003. *Review of Educational Research, 76*(1), 1–62.

Crystal, D. S., & Stevenson, H. W. (1991). Mothers' perceptions of children's problems with mathematics: A cross-national comparison. *Journal of Educational Psychology, 83*(3), 372–376.

Kang, Y., & Cao, Y. (2013) 中英美小学和初中数学课程标准中内容分布的比较研究 [Comparative study of content distribution of mathematics curriculum standards from China, the United Kingdom and the United States]. *Curriculum, Teaching Material and Method, 4.*

Leung, F. K. S. (2001). In search of an East Asian identity in mathematics education. *Educational Studies in Mathematics, 47*(1), 35–51.

Li, J. (2004). A Chinese cultural model of learning. In L. Fan, N. Wong, J. Cai, & S. Li (Eds.), *How Chinese learn mathematics: Perspectives from insiders*. Singapore: World Scientific.

Liu, H. (2007). Influence of China's imperial examinations on Japan, Korea and Vietnam. *Frontiers of History in China, 2*(4), 493–512.

Peng, P. (2008). 城市义务教育阶段学生课外补习的实证研究——基于武汉市洪山区的调查与分析 [An empirical research on supplementary tutoring of compulsory education students: Based on the survey and analysis in Hongshan District, Wuhan] (unpublished master dissertation). Hu Bei, China: Central China Normal University.

Runes, D. D. (1960). *Dictionary of philosophy*. Progress Publishers.

Stein, M. K., Grover, B. W., & Henningsen, M. (1996). Building student capacity for mathematical thinking and reasoning: An analysis of mathematical tasks used in reform classrooms. *American Educational Research Journal, 33*(2), 455–488.

Wang, J., & Lin, E. (2005). Comparative studies on U.S. and Chinese mathematics learning and the implications for standards-based mathematics teaching reform. *Educational Researcher, 34* (5), 3–13.

Wang, L. (2012). 数学教师对学生学业成就的影响研究 A research on mathematics teachers' effect on students' achievement] (unpublished doctoral dissertation)]. Beijing, China: Beijing Normal University.

Wang, L., Li, X., & Li, N. (2014). Socio-economic status and mathematics achievement in china: A review. *ZDM Mathematics Education, 46*(7), 1051–1060.

Wang, Y., & Zhu, S. (2009). 中国城镇居民家庭教育投入的状况和特征 [The situations and characteristics of the educational inputs of Chinese urban families]. *Education and Economics, 4,* 11–16.

Wong, N. Y. (2004). The CHC learner's phenomenon: Its implications on mathematics education. In L. Fan, N. Y. Wong, J. Cai, & S. Li (Eds.), *How Chinese learn mathematics: Perspectives from insiders* (pp. 503–534). Singapore: World Scientific.

Wong, N. Y., Wong, W. Y., & Wong, E. W. (2012). What do the Chinese value in (mathematics) education? *ZDM—The International Journal on Mathematics Education, 44*(1), 9–19.

Wu, L. & Cao, Y. (2013). 初中数学课程内容分布的国际比较研究 [International comparative study of content distribution of junior high school mathematics curriculum]. *Journal of Educational Studies, 2*.

Zeng, X., & Zhou, H. (2012). 北京市四、八年级学生课后补习的代价与收益 [A descriptive analysis on after-school tutoring in Beijing: Its costs and benefits]. *Journal of Educational Studies, 8*(6), 103–109.

Zhang, Y. (2011). *The determinants of national college entrance exam performance in [People's Republic of] China—With an analysis of private tutoring* (unpublished doctoral dissertation). New York, U.S.A.: Columbia University.

Zhou, L. (2010). 数学竞赛在中国的实践质问其教育价值 [Practice of mathematics competitions in china questions their educational value]. *Journal of Mathematics Education, 19*(5), 28–30.

Part II
Mathematical Curriculum and Textbook

Chapter 6
Developments and Changes in the Primary School Mathematics Curriculum and Teaching Material in China

Historical Development of Primary School Mathematics Courses and Teaching Materials

Beibei Ye

Abstract Since the founding of the P.R.C., primary school mathematics has been regarded as a fundamental course, and its curriculum development and implementation have been of major concern to educational administrative departments at different levels from central to local. In the past, decisions about curriculum development were made by officials, but now officials, mathematicians, researchers, and teachers all contribute. Periodically, changes have taken place in length of schooling, course offerings, syllabus, and textbook compilation.

6.1 Succession and Reform of Primary School Mathematics Courses in the Republic of China (1949–1952)

Since the founding of the P.R.C., primary school mathematics has been regarded as a fundamental course, and its curriculum development and implementation have been of major concern to educational administrative departments at different levels from central to local. In the past, decisions about curriculum development were made by officials, but now officials, mathematicians, researchers, and teachers all contribute. Periodically, changes have taken place in length of schooling, course offerings, syllabus, and textbook compilation.

November 1949 saw the founding of the Ministry of Education of the Central People's Government. In December, the Ministry of Education held the first national conference for education and determined a guideline that "education must serve the national construction, and schools must be accessible to workers and peasants"

B. Ye (✉)
Faculty of Education, Guangxi Normal University, Guilin, China
e-mail: ybb_533@qq.com

Y. Cao and F.K.S. Leung (eds.), *The 21st Century Mathematics Education in China*, New Frontiers of Educational Research,
https://doi.org/10.1007/978-3-662-55781-5_6

(CCCPC Literature Research Office, 2011, pp. 74–76). At a national publishing conference held in September 1950, the attendees jointly put forward the proposal that primary and secondary school teaching materials must be supplied uniformly all over the country. Thus, the Publishing Department of China and the Ministry of Education jointly formed the People's Education Press. From then on, the primary school mathematics curriculum has been transformed and explored step by step.

6.1.1 Development of Transitional Primary School Mathematics Curriculum

In 1951, the Government Affairs Council for decisions about the reform of the educational system was established. Based on this Council's recommendations, the Ministry of Education (MOE) implemented a pilot "five-year consistent system" in primary schools (National Academy of Education Administration, 1999, pp. 106–107). In June, 1951, the Department of Basic Education in the MOE selected six primary schools in Beijing to start this academic political experiment. (Wang, 1996, p. 278) Meanwhile, places in the southern and northern parts of China still continued the previous "4+2" educational system, which divided the primary school into junior primary for four years and higher senior primary for two years.

With the release of a document titled *Tentative Standards and General Principles for the Primary School Curriculum,* the education system was changed into a consistent five-year one, with the weekly numbers of class periods for primary school mathematics courses from the first grade to the fifth grade established as 5, 6, 6, 6, and 7, respectively. It was established that the course of abacus calculation would be offered in the first and second semester in grade 3, and in the first and the second semester in grade 4, with two class periods per week for each of these. The total number of mathematics periods was thus 1216, including abacus calculation courses for grade 3 and grade 4. On this basis, the MOE formulated the *Tentative Standard for Primary School Mathematics Curriculum (Draft)* (hereafter referred to as the *50 Outline*), which was launched in July 1950. In accordance with the five-year consistent system, the *50 Outline* syllabus contained the content and numbers of teaching hours for written calculations for five grade levels. The main teaching content was: four basic operations on integers, four basic operations on simple decimals, simple fraction arithmetic, basic calculation and application of percentages, weights and measures in the metric system, simple calculation of the units of time, simple geometry, simple diagrams, the account-keeping methods in cooperatives, calculations on integers and decimals with an abacus, and conversion of jin (500 g equals one jin) and liang (50 g equals one liang) (Curriculum and Teaching Materials Research Institute, 2010, p. 16).

During this period, there was no unified system for textbooks. Their publication and application differed from the north to the south of the country. In some areas of the north, the textbooks were still the same as those in the districts governed by the Communist Party of China, while, in the south, textbooks for junior primary schools (eight volumes) and higher senior primary schools (four volumes) were mainly compiled by Ziyi Yu and published by Dadong Publishing House; these textbooks were in line with the "4–2" education system.

6.1.2 Implementation of Transitional Primary School Mathematics Curriculum

In this period, the elicitation method was the prominent way of teaching mathematics knowledge. In 1950, the Chaha'er Publishing House edited and published *Research on Arithmetic Teaching*. This indicated the state of arithmetic education in North China and puts forward that heuristic education must be implemented in arithmetic teaching and equation teaching must be combined with reality and comply with children's existing cognitive levels. Influenced by Herbart's Five Formal Teaching Steps, the mathematics teaching content of this period focused mainly on the calculation of numbers and quantity, drawing on teachers' experiences, particularly on the discussion of problem-solving teaching. At that time, primary school mathematics teachers began to think about how to permeate mathematics teaching with moral education.

In terms of the development of course resources, the textbooks in common use in the old liberated areas and various regions were adapted. In the meantime, Dewey's pragmatic thinking was also having some influence. Courses developed by Ziyi Yu proved to be a good example and were used widely in South China.

6.2 Imitation of the Soviet Union's Mathematic Courses (1952–1958)

The Central Government of the CPC announced that China would "adopt the Russian ideas" and would draw on the help of the Soviet Union in various aspects. This policy resulted in China's education imitating the Soviet Union's education mode considerably.

6.2.1 Development of Primary School Arithmetic Courses in the Soviet Mode

In June, 1952, the MOE organized a committee to draft a teaching program for all subjects in primary and secondary schools. According to the policy of "sinochem the advanced experience from the Soviet Union after learning about it," and referring to the curricula, teaching materials, and teaching programs of the former

Soviet Union's ten-year school system, our country authorized and established our own mathematic courses for primary and secondary schools. The MOE decided to implement a five-year system in primary schools from 1952. However, after one year (in November, 1953), the Government Administration Council decided to stop the five-year system and to continue to use the "4+2" system in primary schools (junior primary school for four years and higher senior primary school for two years); thus, the arithmetic teaching content that covered four years in the former Soviet Union needed to be expanded to cover six years in our country.

On March 18, 1952, the MOE issued *Tentative Regulations for Primary Schools (Draft)*, which stipulated that the five-year system would be adopted in primary and secondary schools, and that arithmetic courses would be offered from the first grade to the fifth grade. The numbers of weekly periods for primary school mathematics courses from the first grade to the fifth grade would be 5, 6, 7, 7, 7, respectively, with 1216 class periods in total, which was the same as for the period between 1949 and 1952. Arithmetic included abacus calculation, which was taught in the fourth and fifth grades. In September, 1953, the MOE issued the *Tentative Teaching Programme for Primary Schools (4 + 2 Educational System) (Draft)*, according to which the numbers of weekly periods from the first and the sixth grade were 6, 6, 7, 7, 7, 7, respectively, totaling 1520 h, an increase of 304. From the fourth year, arithmetic included abacus calculation, which had one class period every week on average; this was realized through massed learning in various schools in certain periods. In January 1954, the *Tentative Teaching Programme for Primary Schools (4 + 2 Educational System) (Draft)* was revised once more. On September 2, the MOE issued the teaching program for primary schools, in which the six-year system was adopted, and in which the weekly class periods of arithmetic from the first grade to the sixth grade were 6, 6, 6, 7, 6, 5, respectively, with 1224 class periods in total, nearly the same as the original. The abacus calculation was taught in the fourth and the fifth grades, with one lesson per week in each. In June 11, 1957, the MOE issued the *Teaching Programme for Primary School of the School Year from 1957 to 1958*, in which, the total number of arithmetic lessons was kept at 1224, while the six weekly lessons were offered for each grade from the first to the sixth. Abacus calculation was placed in the fourth and the fifth years, but could also be taught in the sixth year.

In 1951, the MOE translated the Soviet Union's ten-year-system teaching program for primary school arithmetic and formulated the first teaching program for abacus calculation in primary schools in new China. In December, 1952, the MOE issued the *Teaching Programme for Primary School Arithmetic (Draft)* and *Teaching Programme for Abacus Calculation for Primary Schools (Draft)* (the two drafts, combined, were abbreviated as the *52 Programme*).

The compiling and publishing of the primary school mathematic textbooks during this period were highly uniform, undertaken by the People's Education Press. The country transferred primary school educational experts to the People's Education Press, to specialize in the compiling and publishing of the mathematic textbooks.

During this period, the primary school mathematic textbooks are based mainly on those of the Soviet Union, although the four-year teaching content had to be expanded to cover five or six years in China. The People's Education Press assigned Deyuan Huo, Feiyu Cao, Runquan Li, Shiyi Sun, and Youpi Xia to be the editors in chief of the primary school arithmetic textbooks ("4+2" educational system) for the 1952 edition. This set of textbooks came off the press successively from April 1952 to August 1954, including *Junior Primary School Textbook—Arithmetic (8 volumes altogether)*, *Junior Primary School Textbook—Abacus calculation (1 volume altogether)*, *Senior Primary School Textbook—Arithmetic (4 volumes altogether)*, and *Senior Primary School Textbook—Abacus calculation (2 volumes altogether)*, based on which, a set of primary school arithmetic textbooks was compiled, and used until 1958.

6.2.2 Implementation of the Primary School Arithmetic Courses in the Soviet Mode

During this period, China's primary school mathematics teaching was based on the Soviet teaching mode of "five links" advocated in Kaiipob's (1893–1978) *Pedagogy*: organizing teaching, reviewing previous lessons, teaching new lessons, consolidating new lessons, and assigning homework. The principle of intuitional instruction was emphasized, and teachers were called on to help their primary school students to understand the basic knowledge of mathematics. The teaching of mathematical problem-solving methods was reformed. In 1953, the People's Education Press translated and published A.C. Пчелко's *Teaching Methodology of Primary School Arithmetic*, and our country's education periodicals published corresponding book reviews, recommending it to the public. In addition, many books related to arithmetic teaching were published, which played a certain role in reforming the old teaching methods.

At that time, the work of developing primary school mathematics courses was focused mainly on the compilation and the translation of the Soviet Union's arithmetic textbooks, and People's Education Press and various provinces and cities published a series of references based on the unified textbooks.

6.3 The First Exploration into a Primary School Mathematics Curriculum with Chinese Characteristics (1958–1966)

In 1958, the Central Committee of the Communist Party of China (CPC) put forward the goal of going all out and striving to build a socialist country in a faster, better, and more economical way, and the Great Leap Forward swept across the country. From the second half of 1960, the national economy entered a difficult three-year period because of a natural disaster. From 1963 to 1965, Chinese education began to advance in the right direction and witnessed steady development and better results by drawing on past experiences and correcting mistakes.

6.3.1 Developing a Full-Time Mathematics Curriculum with Chinese Characteristics

In 1958, *An Instruction to Education from the Central Committee of the CPC and the State Council* was released to call for reforms in the school system. Accordingly, reforms like bringing forward the school starting age, running primary schools in various ways, and adopting a five-year system in primary schools were carried out in various regions, but without obvious effects (National Academy of Education Administration, 1999, pp. 108–109). In May 1959, the Central Committee of the CPC and the State Council issued *The Stipulation on Reforms in the School System on an Experimental Basis*, requiring that reforms in the school system should be carried out in an organized and well-directed way. Reform attempts in some local primary schools were halted, and the management of reform in the school system was regulated and adjusted. As a matter of fact, the six-year-system primary mathematics curriculum was prevalent during that time, while the five-year system was carried out on an experimental basis.

In the first half of 1959, the MOE convened a nationwide forum on Primary and Secondary Mathematics Teaching, aimed at adjusting primary and secondary mathematics curricula and specific teaching content, as well as revising syllabi and compiling national textbooks on the basis of the adjustment. The attendees also "conducted a comparative study of representative course standards and syllabuses issued in China between 1912 and 1956, on primary and secondary mathematics curriculum setting in some eastern European countries, and on mathematics curriculum setting and teaching content before and after liberation, as well as between China and some eastern European countries" (Lv, 2007). In 1963, the MOE rewrote and released a *New Education Program for Full-time Primary and Secondary Schools (Draft)*, in light of the *Central Committee's Draft of Discussion on Trial Implementation of Work Regulations in Full-time Primary and Secondary Schools and Instructions on Issues of Current Primary and Secondary Education*. In this teaching program (six-year system) for primary schools, the numbers of weekly arithmetic classes 6, 6, 7, 8, 9, and 9, respectively, for grades 1–6, totaling 1649, increasing by 129 compared with the prospective total class periods (1520) planned according to the "4+2" system in 1954. On July 14, 1964, the MOE released *Notification on Adjusting and Simplifying Primary and Secondary Curriculum*, taking away one class period weekly from the total class period of arithmetic from grade three to grade six.

On the basis of this teaching program, the MOE entrusted the People's Education Press to draft the *Full-time Primary Mathematics Syllabus*, which was officially released in September, 1963, also known as the *63 Outline*. This outline proposed three capabilities, namely arithmetic capability, logical reasoning capability, and the capability of space concept (Curriculum and Teaching Materials Research Institute, 2001a, p. 82), as the program objectives.

In 1958, the Central Committee of the CPC and the State Council released the *Regulations for Delegating Educational Management Rights*, pointing out that "local schools could revise or replenish all kinds of instructional teaching programs, syllabi, and general textbooks released by the MOE and relevant central ministries in light of specific conditions in local regions and schools; they could even compile their own textbooks" (Party School Theory Research Institute of the Central Committee of the Communist Party of China, 2005, pp. 76–77). In September 1958, the *Notification About Issuing No Textbooks in the Future* was released by the MOE; this pointed out that the MOE would no longer issue a list of textbooks and that regions could compile their own, although the People's Education Press was still entitled by the MOE to release instructional teaching programs and syllabi, and to compile national basic textbooks for every province (General Office the Ministry of Education of the People's Republic of China, 1959).

Mathematics textbooks of this period included provisional ones published by the People's Education Press, arithmetic textbooks for ten-year-system schools, arithmetic textbooks used by the elementary and advanced primary parts of the full-time twelve-year-system schools, and various textbooks compiled in different regions. One of these was a set of *Mathematics Textbooks (Trial Edition) for Nine-year Full-time Schools (including elementary algebra)* compiled by the Beijing Normal University research group in April, 1960. In August, 1961, the research group drew on a variety of local reform experiences and recomposed this set of textbooks, producing *Testing Textbooks (Trial Edition) for Ten-year-system Schools*, which were published by the People's Education Press. This included ten volumes of primary mathematics textbooks for the five-year-system primary schools. The elementary and secondary curriculum innovation research group, with the Department of Mathematics, East China Normal University, compiled a set of *Five-year-system Primary Mathematics Textbooks,* which were examined and approved by the Shanghai Elementary and Secondary Mathematics Course Innovation Committee, and published by the Shanghai Educational Publishing House. In addition, the Jiangsu Teaching Materials Writing Committee compiled the *Five-year-system Primary School Textbooks (Trial Edition) in Jiangsu-Elementary Mathematics*, which were published by the Jiangsu People's Publishing House in 1960. The Zhejiang Primary and Elementary Teaching Materials Innovation Committee compiled the *Five-year-system Primary School Textbooks (Trial Edition)-Mathematics*, which were published by the Zhejiang People's Publishing House (1958–1961). The Fujian Primary and Elementary Teaching Materials Compiling and Examining Committee, together with the Fujian Normal College, compiled *Five-year-system Primary School Mathematics Textbooks (Trial Edition)* (1960–1961). The Jiangxi Nursery, Primary, and Secondary School Teaching

Materials Compiling and Examining Committee compiled the *Full-time Five-year-system Primary School Textbook-Mathematics*, published by Jiangxi People's Publishing House (1960–1961). The Ningxia Hui Autonomous Region Elementary and Secondary School Teaching Materials Compiling and Examining Group compiled the *Five-year-system Primary School Mathematics Textbooks*, which were published by Ningxia Hui Autonomous Region People's Publishing House (1964–1966). The Shandong Normal College compiled the *Five-year-system Primary School Mathematics (Trial Vision)*, published by Shandong People's Publishing House (1960–1962). The Department of Mathematics, Gansu Normal University, compiled a ten-year-system *Primary School Arithmetic Textbook*, published by Gansu People's Publishing House (1960–1961). The Psychology Institute, Chinese Academy of Sciences, compiled the *Nine-year-system Arithmetic Textbooks (Trial Edition)*, which were published by Beijing Publishing House (1960).

6.3.2 Implementation of Full-Time Primary Mathematics Curriculum with Chinese Characteristics

In April, 1960, the Beiguan Primary School in Heishan County, Liaoning Province, initiated the teaching methodology of "teaching only the essential and ensuring plenty of practice," which was originally applied in teaching Chinese, and later in mathematics. During 1962 and 1966, reforms in teaching methodology were carried out across the nation, and the methodology of "teaching only the essential and ensuring plenty of practice" was adopted by all schools. The reforms were mainly about combining mental arithmetic and written calculation, and innovation in abacus calculation teaching. The above-mentioned methodology had greatly expanded the scale of primary school arithmetic teaching and had enabled pupils to improve their arithmetic knowledge and calculation skills significantly. In July 1963, in the notifications about carrying on reforms and experiments in primary and secondary schools teaching, the MOE pointed out that it was confident that teaching tasks for the six-year-system primary schools could be accomplished in schools adopting the five-year consistent system. Pupils in five-year-system schools were taught more in arithmetic classes than those of the same grade in six-year-system schools. Students in five-year-system schools were physically and mentally competent, and they were more likely to make the required effort. In the fall of 1960, the five-year system was introduced. At the same time, in two grade 5 classes in the Beijing Second Experimental Primary School, the Psychology Research Institute, Chinese Academy of Sciences, tried to teach students both arithmetic and algebra; the results showed that after learning algebra, the students could solve arithmetic problems easily. Therefore, it was believed that some content taught in secondary schools could be taught earlier in primary schools. Experimental reforms in arithmetic, like "teaching the essential and ensuring plenty of practice," provided strong data and practice support for removing arithmetic content from secondary schools to primary schools.

During this period, textbooks and teaching reference books (or materials) still made up most of the primary mathematics teaching resources, and developing and using training aids became a new expansion of teaching resources. Physical and visual teaching aids were important course resources during this period. Teachers in local primary schools had developed a variety of visual aids for teaching arithmetic, geometry, and the like. For example, the Xianghanjiang Experimental Primary School in Putian County developed 25 arithmetic teaching aids (Xianghanjiang Experimental Primary School in Putian County, 1960). Primary school teachers introduced easy-to-make and simple arithmetic teaching aids (Author unknown, 1957; Huang, 1957); they also encouraged students to make visual mathematics aids in labor skill classes in such a way as to enhance the students' understanding of knowledge. Some teachers shared their experiences of using teaching aids in their teaching practice through articles in periodicals (Zheng, 1963; Xu, 1960; Hu, 1957).

6.4 Implementation of Regional Independence in Primary Mathematics Curriculum (1966–1976)

The release of the *May 7 Directive of 1966* and *The Notice of the Central Committee of the CPC* (known as *the May 17 Notice of 1966*) marked the beginning of the Great Proletarian Cultural Revolution, from which China entered a decade of political and economic turmoil until October 1976. After the Central Committee of the CPC issued The *Comments on the Great Proletarian Cultural Revolution in Secondary Schools* in February 1967, the primary mathematics curriculum was retained as essential content. During this period, the education field had no unified teaching plan or teaching program. Therefore, the government encouraged the faculty at all school levels to develop their own teaching plans, teaching curriculum, teaching content, and textbooks.

6.4.1 Development of Regionalized Primary Mathematic Curriculum

During this period, the flexible primary education system was implemented nationwide. The five-year primary mathematics education system coexisted with the six-year system, and many places canceled the junior and senior primary school model, putting the five-year consistent system into trial use. There were 14 provinces and autonomous regions trying out the nine-year education system (five years for primary school, two years for secondary school, and two years for high school); seven provinces, cities, and autonomous regions tried out the ten-year system (five years for primary school, three years for secondary school, and two years for high school, or six years for primary school and four years for secondary school); nine provinces and autonomous regions tried out the nine-year system in rural schools and the ten-year system in urban schools. The Tibet Autonomous Region tried out the coexistence of the five-year and six-year systems in the primary school and the three-year secondary school system (National Institute of Education Sciences, 1983, p. 475).

The main contents of the mathematic curriculum were knowledge and operations with integers, decimals, percentages, and proportion as well as measurement, geometry, and simple statistical diagrams. In some provinces, students learned more about the basic knowledge of geometry than in other places, but in all provinces the primary school mathematics content did not go beyond ratio and proportion. A few provinces set up calculation with the abacus as a separate course, while most integrated it into the arithmetic course. In setting the primary school mathematics curriculum, the emphasis was laid on the simplification and condensation of knowledge, specifically arranging the topics of integers, decimals, and fractions alternately and combining the knowledge of geometry, algebra, and formulae scientifically.

As there was no unified textbook system nationwide, groups of primary and secondary schools in various regions developed their own and issued them via the New China Book Store after approval and regulation from the local education administrative departments. From 1968, all provinces, autonomous regions, and municipalities had established their local textbook-compiling groups for primary and secondary schools. Some places with insufficient experience adopted the textbooks of Beijing or Shanghai. In the second half of 1972, some provinces, autonomous regions, and municipalities began to collaborate on the basis of self-compilation of textbooks so as to solve the difficulties encountered and ensure the quality. From 1973, some provinces, autonomous regions and municipalities directly under the Central Government also compiled (mainly done by their subordinate regions and counties) a primary mathematics textbook of "three calculations" (namely abacus, oral, and calculations) as an option for schools.

6.4.2 Regional Independence in Primary Mathematic Curriculum

During this period, all provinces carried out large-scale teaching experiments on the combination of three types of calculations, oral, abacus, and written. This occurred in 29 provinces, cities, and autonomous regions. Their methods were varied. Some provinces put oral calculation as the fundamental method while emphasizing written calculation and using abacus calculation as a tool; some used abacus calculation as the basis of their teaching, reforming written calculation, and promoting oral; some laid emphasis on the written, with abacus calculations playing a supportive role in the hope that the two would support each other. These "three calculations" experiments were still in practice until the 1980s in the form of small experimental classes in some provinces. This experiment effectively promoted the combination of oral calculation and writing calculation in the primary mathematics curriculum, and the role and significance of abacus calculation as a calculation tool were also taken into consideration.

Valuable resources were left behind by the "three calculation" textbook for the development of primary mathematics curricula during this period.

As stated above, this development started with adopting and reforming the curriculum of the Republic of China, then came the large-scale adoption of the Soviet model, after which came the first attempt to develop a primary school curriculum for New China, and finally regionalization due to the impact of the Cultural Revolution. The primary mathematics curricula were influenced heavily by the political movements in New China; for example, the *52 Syllabus* and its corresponding curriculum and textbooks were the result of learning from the Soviet Union. During the Cultural Revolution, local curricula, textbooks, and curriculum reforms were branded deeply by political movements. Under the influence of pragmatism and the Soviet education theories, the combination of teaching and practice was emphasized, and the focus on two fundamentals and three abilities was introduced as well.

6.5 Reexploration of the Chinese Characteristics Primary Mathematics Curriculum (1976–1986)

After the end of the Cultural Revolution, the MOE began to issue teaching plans for primary and secondary schools to standardize the primary school mathematics curricula. In 1977, under the direct leadership of the MOE, resources were organized to redraft the mathematics teaching program for primary schools in the ten-year school system, and to compile national mathematics textbooks. In addition, the teaching experiments in the five-year and six-year systems in various regions were carried out in a scattered but orderly way, promoting curriculum reform and development.

6.5.1 Research and Development of Diversified and Regional Mathematics Curriculum of Primary Schools

On January 18, 1978, the MOE issued the *Trial Draft Teaching Plan for Full-Time Primary and Secondary Schools*, which stipulated that the full-time systems for primary and secondary schools would be ten years, namely five years for primary schools and five for secondary schools. Since 1980, a few provinces and municipalities had changed to the six-year primary school system. Soon afterward, in 1980, the *No. 84 Document* from the Central Government stipulated that primary schools should implement the coexisting five-year and six-year school systems; those in urban areas could implement a trial six-year school system, and those in rural areas could remain with the existing system for the time being. Beijing, Shanghai, and Tianjin implemented the six-year system, at first trialing the plans, preparation, and processes in the urban areas.

On September 22, 1978, the MOE issued the *Notice of Regulations and Provisional Rules for the Trial Implementation in Full-Time Secondary Schools (Trial Draft),* and *Regulations and Provisional Rules for the Trial Implementation in Full-Time Primary Schools (Trial Draft).* The second chapter of *Teaching Work* pointed out that: "mathematics in primary schools should strengthen the teaching of mathematics basic knowledge and the training of basic skills." Students should be familiar with the formulae and able to do arithmetic calculations correctly and rapidly. They should be trained to develop the abilities of calculation, logical thinking, and problem solving. The written form should be in accordance with the regulations (Curriculum and Teaching Materials Research Institute, 2001b, p. 315). Later, the MOE issued the *Trial Implementation Draft of the Teaching Plan for Full-Time Ten-year-school-system Primary and Secondary Schools.* It stipulated that mathematics for primary schools was for five years, with 7, 7, 6, 6, and 6 weekly classes for grades 1–5, respectively. There were 42 weeks for grades 1–3, and 39 weeks for grades 4 and 5, with 1302 class periods in total. On March 13, 1981, the State Education Commission released the revised draft of this document. There were, respectively, 6, 6, 6, 7, and 7 glasses for grades 1–5, with 150 class periods in total. On August 15, 1984, a six-year plan was issued, with separate versions for urban and rural school systems. The urban version stipulated 5–6, 5–6, 6, 6, 6, and 6 lessons per week for the six grades, respectively, with 1156–1224 class periods in total, and rural document 6, 6, 6, 6, 6, and 6, with 1224 in total. Grades 1 and 2 in urban primary schools were expected to have 5 or 6 periods of mathematics every week, and schools with better conditions were advised to have 5 periods. Grades 4 and 5 in rural primary schools had separate abacus calculation courses, hence did not cover this content in the general teaching materials. The mathematics teaching for senior grades in rural areas was expected to supplement proper knowledge of measurement, statistics, and accounting. Mathematics courses for grades 4, 5, and 6 in rural primary schools taking these measures could add up to seven class periods (Curriculum and Teaching Materials Research Institute, 2001b, p. 347). During this period, our country basically achieved full-time primary school education, and the mathematics curriculum in primary schools implemented two kinds of full-time curricula, one of five years and the other of six.

The MOE organized a national mathematics textbook-compiling team for primary and secondary schools to draft primary and secondary school teaching plans from September, 1977, on the basis of primary school mathematics textbooks and teaching programs at home and abroad, and mimeographed these into books in order to seek advice from educational administrations, teaching and research departments, and normal colleges and universities. In February, 1978, the MOE issued the *Teaching Program for Mathematics in the Full-time Ten-year-school-system Primary Schools (Trial Draft)* (briefly referred to as the *78 Program*), announcing that primary schools should implement a consistent five-year school system. The *78 Program* puts forward for the first time that "we should simplify the traditional arithmetic content and appropriately imbue the thinking methods of modern mathematics." After publishing the *Compulsory Education Law*, the State Education Commission organized resources to revise the *78 Program* under the principle of "lowering difficulty, reducing burden, and specifying teaching requirements." The revision of the program was to be based on the actual situation of mathematics teaching in the primary schools of our country at that time, to summarize the teaching practice experiments between 1978 and 1986, and to remain basically unchanged in the mathematics textbooks in current primary schools. The *Revised Draft* was issued by the State Education Commission in December, 1986; its full name was the *Teaching Programme for Mathematics in Full-time Primary Schools* (briefly referred to as the *86 Program*). The *86 Programme* adopted both the five-year and six-year systems and was richer in course target levels and contents.

In September, 1977, the MOE decided to rely on the textbook-compiling staff for primary and secondary schools in the People's Education Press and appointed a larger number of teachers and textbook-compiling staff from primary, secondary schools, and universities in 18 provinces, autonomous regions, and municipals. Using the format of a working conference, it established teams to compile general textbooks for each discipline and started to draft teaching programs. At the same time, it employed some mathematics experts and educators as consultants and set up textbook-compiling, editing, and leading groups.

According to the *78 Program* issued by the MOE, the *First Volume of Primary School Mathematics Textbooks for Full-time Ten-year-system Schools (Trial Edition)*, edited by the People's Education Press, was released nationally in the autumn semester of 1978, and the other volumes were published in succession in 1983. This series of textbooks emphasized the "two bases" of teaching and the development of "three abilities," arithmetic, logical thinking, and initial spatial ideas. The textbooks were in use until the end of the 1990s and were regarded as being of the highest quality, the longest used, and the most stable after the founding of the country. In order to be consistent with the actual situation of the coexisting five-year and six-year school systems, and especially to meet the needs of some areas implementing the latter, the People's Education Press edited and published a series of *Six-Year Primary School Mathematics Textbooks* after 1980, with 12 volumes in total. They were all published, and the contents were the same as those in the five-year-school-system textbooks; to adapt to the characteristics of the six-year schools, the distribution of the knowledge and focus was adjusted and slightly more practice examples were included.

Various regions had also compiled experimental textbooks with distinct features on the basis of teaching reform experiments. These included the *Primary School*

Experimental Mathematics Textbooks produced in collaboration between the Education College of Beijing Normal University and Beijing Jingshan School; the *Primary School Experimental Mathematics Textbook (Trial Edition), Volumes 1–4,* edited by the National Institute of Education and Science, Educational Science Publishing House (1981); the *Primary School Experimental Mathematics Textbooks* by the Office of Educational Research of Hangzhou Normal University and Heilongjiang Education Institute; the *Modern Primary School Mathematics Experimental Textbook* edited by Liu Jinghe; the *Primary School Textbook of Experimental Teaching* compiled by Jiang Leren; the *Primary School Mathematics Experiment Textbook for the Comprehensive Construction Method* compiled by Zhao Songguang; the *Primary School Mathematics Experimental Textbook* compiled by the Curriculum and Teaching Material Research Institute; the *Trial Primary School Mathematics Textbook* edited by the Teaching and Research Division of the Shanghai Municipal Bureau of Education; and the newly compiled *Primary School Experimental Mathematics Textbook* edited by Wang Jizhen. The People's Education Press decided to compile a series of five-year-system primary experimental mathematics textbooks.

6.5.2 Implementation of Diversified and Regional Mathematics Curricula in Primary Schools

The process of teaching mathematics in primary schools after the reform opened up concerns about students. Using their development as the starting point, further experiments were conducted:

- The "three calculation" combination in Hunan province,
- The teaching reform of "student orientation' conducted by Ma Xinlan, a teacher of the Central Primary School of Happy County, Chaoyang District of Beijing,
- Experimental teaching method research done by Qiu Xuehua,
- Reform and experiment on methodology in Changchun Second Experimental Primary School,
- The whole-teaching experiment in mathematics in primary schools, in Fujian Province,
- Reform and experimentation on textbooks and teaching methods for primary school mathematics in the Department of Mathematics of Shanghai Normal University, and
- The tracking-of-teaching experiment in primary school mathematics by the College of Education, Shanxi University.

These studies laid the foundation for primary school curriculum development in the 1990s, particularly for the establishment of primary school mathematics curricula in compulsory education.

This was the recovery period for mathematics education in primary schools. Primary schools from all over the country started mathematics journals in succession, which provided platforms for the primary school mathematics educators to engage in research and exchanges. For example, the journals included *Mathematics Teachers in Primary Schools* by Shanghai Education Press in 1978, *Curriculum Teaching: Material Teaching Method,* sponsored by the Curriculum and Teaching Material Research Institute of the People's Education Press in 1981, *Mathematics in Primary and Secondary Schools,* directed by the MOE and sponsored by the Chinese Education Society in 1982, *Research on the Teaching Methods in Primary Schools* by Jiangxi Education Press in 1980, and *Primary School Teaching (Mathematics Edition)* (formerly *Young Teachers of Primary School*). Moreover, as a result of the experiments, the textbooks, teaching reference materials, and teaching aids had become richer. In particular, some foreign teaching aids were introduced into our country, such as the geoboard for plane geometry, and Cuisenaire used for the four arithmetic operations on integers.

At the same time, further experiments were conducted to explore the Chinese characteristics of the mathematics curriculum. In the 1980s, Cao Feiyu from the People's Education Press and his graduates conducted a series of experiments on the textbook contents of primary school mathematics and the psychology of mathematics, including tests of preschool children's mathematical concepts and calculation abilities, and experiments on the teaching of topics such as addition and subtraction to 20 and simple addition and subtraction applications, research on developing grade 2 students' abilities to answer two-step questions, and on developing students' abilities to answer fractional application questions and simple equations.

6.6 Establishment of Primary School Mathematics Curriculum for Compulsory Education (1986–2001)

On April 12, 1986, the *Compulsory Education Law of the People's Republic of China* was enacted; in February 1993, the Central Committee of the CPC and the State Council published the *Outline for Reform and Development of Education in China*. In June 1999, the *Resolutions of the Central Committee of the CPC and the State Council on Deepening Education Reform and Pushing Forward All-round Quality Education* was issued, and in the same year the third national education conference was held. In 1999, the Central Committee of the CPC and the State Council circulated the State Education Commission's *Action Plan for Invigorating Education in the twenty-first Century*. All of these documents were proposed in order to reform the elementary education curriculum system and to speed up the construction of a new system adapted to the requirements of the time.

6.6.1 *Development of Primary Mathematics Curriculum for Full-Time Compulsory Education*

In 1988, the MOE issued the *Notice on the Issuance of Teaching Plans for Primary and Secondary Schools in Full-time Compulsory Education (Trial Draft)* and *Syllabus for 24 Subjects (Preliminary Draft)*. This was followed, in 1992, by the *Notice on the Issuance of Curriculum Plans for Primary and Secondary Schools in Full-time Nine-year Compulsory Education (Trial Draft)* and the *Syllabus for 24 Subjects (On Trial)*. Both of these documents were fit for the "six-three" mode of six-year primary and three-year secondary schooling, the "five-four" mode of five-year primary and three-year secondary schooling, the "nine-year consistent" system, and the transitional education system of five-year primary and three-year secondary schooling. The document supported carrying out the experimental work of the "five-four" mode in full swing. In fact, this was a period when the five-year and six-year primary school modes both existed.

The purpose of the mathematic curriculum was mainly to help students to grasp the most basic knowledge of integers, decimals, and fractions, use the four arithmetic operations on integers, decimals, and fractions correctly and swiftly, learn simple geometric figures and abacus calculation knowledge, develop early logical thinking and spatial concepts, and make use of prior knowledge to solve some simple practical problems (Curriculum and Teaching Materials Research Institute, 2001b, p. 353). From the first to the sixth grades, there were 4, 5, 5, 5, 5, and 5 mathematics lessons, respectively, per week, totaling 986; from the first to the fifth grades, there were 5, 6, 6, 6, and 6, respectively, totaling 986. This was a reduction of 20%.

In the *Notice on the Issuance of Curriculum Plans for Primary and Secondary Schools in Full-time Nine-year Compulsory Education (Trial Draft)* and the *Syllabus for 24 Subjects (On Trial)* issued by MOE in 1992, it was stated that the primary and secondary mathematics curricula should help students to grasp the basic skills in the four arithmetic operations of integers, decimals, and fractions, learn some simple geometric figures, equations, and abacus calculation knowledge, learn some simple and basic statistical knowledge, develop early logical thinking and spatial concepts, and make use of learned mathematic knowledge to solve some simple practical problems (Curriculum and Teaching Materials Research Institute, 2001b, p. 374). From the first grade to the sixth grade, there were 4, 5, 5, 5, 5, and 5 weekly classes, respectively, with 986 class periods in total; from the first grade to the fifth grade, there were 5, 6, 6, 6, and 6, respectively, with 986 in total. In 1994, with the publication of a new work schedule by the State Council, curriculum (teaching) planning for primary and secondary schools was adjusted. On July 4, the MOE issued further adjustments to full-time primary, secondary, and senior high school teaching plans. There were no changes to the numbers of lessons in the six-year system but in the five-year system it was changed to 5, 6, 6, 5, and 6, and the total class periods were reduced from 986 to 952.

After the publication of the Nine-year Compulsory Education Law, the State Education Commission authorized the People's Education Press, the Education Commissions (Education Department, Bureau of Education) of Beijing, Shanghai, Liaoning, Shandong, Guangdong, and Xi'an, and Beijing Normal University and other departments as a committee to compile teaching plans for each subject. In 1987, according to the curriculum plan for primary and secondary schools in full-time compulsory education, the committee drafted the *Teaching Plan for Primary Mathematics for Full-time Nine-year Compulsory Education (Draft for Comment)*; in August 1987, the State Education Commission summoned a special meeting to study and discuss the teaching plans for each subject. In November 1988, the State Education Commission issued an updated draft, which became the basis for the compilation of a trial pack for primary mathematics in nine-year compulsory education, further teaching reform experiments, and the promotion of compulsory education textbooks in our country.

After the publication of the *88 Plan*, the People's Education Press compiled the corresponding *Experimental Textbook of Primary Mathematics*. After three years of application in experimental zones and soliciting opinions from the whole nation, the State Education Commission completely overhauled the *88 Plan*. After being authorized by the Primary and Secondary School Textbook Review Committee of the State Education Commission, the State Education Commission issued the trial version of the *Teaching Plan for Primary Mathematics of Full-time Nine-year Compulsory Education* (referred to as the *92 Plan*). From 1994, adjustments were made that included curriculum changes and content adjustment, and in March 2000, the final version (referred to as the *00 Plan*) was issued.

In this period, a system of textbook authorization started to take shape. In September 1986, a national committee was established to consider the nation's teaching plans and authorizing textbooks for primary and secondary subjects, which helped to promote diverse and formal development of textbooks. Following a conference held in 1988, it was decided that textbooks should be developed to reflect different cultures, educational systems, regions, and levels. The People's Education Press and other departments all contributed to the compilation and publication of primary mathematics textbooks, which created a boom for textbook development.

From 1990 to 1995, the People's Education Press successively compiled and published two sets of mathematics experimental textbooks for five-year and six-year primary schools in full-time compulsory education, providing for experimental zone in the nation and soliciting their opinions. Various sets of five-year and six-year primary mathematics textbooks were published nationwide, by the People's Education Press, Beijing Normal University, Shanghai, Zhejiang, Northeast Normal University with eight other normal universities, the Department of Education of Guangdong Province, State Education Commission of Fujian Province, State Education Commission of Hainan Province, and South China Normal University, the State Education Commission of Sichuan Province and Southwestern Normal University, and Henan Province. The National Primary and Secondary School Textbook Authorization Committee authorized all of these sets, and they were used in experimental zones nationwide; from the autumn semester of 1993, they were used by primary schools nationwide.

6.6.2 Implementation of Primary Mathematics Curriculum for Full-Time Compulsory Education

In order to focus on the development of thinking abilities, frontline teachers used pedagogical and psychological theories, paid attention to analyzing students' work, and brought in diverse classroom-teaching methods. Visual teaching was used widely, introducing visual objects, teaching tools, and images which connected mathematics with children's lives, creating a learning context, and encouraging them to "use their hands, open their mouths, and work their brains."

During this period, visual teaching tools from abroad were introduced. Traditional visual teaching tools also sprang up in diverse textbooks, for instance: Traditional tangrams and kaleidoscopes from ancient China appeared in primary textbooks; images increased, paper teaching tools were provided in appendices, and corresponding teaching tool boxes were also produced by textbook publishing houses.

Electronic education instruments came into primary mathematic classes, including projectors, televisions, computers, and other multimedia devices that enriched teaching content with sound, light, and increased dynamic and generative properties. In the 1990s, a meeting of the Professional Commission for Mathematics Teaching in primary schools in Chinese Society invited three teachers to deliver observation lessons. This was the forerunner for large classroom-teaching observation and communication meetings to be held every two years and recorded on video.

References

Author unknown. (1957). 简单易制的算术教具 [Simple and easy-to-make arithmetic training aids]. *Elementary Education Journal, 73*, 11.

CCCPC Party Literature Research Office. (2011). 建国以来重要文献选编*(*第一册*)* [Anthology of major documents since the founding of P.R.C.] (Vol. I). Beijing: Zhongyangwenxian Press.

Curriculum and Teaching Materials Research Institute. (Ed.). (2001a). *20* 世纪中国中小学课程标准*:* 教学大纲汇编 [Collection of primary and secondary school curriculum standards and syllabus of the twentieth century China]. Beijing: People's Education Press.

Curriculum and Teaching Materials Research Institute. (Ed.). (2001b). *20* 世纪中国中小学课程标准•教学大纲汇编*(*课程*(*教学*)*计划卷*)* [Curriculum standard for primary and secondary schools in China in the 20th century syllabus compilation (curriculum teaching) plan volume)]. Beijing: People's Education Press.

Curriculum and Teaching Materials Research Institute. (2010). 新中国中小学教材建设史*1949-2000*研究丛书 数学卷 [Research series on the construction history of the textbooks in primary and secondary schools from 1949 to 2000 mathematics volume]. Beijing: People's Education Press.

General Office the Ministry of Education of the People's Republic of China. (Ed.). (1959). 教育文献法令汇编*(1958年)* [Education literature and decree compilation (1958)]. Beijing: Ministry of Education of China.

Hu, H. W. (1957). 在应用题教学中运用直观教具的初步体会-谈运用直观教具的适当性 [Experience of applying visual aids in teaching students how to solve mathematics word problems]. *Anhui Education, 6,* 24.

Huang, J. L. (1957). 小学算术教具箱 [Primary school arithmetic training aids box]. *Elementary Education Journal, 87,* 27.

Lv, S. H. (2007). 20 世纪中国中学数学课程发展 (1950–2000) [Development of China's mathematic courses in the 21st century (1950–2000)]. *Mathematic Bulletin, 7,* 8–15.

National Academy of Education Administration. (1999). 新中国教育行政管理五十年 [Fifty years of the education administration in New China]. Beijing: People's Education Press.

National Institute of Education Sciences. (1983). 中华人民共和国教育大事纪*(1949–1982)* [Big Events in the education of the People's Republic of China (1949–1982)]. Beijing: Education Sciences Press.

Party School Theory Research Institute of the Central Committee of the Communist Party of China. (2005). 历史的丰碑:中华人民共国国史全鉴 教育卷 [Monument of history: Review on the history of People's Republic of China, education volume]. Beijing: Central Party Literature Press.

Wang, Q. (1996). 中国小学数学教学史 [The teaching history of Chinese primary school mathematics]. Jinan: Shandong Education Press.

Xianghanjiang Experimental Primary School in Putian County. (1960). 介绍25件算术教具的制法和用法 [Introduction to the preparation and usage of 25 arithmetic training aids]. *Fujian Education (Primary Education Edition), 3,* 26.

Xu, S. Z. (1960). 直观教具在小学算术教学上的应用 [Application of visual training aids in elementary arithmetic teaching]. *Fujian Education (Primary Education Edition), 3*, 25.

Zheng, Z. X. (1963). 试论直观教具与运动分析器结合的教学方式对小学低年级学生掌握算术知识的作用 [On how the combination of visual aids and motion analyzer affect primary school low-grade students in mastering arithmetic knowledge]. *Academic Research (Guangzhou), 5,* 27–31.

Chapter 7
The Development of Mathematics Curriculum and Teaching Materials in Secondary Schools in the Second Half of the Twentieth Century

Shi-hu Lv and Chun-yan Cao

Abstract The evolution of the secondary school mathematics curriculum in China has been ongoing for more than sixty years, since 1949. To harness the history of these sixty years' development, it is necessary to select the appropriate focus, and this depends on how to comprehend the curriculum.

The evolution of the secondary school mathematics curriculum in China has been ongoing for more than sixty years, since 1949. To harness the history of these sixty years' development, it is necessary to select the appropriate focus, and this depends on how to comprehend the curriculum.

In China, the term "curriculum" usually refers to the educational content and processes which are set to achieve a school's training objectives (Lv, 1999). In other words, it refers to the plans, aims, and guidance, which are embodied in the Curriculum Plan (Teaching Plan), Syllabus (Curriculum Standard), and Teaching Material, these three together forming the curriculum.

Well-known American curriculum scholar J.I. Goodlad proposed that a curriculum has five different forms, the ideological curriculum, the formal curriculum, the perceived curriculum, the operational curriculum, and the experiential curriculum (Shi, 1996). The ideal curriculum is the one put forward by research institutions, academic groups, and curriculum experts. This curriculum is in the stage of theoretical research, representing the theoretical level of national curriculum research. The influence of this curriculum depends on the extent to which it is adopted by the authority. The formal curriculum is that promulgated by the administrative department of education, including curriculum plans, standards (syllabus), and textbooks;

S. Lv (✉)
College of Education, Northwest Normal University,
Lanzhou 730070, Gansu, China
e-mail: shihulv@163.com

C. Cao
College of Education Science, Shaanxi Xueqian Normal University,
Xian 710061, Shaanxi, China

Y. Cao and F.K.S. Leung (eds.), *The 21st Century Mathematics Education in China*, New Frontiers of Educational Research,
https://doi.org/10.1007/978-3-662-55781-5_7

this is also known as the official curriculum. The curriculum at this level has the significance of law and represents the educational administrative level of a country. The perceived curriculum is the one comprehended by teachers; it is determined by the teachers' own understanding, explanations, and their own subjective desire to follow the formal curriculum. The operational curriculum is that put into practice in the classroom; it is also called the practice curriculum, reflecting the teaching and learning activities of teachers and students. The experiential curriculum is the one experienced by the students, constituted by what the students obtain from the operational curriculum and their opinions of this.

Based on this understanding of the curriculum, the mathematics curriculum in this study refers to the formal curriculum, which is promulgated by officials and includes the Curriculum Plan, Mathematics Curriculum Standards (Mathematics Syllabus) and Mathematics Teaching Material. The above understanding of the curriculum is in accord with China's actual situation and habits. Since the Qing Dynasty, the school curriculum has adopted a centralized management mode, thus the curriculum design, curriculum objectives,curriculum content, and teaching demands are established by the government. Therefore, this chapter will take the Mathematics Curriculum Standards (Mathematics Syllabus) and Mathematics Teaching Material (Mathematics textbook) as reference points to trace the history of the middle school mathematics curriculum during these sixty years.

The Mathematics Curriculum Standards (Mathematics Syllabus), as the official curriculum documents, mainly set the curriculum objectives, curriculum content, and organization and are put into practice through the Mathematics Teaching Material. Thus, the purpose of this study was to analyze the characteristics of the middle school mathematics curriculum in relation to these three aspects and to describe the inspirations that led to the current mathematics curriculum reform.

In the second half of the twentieth century, the development of the Mathematics Curriculum and Teaching Material was characterized by six stages. The first stage was choosing the directions for developing the mathematics curriculum (1949–1957), and the second was exploring the development of China's mathematics curriculum for the first time (1958–1961). The third stage involved reverting to the mathematics curriculum model of the Soviet Union (1961–1966). In the fourth stage, some setbacks were encountered in the process of mathematics curriculum development (1966–1976). The fifth stage explored the developmental path of the mathematics curriculum for the second time (1977–1991), and the sixth addressed trying to establish China's mathematics curriculum system.

7.1 Choosing the Directions for Developing the Mathematics Curriculum (1949–1957)

When the People's Republic of China was established, middle school mathematics educators faced the choice of inheriting the existing mathematics curriculum system of the former Republic of China or establishing a new one. Affected by the political situation at that time, the focus turned to inheriting and reforming China's existing middle school mathematics curriculum, but after a short time, a comprehensive study of the Soviet Union mathematics curriculum was chosen.

7.1.1 Inheriting and Reforming the Existing Middle School Mathematics Curriculum

In the first half of the twentieth century, the middle school mathematics curriculum in China had experienced two stages of development, learning from foreign systems and exploring localization. During the years of 1929–1949 in particular, China had systematic mathematics curriculum standards and mathematics textbooks. At the beginning of the establishment of the People's Republic of China came a more systematic development of mathematics curriculum standards and textbooks prepared in accordance with the curriculum standards, to form a relatively complete mathematics curriculum. Most of the country still had mathematics teaching materials from the former Republic of China, and one option at that time was to inherit this middle school mathematics curriculum.

On November 1, 1949, the Ministry of Education of the Central People's Government was established. After its foundation, the Ministry of Education recognized that students were overloaded with schoolwork, and one reason was that the teaching content of mathematics, physics, and chemistry was too much and the arrangement was unreasonable. To solve this problem, in 1950 the Ministry of Education convened a forum to adjust the teaching materials for these three subjects. In this forum, twelve experts, Z. Fu, T. Cheng, Q. Wei, G. Han, J. Wang, C. Liu, C. Zhou, Z. Cao, S. Zhong, Z. Zhao, S. Guan, and M. Wang, drafted the *Simplified Mathematics Syllabus* (hereafter referred to as the *Syllabus*), which became the popular mathematics reference for teaching all over the country. On December 1, 1950, the People's Education Press was formally established and began to adapt the mathematics teaching material of the Republic of China according to the Syllabus. A set of twelve-year secondary schools mathematics textbooks (known as *Simplified Textbooks*) was published before the fall semester of 1951. The Ministry of Education required middle schools in national major administrative regions to adopt the *Simplified Textbooks*. Thus, this was the first set of mathematics teaching materials in China.

At the same time, before the first national Secondary Education Conference in March 1951, the Ministry of Education organized relevant personnel (mainly Beijing Normal University teachers, the editor of the People's Education Press, and 16 middle school teachers from Beijing and Hebei) to draft the middle school mathematics curriculum standards for the *Syllabus*. Modifications were made according to suggestions made during the conference, then submitted to the Ministry of Education. At that time, the Ministry of Education decided to do a comprehensive study of the experiences of the Soviet Union. It proposed basing the middle school mathematics syllabus on the ten-year school syllabus of the Soviet Union. Therefore, the curriculum standard draft was not formalized, and in the end the decision to inherit the mathematics curriculum of the former Republic of China was not realized.

7.1.2 A Comprehensive Study of the Soviet Union Mathematics Curriculum

From the second half of 1952, affected by the One Side Policy of the Communist Party of China, the learning from the Soviet Union's educational experience became more comprehensive and systematic after 1952. The proposed strategy was first to draw on the Soviet Union's advanced experience and then to adapt it to the Chinese context (Wei & Zhang, 1996). From this time on, the middle school mathematics curriculum was based on a comprehensive study of the Soviet Union.

The People's Education Press put forward a guideline for compiling a suitable new Chinese textbook based on the Soviet Union's middle school one, and then published a set of twelve-year textbooks for middle schools during 1952–1953, first published by the Northeast Region. In 1950, the Ministry of Education of the Northeast People's Government had translated the Soviet Union junior middle school textbook. During the translating process, they modified some content to suit the context of China and this why it was referred to as "compiled." The compiled textbooks of arithmetic, algebra, and plane geometry for the junior middle school and algebra, plane geometry, solid geometry, and plane trigonometry for the senior middle school were all published by the Northeast People's Publishing House. Some additional books of exercises were developed and, for the convenience of the publishing, these were attached to the textbooks. This was the second set of mathematics teaching materials for middle schools.

In July 1952, the Ministry of Education established syllabus drafting committees for various disciplines in primary and middle schools. There were subject groups within the committee and the middle school mathematics group consisted of 12 people. This committee proposed taking the latest ten-year Soviet Union syllabus (four years primary school and six years middle school) as a basis for compiling the *Chinese Mathematics Syllabus*. The modifications were made gradually, and the result was that the middle school of our country is similar to that of the Soviet Union. The advice was approved by the Ministry of Education and published by the People's Education Press in December 1952 (1952 Syllabus).

The 1952 Syllabus consisted of two parts, commentary and outline. The commentary part included general commentary and subjects commentaries. The general commentary stated the purpose and principles of mathematics teaching in middle school, and the subject commentaries stated the teaching purpose, orientation, methods, and content in every subject. In the outline, the key teaching content and class hours for the curricula of every subject were stated by grades (Curriculum and Teaching Materials Research Institute, 2001a). This syllabus was concerned with basic knowledge and skill training focused on systematic and logical mathematics content and adopted the style of the Soviet Union' curriculum at that time.

The 1952 *Syllabus* was revised twice, in 1954 and 1956, but the structure and content were not changed substantially. Therefore, the 1952 *Syllabus* was actually the basis for the mathematics textbook teaching for middle schools in the ensuing time period.

Because the first edition textbook published by the People's Education Press had not used the 1952 *Syllabus* as a reference, except for solid geometry in the senior middle school, the two were not consistent in teaching practice. For this reason, the People's Education Press began to write a new twelve-year textbook (referred to as *RMT*) during 1954–1957; this took the Soviet Union's textbook and exercises for reference. From 1955 to 1959, the Ministry of Education issued a document recommending the use of the RMT, as the third set of mathematics textbooks.

In conclusion, during this period, China had a national unified mathematics syllabus and textbook. The syllabus middle school syllabus was translated from that of the Soviet Union, and the mathematics textbooks were compiled from the Soviet Union syllabus and textbooks. Both the syllabus and textbooks emphasized basic knowledge and skills and paid attention to the scientific, systematic, and ideological nature of the content. The content arrangement was systematic and considered students' receptivity, which led to a significant increase in student achievement. However, in adapting the Soviet Union's curriculum some middle school content was removed, including statistics, probability, and determinant and analytic geometry. This resulted in the teaching material covering less knowledge and at a low level, which did not address students' needs for further study and life in general.

7.2 Exploring the Development of China's Mathematics Curriculum for the First Time (1958–1961)

From 1958, China tried to break away from the Soviet Union's education and began to develop its own system. Although this was a period of transition and some of the decisions may even have been rash, there were still some "original" ideas and practices in curriculum reform which laid the foundation for further development of the mathematics curriculum of China.

7.2.1 Reflections on the Shortcomings of the Soviet Union Curriculum and Adjustments to the Curriculum Content of Mathematics

In 1958, the Ministry of Education decided to adjust the curriculum and teaching content in order to move forward from the Soviet Union experience. In the first half of 1959, the Ministry of Education decided to hold a mathematics teaching forum for primary and middle schools, with the purpose of adjusting the curriculum and teaching content and revising the syllabus and textbook writing. The purpose of the forum was to draw on the suggested syllabus revisions for primary and secondary schools. The process included obtaining advice about the mathematics textbook and mathematics needs from factories, universities, technical secondary schools, and primary and middle schools, as well as some comparative studies of the representative curriculum standards and syllabus during 1912–1956 and a comparative study, between China and Eastern Europe, of curriculum arrangements and teaching content before and after liberation. The revised version (referred to as the *Report*) was submitted to the Culture and Education Office of the State Council in January 1960 and became an important document to guide mathematics textbook writing and teaching in middle schools at this time.

The curriculum adjustments proposed by the *Report* included putting the teaching content of arithmetic into the primary school completely, moving the teaching content of plane geometry and linear equations of two unknowns from the senior middle school to the junior middle school, and increasing plane analytic geometry (Curriculum and Teaching Materials Research Institute, 2001a).

Based on the *Report*, the People's Education Press edited and published a set of a twelve school mathematics textbooks for temporary use during the transition period. This temporary set of textbooks was revised, reorganized, and refreshed in part. The plane geometry section in the junior middle school and the solid geometry and plane analytic geometry sections in the senior middle school were newly written, while the plane geometry section in the senior middle school remained as the original. This set of textbooks was introduced in 1959 and continued until 1966. This was the fourth set of mathematics textbooks for middle schools.

7.2.2 A Test Program to Reform the Education System and Modernize Mathematics Education

In May 24, 1959, The CPC Central Committee and State Council issued a document titled *Provisions for the Experimental Reform of the Education System*, which was used in all provinces, municipalities, and autonomous regions in China. Until March 1960, this reform was introduced in various regions. At this time the sentiment was, "Go all out, aim higher, better and more to build Socialism." The "Education Revolution" and the "Great Leap Forward Movement" included a proposal to carry out a program of mathematics education modernization and a corresponding curriculum reform in the spirit of "better and more."

By the end of 1959, the Science Department of the Central Propaganda Department allocated the task of modernizing mathematics education to the Beijing Normal University, Institute of Mathematics, People's Education Press, and other units, and designated Beijing Normal University to put forward a proposal as soon as possible. The Beijing Normal University Research Group for Mathematics Education Reform in Primary and Middle Schools carried out extensive research in early 1960. They visited factories, enterprises, schools, the Scientific Research Unit and the Design Institute of the Manufacturing Department, and investigated their mathematics requirements, conducted interviews with a number of senior mathematics scholars, collected data related to home and abroad in mathematics education, and analyzed the situation and problems in mathematics teaching. On this basis, another nine-year scheme was set up, with five courses: algebra (including primary school arithmetic, junior middle school algebra, and plane geometry for Grades 1–6); elementary functions (including senior middle school algebra, triangle and plane analytic geometry for Grades 6–8); calculus (including limit differentiation, integrals, and first-order differential equations, for all nine Grades); probability theory and mathematical statistics (for all nine Grades); and map-making (including plane geometry and solid geometry, basic drawing, orthographic views, and solid figure, for Grades 7–9) (You, 1993). The basic spirit of this program was to take the function as the guiding principle, highlight the "Combine Number with Geometrical Figure" and "Combine Calculation with Concept". Meanwhile, let function command the algebra and break the Euclidean geometry system, and then insert necessary graphic knowledge into algebra and map-making textbooks. The People's Education Press proposed a program of reform called *Preliminary Opinions for Textbook Reform*, which suggested finishing arithmetic and introducing algebra in the five years of primary school and continuing with algebra, geometry, trigonometry, analytic geometry, and calculus in the five-year middle school (Wei & Zhang, 1996).

In this period, more than 3400 secondary schools nationwide tested the new educational system, as did about 18% of the secondary schools (Mao & Shen, 1989). Various regions began to edit some teaching materials experimentally. Of these, the nine-year system experimental textbook, edited by the Beijing Normal University Department of Mathematics, had the greatest influence. This set of teaching materials included ten volumes of algebra, one of elementary functions, one of calculus, one of probability, and one of graphics, and it was trialed first in the Beijing Jingshan School (founded in 1960). Half a year later, it was evident that it was difficult to complete the scheduled tasks, so it was modified to a ten-year textbook series. This was titled the *General Education Reform Group of the Mathematics Department of Beijing Normal University* and published by the People's Education Press in 1960.

In October 1960, in order to meet the demands for the new education system, the Ministry of Education decided to edit new textbooks for ten-year schools and submitted the *Report on Meeting the Teaching Reform and Adapting the Teaching Materials.* The Ministry of Education established a leading group of textbooks for primary and secondary schools, choosing a number of excellent teachers from many provinces and universities to contribute, including three experts, L. Hua, Z. Guan, and E. Ding, who acted as consultants. The editing group held a seminar to discuss issues such as the teaching objectives, teaching content, and system of the textbooks. They specifically discussed two main controversial issues, whether to add plane analytic geometry content and whether to arrange the material according to subject or to arrange the content synthetically. Based on this, the draft version of the

Editing Program for Mathematics Textbooks for Ten-Year Schools (*Editing Program* for short) was produced (Curriculum and Teaching Materials Research Institute, 2001a). After this, the editorial group began a set of mathematics textbooks for ten-year schools. This included three volumes of algebra for junior high school, two of plane geometry for junior high school, two of algebra for senior high school, and one of solid geometry for senior high school. Of these, algebra for secondary school was new, while plane geometry for junior high school and solid geometry for senior high school were unchanged. The People's Education Press started to publish this set of textbooks in 1961, and it was used until 1966. This was the fifth mathematics textbook series.

In conclusion, this period witnessed a movement away from the mathematics curricula of China before the People's Republic and away from curricula adapted from the Soviet Union, to China developing its own new mathematics curricula. This period saw the twists and turns of the education management authority and was noted as the "Education Great Leap Forward" and "The Education Revolution." All provinces performed reform experiments on the length of schooling and on locally written textbooks to different extents, which broke down the national unified situation. In the mathematics curriculum, the transfer of middle school arithmetic to the elementary school and the transfer of secondary school plane geometry to primary school changed the decades-old curriculum pattern. The mathematics curriculum content for secondary schools gained plane analytical geometry and also expanded content about preliminary probability, determinants, cartography, and measurement, basically bringing the level of mathematics back to previous levels found in China. In particular, during the test of the Chinese mathematics curriculum system modernization, it was the first time to discuss the content system and produce a new idea. For example, the thought of "take the function as the guiding principle, highlight the 'Combine Number with Geometrical Figure' and 'Combine Calculation with Concept'". The fruitful results of the mathematics curriculum content update, system construction, and reform implementation included key ideas and reform practices that have had important influences on the subsequent development of the mathematics curriculum.

7.3 Reverting to the Soviet Union Mathematics Curriculum (1961–1966)

From 1961, there was reflection on the education system reform and the modernization of mathematics. After drawing on the lessons learned from the curriculum of the Republic of China Period and the Soviet Period and making a comparative study of home and abroad, a new mathematics syllabus was developed, which strengthened the basic knowledge and skill and reverted to the mathematics curriculum model of the Soviet Union.

7.3.1 Reverting to Basic Knowledge and Skills Training, Based on Lessons Learned from the Curriculum of the Republic of China Period and the Soviets Period

In the second half of 1960, there was significant evidence of harm caused by "The Education Revolution." In February 1961, the Ministry of Education held a general education forum for the new education system, which pointed out that system tests should not be too large, changed the length of time from ten-year to twelve-year schools, and no longer offered the nine-year coherent style school test. In 1961, the Culture and Education Group of the CPC Central Committee proposed rewriting better quality full-time primary and secondary school textbooks for twelve-year schools on the basis of past experience in textbook writing. According to this instruction, the Ministry of Education started the preparation work in June 1961 and the People's Education Press took up this responsibly. The mathematics editors of the People's Education Press consulted information from all time frames and from all over the world, and did a comparative study of mathematics curricula and teaching for primary and secondary schools before and after the liberation (the period of the Northern Warlords Government and the National Government) and across countries, including the Soviet Union, the German Democratic Republic, America, and Japan. Based on this, the draft version of the *Mathematics Syllabus for Full-Time Primary and Secondary Schools* was issued in October 1961. The Ministry of Education held a forum in Beijing and conducted research in several provinces to collect feedback and advice from teachers and experts, as well as a monographic study to collect feedback from all provinces.

In 1963, the People's Education Press drafted the *Mathematics Syllabus for Full-Time High Schools* based on the *Provisional Regulations of Full-Time Secondary Schools* issued by The CPC Central Committee in March 1963 (Curriculum and Teaching Materials Research Institute, 2001a), the *Notes on Performing The New Teaching Plan For Full-Time Primary And Secondary Schools* published by the Ministry of Education in July, 1963 (Curriculum and Teaching Materials Research Institute, 2001b), and the results of the comparative research and monographic study.

The 1963 Syllabus had five parts: teaching aims and requirements, teaching content, teaching content and arrangement, teaching precautions, teaching requirements in every subject. The proposed aim was to foster calculating ability, logical thinking ability, and spatial imagination; this was the first time these three abilities had been included. It was proposed that teachers should pay attention to concepts, rules, theorems, formulae and solving problems, methods and steps of proof. The teaching process needed to strengthen the students' calculating ability, logical thinking ability, and spatial imagination. There was also a need to connect it to real life while still acknowledging the learning of basic knowledge (Curriculum and Teaching Materials Research Institute, 2001b). The style of the 1963 *Syllabus* was similar to that of 1952.

7.3.2 *Trialing and Revisions of the New Mathematics Curriculum for Twelve-Year Schools*

In October 1961, after the People's Education Press drafted the *Mathematics Syllabus for Full-Time Primary and Secondary Schools,* it began to edit and write Mathematics Textbooks for full-time primary and secondary schools (often called *New Textbooks for Twelve-Year Schools*). The draft was trialed in schools in Jingshan, Fengsheng, Erlonglu in Beijing and other provinces. This set of textbooks, first published in autumn of 1963, had four volumes of junior high school algebra, two of junior high school plane geometry, for one of senior high school solid geometry, one of senior high school plane triangles and one of senior high school plane analytic geometry. This set of books was used until the *Cultural Revolution* in 1966. It was the sixth set of mathematics teaching materials for secondary schools.

After being used for less than a year, this set of textbooks was marked for revision by the Ministry of Education because of feedback regarding problems such as difficult content, burdensome tasks, and too much burden for students from some regions and schools. The Ministry of Education ordered the People's Education Press to modify the textbook and issued the *Notes on the Textbook for Primary and Secondary Schools.* In the spring of 1964, The Ministry of Education then requested the People's Education Press to modify the textbook for twelve-year schools, with a completion date of 1965. Shortly afterward, due to the outbreak of the Cultural Revolution, the Ministry of Education only published the algebra textbooks for junior high schools and the plane geometry textbook for senior high schools, labeled as *The Undetermined Edition*; the other textbooks were never officially published.

In conclusion, after the discussion and research on the problems of the mathematics curriculum reform during this period, the 1963 *Syllabus* was an important advancement of the mathematics curriculum. The *Three Abilities* were proposed as the teaching aim for mathematics. The structure of the 1963 *Syllabus* offered a paradigm for the future. From the process of developing the 1963 *Syllabus* and the subsequent teaching material, we can see that the procedure of the mathematics curriculum research was scientific and normative, and that it accumulated experience for constructing the mathematics.

7.4 Encountering Setbacks in the Process of Mathematics Curriculum Development (1966–1976)

From May 1966 until the end of the Cultural Revolution in October 1976, the mathematics curriculum development encountered setbacks. At the beginning of the Cultural Revolution, previous teaching material was denied completely, and the People's Education Press and their higher authorities of the Ministry of Education were paralyzed.

During this period, mathematics curricula all over the country died out. There was no unified teaching plan or syllabus, and the entire pre-Cultural Revolution textbook system was stopped. Mathematics was limited to basic Industrial and Agricultural knowledge, and some provinces edited and wrote their own teaching materials. Later, a number of provinces, autonomous regions, and municipalities edited textbooks in collaboration, as self-compiled teaching material.

During this period, there was no unified mathematics syllabus and no unified teaching material used throughout the country. Textbooks which were written or chosen locally were basically collections or assemblies of previous ones. The textbooks of this period could be divided into three types: streamlined, practical, and intermediate. The streamlined type was streamlined compared to previous texts, but the contents of the applications from the 1963 textbooks had been added. These textbooks were used more in the later period of the Cultural Revolution (1966–1976). The practical type were textbooks that connected the calculations of industrial production, graphing, measurement, etc., in the city, and accounting, measurement with abacus calculations, etc., in the countryside. The *Temporary Mathematics Textbooks for Shanghai Middle Schools*, edited by the Shanghai Middle and Primary Schools Textbooks Editing Group in 1968, were an example of the practical type. These textbooks emphasized the prominent political, one-sided emphasis on the practical, but weakened the basic knowledge. The intermediate type of textbooks lays somewhere in between the streamlined and practical types. For example, the middle school trial textbooks *Mathematics* (four books), edited by the Shandong Middle and Primary Schools Textbooks Editing Group in 1970, followed a policy of combining basic, useful mathematics while also introducing some agriculture accounting and common sense. Nationally, the majority of mathematics textbooks were of the intermediate type.

7.5 Exploring the Development of the Mathematics Curriculum for the Second Time (1977–1991)

In October 1976, the Cultural Revolution came to an end and all the country entered into the period of Overall Rectification. In the education area, the teaching order began to be rectified and restored and the essence of education was discussed; this preceded the educational reform by addressing adjustment, reform, rectification, and improvement.

7.5.1 The Experiment of the Unified Comprehensive Mathematics Curriculum

In September 1977, the Ministry of Education decided to edit the teaching material in the form of a national working conference for preparing primary and secondary materials. To enrich the power of primary and secondary school mathematics, the editors of the People's Education Press assigned to all provinces were recalled and sixteen personnel were loaned from all provinces to found the group to edit and write textbooks for primary and secondary schools. As well, eight experts, B. Su, Z. Guan, X. Duan, Z. Jiang, W. Wu, L. Yang, G. Zhang, and E. Ding, were invited by the Ministry of Education as consultants for primary and secondary mathematics textbooks. The team was responsible for drafting the *Teaching Plan for Full-time Primary and Secondary Education and* editing the *Mathematics Textbook for Full-time Primary and Secondary Education.*

In September 1977, the Mathematics Writing Team began drafting the *Teaching Syllabus for Full-time Primary and Secondary Education.* During this process, the team responsible for the primary and secondary textbooks analyzed the teaching materials and syllabi of Japan, Britain, France, and America. By the end of the year, they had completed a draft of the *Syllabus for Primary and Secondary Mathematics*, and sent it to administrative departments for education, the Teaching and Research Group, normal colleges and universities, and experts, consultants, teachers of provinces and autonomous regions for advice. Comments on the draft outline focused on the mathematics content, mathematics curriculum modernization, and arrangement. To modernize the mathematics curriculum content, one point of view was that the content should be increased to enable students to understand and be in touch with the basics of modern mathematics. Another point of view was that the secondary school content laid a solid foundation for students and that in the current condition teachers could not finish the existing tasks, so it was unnecessary to add any new content. There were two viewpoints about the arrangement of the curriculum content of mathematics curriculum. One was to cancel the division of algebra, geometry, and triangles and make it a comprehensive mathematics curriculum. The other viewpoint was that the study of these three topics had different research objectives, methods, and teaching requirements, so it was hard to form a comprehensive curriculum, and that it should be divided. The editing team did some research and proposed that the curriculum content be modernized by using the principles of "selection, increase, and penetration." In January 1978, the team responsible for primary and secondary mathematics textbooks completed the drafting of the curriculum. In February 1978, the *Mathematics Syllabus for Ten year Full-time Primary Schools* (*1978 Syllabus*) was issued by the Ministry of Education, and it was implemented from autumn of 1978.

The content of the *1978 Syllabus* included confirmation of the teaching purpose and content arrangement, teaching notes, teaching requirements, and teaching content (which was similar to that of the *1963 Syllab*us). The principles of "selection, increase, and penetration" were applied: selecting and simplifying the traditional mathematics content for high schools, increasing the preliminary knowledge of calculus, probability, statistics, and logical algebra, and penetrating the curriculum with modern mathematical ideas such as sets and correspondence. This became a mixed-content curriculum rather than a divided one, and the content of algebra, geometry, and triangles and calculus was added to the comprehensive curriculum (Curriculum and Teaching Materials Research Institute, 2001b).

During this period of editing, *Mathematics Textbooks for Full-time Ten-year Secondary Schools* were developed, completed in April 1980. This set of teaching materials included six volumes for junior high school and four for senior high school. It was the seventh common set of mathematics textbooks for secondary schools.

In this set of teaching materials, there were great changes in the content: some traditional content was removed, and increases were made in necessary basic knowledge of technology, such as probability and statistics, logic and algebra. As well some modern basic concepts and content such as sets, correspondence, and calculus were added.

7.5.2 The Experimental Selective Discipline Mathematics Curriculum

In December 1980, the CPC Central Committee and State Council issued *Several Issues Concerning Primary and Secondary Education.* This proposed that the education system would change gradually to twelve years. On April 17 of 1981, the Ministry of Education issued *Notes on the Teaching Plan for Six-year Full-time Key Secondary Schools*, in which it ruled that the education system for secondary schools would be six years (three years each for junior high school and senior high school), to be developed and implemented before 1985. The People's Education Press developed the *Mathematics Syllabus for Six-year Full-time Key Secondary Schools* (*1982 Syllabus*) and the *Teaching Plan for Six-year Full-time Key Secondary Schools*. In the *1982 Syllabus*, the senior high school curriculum was set up to follow three different strands: the first was a single elective course, the second was an arts strand, and the third was a science strand. Following this syllabus, the People's Education Press compiled several textbooks for six-year key secondary schools. This set included four volumes of algebra and two of geometry for junior high school, and, for senior high school, three volumes of algebra, two of algebra with geometry, and one each of solid geometry, plane analytic geometry, and calculus. The textbooks for junior high schools were used for general, and key secondary schools were put into use in the autumn of 1983. This was the eighth set of mathematics teaching materials.

During the use of the ten-year mathematics textbooks for secondary schools, which appeared in the *1978 Syllabus*, some problems emerged including difficult content, high requirements, heavy burdens, large differentiation, and low passing rates. Therefore, the Ministry of Education decided to adjust the teaching content and implement two kinds of teaching requirements, so it issued the *Mathematics Syllabus for Senior High Schools* in November 1983, which contained both basic and higher requirements for mathematics teaching content in senior high schools (Curriculum and Teaching Materials Research Institute, 2001b). Accordingly, the mathematics editing group of the People's Education Press adjusted the mathematics textbooks of senior high schools as follows: edited a new senior high textbooks for algebra (B version), solid geometry (B version), plane analytic geometry (B version) and adapted the original mathematics textbooks for the science program keyed to six-year secondary schools: algebra for senior high schools (A version), solid geometry for senior high schools (A version), plane analytic geometry for senior high schools (A version), and calculus for senior high schools (A version). This was the ninth set of mathematics teaching materials for senior high schools, and they were put into use in the autumn of 1984.

7.5.3 Implementing a Unified Mathematics Curriculum

In May 1985, the Central Committee of the Communist Party of China issued the *Decision on the Reform of the Education System.* In June 1985, the Standing Committee of the National People's Congress for the Sixth Session decided to establish the National Education Commission and repeal the Ministry of Education. In September 1986, the newly established National Examination Committee for Teaching Materials for Elementary and Secondary Schools decided to reform the textbook system. This produced a flourishing situation for textbooks under the one syllabus, with a multiple textbook systems. The conference determined the basic steps of the textbook reform and construction as follows. The first step was that the current syllabus would be revised, but that there would be no major changes to the existing arrangement, main system, or content. This syllabus would become the basis for the teaching, examination, evaluation, and revising of the teaching material. The second step was to design a new syllabus and teaching plan for nine-year compulsory education and write teaching material and teachers' books for the new syllabus. Following these steps, the National Education Commission modified and reviewed the current syllabus by reducing the difficulty of the teaching content appropriately, thus decreasing the students' burdens and specifying the teaching requirements. The *1978 Syllabus* was revised according to requirements of the Ministry of Education. In February 1987, the National Education Commission issued the *Mathematics Syllabus for Full-time Secondary Schools* (*1987 Syllabus*).

The text structure of the *1987 Syllabus* was exactly the same as the 1978 one. The focus was on simplifying the traditional mathematics, increasing the statistics content for the junior high schools, and increasing the initial application of limits and preliminary knowledge of probability for senior high schools, and permeating mathematical ideas of sets and correspondence properly. The content was arranged by subject (Curriculum and Teaching Materials Research Institute, 2001b).

7.5.4 Exploring the New System of Mathematics Content for Secondary Schools

In 1978, W. Xiang, a professor in the Mathematics Department at the University of California at Berkeley, returned home to China to give lectures. After he reviewed the syllabus and teaching material of mathematics, he put forward a proposal about the middle school mathematics experiment textbooks to S. Kang (Vice Premier of the State Council) and N. Jiang (Secretary of Education). He wanted to write a set of mathematics experimental textbooks. Kang approved the idea and appointed Jiang to implement this experimental study.

In November 1978, a group was established to develop the *Mathematics Experimental Textbook for Secondary Schools*, with representatives from Beijing Normal University, the Institute of Mathematics of the Chinese Academy Society, the People's Education Press, Beijing Teachers' College, and Beijing Jingshan Middle School. In July 1979, the *Mathematics Experimental Textbook for Secondary Schools* was completed and printed in trail version. In July 1979, the Ministry of Education issued the *Notice of Organizing the Experimental Work of Mathematics Experimental Textbook for Secondary Schools,* and then founded a group for *Experimental Study* in the *Ministry of Education* and other provinces. Under the guidance of the *Experimental Study Group*, some schools started to experiment in September 1985. During the process of experimentation, the editing group absorbed advice from teachers in experimental schools and modified the trail version, finally forming a set of Mathematics Experimental Textbooks for Secondary Schools. This was printed by the Beijing Normal University Printing Group and six versions of mathematics textbooks for junior and senior high schools were published in succession from 1981 to 1986. The first edition for junior high schools was modified and published by the People's Education Press six years later.

The Mathematics Experimental Textbook for Secondary Schools was a new attempt to systematize mathematics teaching content, which was proposed by W. Xiang. He proposed that the guiding ideology of the teaching material should be concise and practical, return to nature, explain profound theories in simple language, and then, follow a logical train of thought. Concise and practical referred to as the use of real-life illustrations in order to handle complexity through simplicity. Teaching material should begin with practical problems and then give the information step-by-step, and then clarify how to use the simple theory to illustrate complex concepts rather than focusing only on the abstract (The State Education Commission Middle School Mathematics Research Group Experimental Textbook, 1994).

The content of the Mathematics Experimental Textbook was selected from algebra, geometry, and analysis and organized in a spiral arrangement. It consisted of two levels each for junior and senior secondary schools. In summary, during this period after the Cultural Revolution, the levels of mathematics curriculum content and requirements were too low and gave rise to the need for mathematics textbook content modernization.

In the beginning, the *1978 Syllabus* had increased the amount of modern mathematics curriculum content, using the principle of selection, increase, and penetration, which leads to the content being too hard and not meeting the needs of teachers and students. After more than ten years of adjustments and revisions, the *1987 Syllabus* resulted.

From 1982, high schools adopted different types of courses, such as elective courses and those with two different kinds of requirements. However, due to the increased content in the textbooks, some teachers could not meet the required standards and generally found the content difficult. Students had heavy burdens and problems surfaced in practice. The *1987 Syllabus* reflected adjustments to address these issues. Although, in 1978, a synthesized approach was used for the arrangement of teaching content, it was phased out because teachers did not adapt well to this new arrangement.

7.6 Trying to Establish China's Mathematics Curriculum System (1992–2000)

7.6.1 Diversified Compulsory Education Mathematics Curriculum

In 1986, in order to implement the Compulsory Education Law, the National Education Commission developed the *Teaching Plan for Full-time Compulsory Education*, both for primary and secondary schools (*Teaching Plan*), and authorized the People's Education Press, Shanghai Education Bureau, Liaoning Institute of Education, and the Mathematics Department of Beijing Normal University to draft the *Mathematics Syllabus for Full-time Nine-year Compulsory Education* for junior high school. In 1992, after the usual process of testing and revision, this syllabus was promulgated by the National Education Commission as the *Mathematics Syllabus for Nine-year Full-time Compulsory Education in Junior High Schools* (*1992 Syllabus*).

Compared with the *1978* and the *1987 Syllabi*, the structure of the *1992 Syllabus* was similar to the former except that it combined the teaching content and arrangement as one part. This syllabus focused on basic knowledge, basic skills, operational ability, logical thinking ability, solving simple actual problems, and developing a good personality. It supplied two choices of a six + three year system or a five + four year system, with similar teaching content in junior high schools and additional elective content (Curriculum and Teaching Materials Research Institute, 2001b).

The *1992 Syllabus* was revised in 2000. In March 2000, the Ministry of Education issued the *Mathematics Syllabus for Nine-year Full-time Compulsory Education in Junior High Schools* (*2000 Syllabus*). Since 1988, different textbooks had been developed for four different conditions: areas with a system of six years of primary school and three years of secondary school, areas that were relatively better developed economically and culturally with better conditions, and remote areas which were weak in economy and culture, such as the pastoral and mountainous areas and poorer primary and secondary schools with less teaching equipment available.

Three syllabi eventuated for nine-year full-time compulsory education, with different types of teaching materials and multiple textbooks. There were two types of teaching materials: one was developed by the National Education Committee, and referred to as one syllabus with multiple textbooks, and the other was produced in Zhejiang and Shanghai, with a self-syllabus (curriculum standards), and referred to as multiple syllabi with multiple textbooks.

The "one syllabus with multiple textbooks" model consisted of sets of Mathematics Textbooks for the six + three system and the five + four system (People's Education Press), a set for the five + four system prepared by Beijing Normal University, a set for the six + three system prepared by the Provincial Department of Education and Huanan Normal University (Guangdong Education Press), a set for the six + three system prepared by the Sichuan education committee and the Southwest Normal University (Southwest Normal University press), a set for lower level students in rural primary schools prepared by the Research Institute of Education Science in Hebei (Hebei Education), a set for the nine-year system in developed cities prepared by the Shanghai Education Bureau (Shanghai Education Press), and a set for the five + three system and Six + three system for rural areas.

As well as the above eight sets of mathematics textbooks, the *Experimental Mathematics Textbook for Secondary Schools* in junior high school and a self-counseling mathematics textbook, proposed by W. Xiang, were approved by the National Examination Committee for primary and secondary schools and as mathematics textbooks for compulsory education in junior high school.

All of these textbooks were modified and the Subject Review Committee of the National Examination Committee for Teaching Materials for Elementary and Secondary Schools began to use them in 1993.

7.7 Testing the Comprehensive Curriculum in Senior High School

After the implementation of the mathematics curriculum for compulsory education in 1993, the National Education Committee Working Group for the New Curriculum for Senior High School drafted the *Mathematics Curriculum Plan for Full-time Ordinary Senior High Schools* and authorized the People's Education Press responsible for drafting the corresponding syllabus. Teachers and teaching–research staff of secondary schools participated in the syllabus drafting. This syllabus was finally issued by the National Education Committee with the name of

Mathematics Syllabus for Full-time Ordinary Senior High Schools (*1996 Syllabus*). This was linked to the *1992 Syllabus*.

Compared with the *1987 Syllabus*, there was some new content for senior high schools. These additions mainly covered simple logic, plane vectors, preliminary knowledge of probability and statistics, and preliminary knowledge of calculus. There were two schemes for solid geometry, one comprehensive and the other focusing on vector geometry methods and provided for compulsory, required elective, and arbitrary elective courses. It also stated that the compulsory courses should start in grades 10 and 11, and that the required elective courses should start in grade 12, targeting three programs, science, liberal arts, and practice. This syllabus required a mixed arrangement of mathematics teaching content for junior high schools, for the first time after the foundation of New China (Curriculum and Teaching Materials Research Institute, 2001b).

The *1996 Syllabus* was revised in 2000, after three experimental phases. In March 2000, the Ministry of Education issued the *Mathematics Syllabus for Full-time Ordinary Senior High Schools* (2000 Syllabus).

The People's Education Press wrote the *Mathematics Textbook for Full-time Ordinary Senior High Schools*, which had seven volumes. Two of these were compulsory textbooks for grades 10 and 11. The second semester of grade 11 was divided into two volumes corresponding to the two schemes of the syllabus. The third volume was an elective textbook for grade 12, including two elective versions for the science and liberal arts programs and the practice program. This set of mathematics textbooks was the tenth one for ordinary senior high schools.

This set of textbooks changed the focus on algebra, solid geometry, plane analytic geometry, and calculus, making them into a comprehensive course with selective content. The teaching material addressed logical and systematic mathematics content and the teaching sequence was arranged by ideology of knowledge and cognition. This set of textbooks was introduced in autumn of 1997 in Tianjin city and Jiangxi and Shanxi provinces, and then gradually popularized all over the country. In 2000, the textbook was revised according to the *2000 Syllabus* and had a new graphic format, but the content and structure were not changed.

In summary, this period saw explorations and new attempts to modify the content of the mathematics course system to address the different resources, talents, and qualities present in different areas. In order to support compulsory education, a diversity of mathematics curricula was developed and demonstrated, establishing three programs with different requirements for different areas. Eight half-sets of materials for compulsory education in mathematics were produced. The policy of multiple syllabi with multiple textbooks was thus implemented in compulsory education. The mathematics curriculum in compulsory education was organized according to separate subjects. From 1997, links were created between compulsory education and ordinary senior high education, with ordinary senior high schools using one syllabus and unified textbooks, arranged in a comprehensive style. Thus, during this period, the explorations of ways to organize the mathematics course system and the new rounds of basic education curriculum reform helped to accumulate experience and lay a foundation for future work.

References

Curriculum and Teaching Materials Research Institute. (Ed.). (2001a). *20 世纪中国中小学课程标准: 教学大纲汇编* [Collection of primary and secondary school curriculum standards and syllabus of the twentieth century China]. Beijing: People's Education Press.

Curriculum and Teaching Materials Research Institute. (Ed.). (2001b). *20 世纪中国中小学课程标准•教学大纲汇编(课程(教学)计划卷)* [Curriculum standard for primary and secondary schools in china in the 20th century syllabus compilation (Curriculum (Teaching) plan volume)]. Beijing: People's Education Press.

Lv, D. (1999). 课程史论 [The theory of curriculum history]. Beijing: People's Education Press.

Mao, L., & Shen, G. (1989). 中国教育通史(第六卷) [The general history of China's education (Vol. 6)]. Jinan: Shandong Education Publishing House.

Shi, L. (1996). 课程理论—课程的基础、原理与问题 [The theory of curriculum: The foundation, principles and issues of curriculum]. Beijing: Educational Science Publishing Company.

Wei, Q., & Zhang, Y. (1996). 中国中学数学课程教材演变史料 [The historical evolution of middle school mathematics curriculum material in China]. Beijing: People's Education Press.

You, M. (1993). 北京景山学校数学教学改革的理论和实践 [The theory and practice of the mathematics teaching reform in Jingshan Hill school in Beijing]. Beijing: Huaxia Publishing House.

Chapter 8
The Evolution of Mathematics Curriculum and Teaching Materials in Secondary Schools in the Twenty-First Century

Shi-hu Lv and Chunyan Cao

Abstract At the end of the twentieth century, a basic education reform was initiated in China, and this led to the development of a new curriculum system. This was a new breakthrough, different from the previous mathematics curricula that had been in place since 1949, both in design and implementation. For example, the new features included the orientation of curriculum goals with student-centered principles, the highlighting of a curriculum structure with selective characteristics, and curriculum organization with a comprehensive block-based design, integrated curriculum content, and the development of basic knowledge and basic skills.

At the end of the twentieth century, a basic education reform was initiated in China, and this led to the development of a new curriculum system. This was a new breakthrough, different from the previous mathematics curricula that had been in place since 1949, both in design and implementation. For example, the new features included the orientation of curriculum goals with student-centered principles, the highlighting of a curriculum structure with selective characteristics, and curriculum organization with a comprehensive block-based design, integrated curriculum content, and the development of basic knowledge and basic skills.

The first phase in the development and trialing of the *Mathematics Curriculum for Basic Education* was for compulsory education, followed by senior high schools.

S. Lv (✉)
College of Education, Northwest Normal University,
Lanzhou 730070, Gansu, China
e-mail: shihulv@163.com

C. Cao
College of Education Science, Shaanxi Xueqian Normal University,
Xian 710061, Shaanxi, China

Y. Cao and F.K.S. Leung (eds.), *The 21st Century Mathematics Education in China*, New Frontiers of Educational Research,
https://doi.org/10.1007/978-3-662-55781-5_8

8.1 Developing and Testing the New Mathematics Curriculum for Compulsory Education

8.1.1 The Development Process

The basic mathematics curriculum reform (referred to hereafter as the new curriculum) was initiated in 1989. In 1989, X. Zhang, board chairman of the *Mathematics Teaching Association of Chinese Education Society* of the People's Education Press, proposed the *21st Century Mathematics Education Project* and this was accepted.

This research project had five main aspects. The first was to consider the evolution of the mathematics curriculum in China; the second was a reflection on the current status of mathematics education; the third was to understand the trend of international mathematics curriculum development; the fourth was to study how to learn mathematics; and the fifth was to explore a new way of evaluating mathematics learning.

This research project was not limited just to the theoretical level. From 1994 to 1999, some young scholars in the research project group cooperated with the Ministry of Education to write a set of primary mathematics textbooks. This was tested in dozens of primary schools in Beijing and one county of Jilin province, a process that contributed useful experience and theory to inform the twenty-first century mathematics education and impact upon the decisions relating to the new basic curriculum reform (Sun, 2007).

In order to implement the plan of action for the twenty-first century and develop a modern curriculum system for basic education, the Mathematics Curriculum Standard Development Team for Compulsory Education was founded in May 1999, with J. Liu as the team leader. The team's work resulted in the *Mathematics Curriculum Standards for Compulsory Education.*

After the Ministry of Education officially launched the curriculum research project in June 2000, the curriculum standard developing team continued to collect opinions about this document from various circles of society and held several research seminars to discuss its structure, content, and format. This led to a revised document, known as the *2001 Standard*, which was issued in July 2001 and tested nationwide.

During the development process, some controversial issues arose, for example, whether the curriculum goals were proper, how to address the basic operation skills, including the speed and accuracy of mental arithmetic and vertical form, when the calculator should be introduced, how to treat arithmetic application problems, how to select and arrange the content of plane geometry, including how to understand the value of Euclidean geometry, how to cater for different students learning different mathematics, how to understand the relationship between proof and graphics understanding, and how to explain formal proof and the relationship between it and exploration. Other issues were concerned with the relationships between project learning and other fields, the content and requirements of statistics and probability, and the effective implementation of evaluation. This led to establishing the goals, structure, content capacity, and requirements of the mathematics curriculum, which played an important role in the *2001 Standard.*

8.1.2 *The Content of* Mathematics Curriculum Standards for Compulsory Education (Experimental Version)

The *2001 Standard* had four parts: Foreword, Curriculum Objectives, Content Standards, and Implementation Recommendations. The Foreword elaborated the Basic Rationale and Design Considerations. The Basic Rationale described six aspects: curriculum, mathematics, learning, teaching, evaluation, and technology. The Design Considerations divided nine school years into three learning stages (grades 1–3, 4–6, and 7–9) based on the physiological and psychological characteristics of children's development. It also listed and described action verbs for the knowledge, skills and processes, and explained the core concepts such as number sense, symbol sense, space concept, statistical concept, application awareness, and inferential ability.

Three aspects of Curriculum Objectives were addressed: knowledge and skills, processes and methods, and affect, attitudes, and values. The overall objectives and those for each phase were described in relation to these three aspects as well as problem solving. For example, the overall objectives were elaborated as follows.

Through mathematics learning at the obligatory stage of schooling, students are able to:

- Acquire important mathematical knowledge (including mathematical facts, mathematical activity experiences), basic mathematical thinking methods, and application skills that are essential for adapting to future social life and further development;
- Begin to know how to deploy various ways of mathematical thinking to observe, analyze, and solve problems encountered in daily living and studies in other disciplinary areas;
- Realize the intimate relationships between mathematics and nature, as well as between mathematics and human society, know the worth of mathematics, increase understanding of mathematics, and gain confidence in learning mathematics with good results;
- Possess some degree of creative spirit and practical abilities, and develop sufficiently in areas of general abilities, affect, and attitudes.

The overall objectives are stated in detail below.

Knowledge and skills
• Abstract authentic problem situations as number and algebra problems; master fundamental knowledge and basic skills pertaining to numbers and algebra; solve simple problems
• Explore how shapes, sizes, and positions of objects and figures are related and transformed; master fundamental knowledge and basic skills pertaining to space and figures; solve simple problems
• Engage in the processes of problem posing, data collection and processing, decision making and prediction; master fundamental knowledge and basic skills pertaining to statistics and probability; solve simple problems

(continued)

(continued)

Mathematical thinking
• Apply mathematical symbols and figures to describe the phenomenal world; establish initial number sense and symbol sense; develop abstract mathematical thinking • Enrich knowledge of space and objects in the phenomenal world; establish initial space concepts; develop iconic mathematical thinking • Explore how information is described by data processing and organization, as well as how inferences are made; develop statistical thinking • Explore how observation, experimentation, guessing, and proving are done in mathematical activities; develop reasonable analogical and induction abilities and initial mathematical deduction ability; present ideas systematically and clearly
Problem solving
• Begin to learn how problems can be posed and comprehended from mathematical perspectives; apply knowledge and skills acquired for problem solving in an integrated manner; develop application awareness • Formulate some strategies for problem solving; experience that problems can be solved in a variety of ways; develop practical abilities and creative spirit • Learn how to cooperate with others; communicate with others about processes and products of thinking • Begin to form an awareness of evaluation and reflection
Affect and attitudes
• Participate positively in mathematics learning activities; demonstrate curiosity and eagerness in mathematics learning • Experience success in mathematics activities; develop strength to overcome difficulties; develop confidence in mathematics learning • Begin to know the intimate relationships between mathematics and human lives, as well as the values and influences of mathematics on human civilization; experience that mathematical activities are full of explorations and creative productions; sense the rigor of mathematical knowledge and appreciate the certainty of mathematics conclusions • Form an attitude that is pragmatic and realistic; develop habits of querying and independent thinking

The *2001 Standard* strengthened the four objectives of knowledge and skills, mathematical thinking, problem solving, and affect and attitudes, which are interrelated and very important to children's development. For example, the development of mathematical thinking, problem solving, and affect and attitudes cannot happen without the learning of knowledge and skills, which, in turn, require these three preconditions.

The Content Standards adopted the expression of "Learning + Field", and each stage covered four learning areas, numbers and algebra, space and figures, statistics and probability, and practical and integrated applications. The learning objectives were elaborated by using action verbs that were clear, understanding, and operational and by focusing on students as main target, rather than teachers.

The structure of specific objective in four areas was as follows (Table 8.1).

The mathematics curriculum content of the *2001 Standard* was increased to include statistics and probability, practical and integrated applications, and a new treatment of the traditional algebra and geometry. Overall, there was a great change in geometry and statistics and probability.

Table 8.1 Structure of mathematics curriculum contents for compulsory education

Stage of schooling	First stage (Grade 1–3)	Second stage (Grade 4–6)	Third stage (Grade 7–9)
Numbers and algebra	• Knowing numbers • Number operations • Common quantities • Exploring patterns	• Knowing numbers • Number operations • Expressions and equations • Exploring patterns	• Numbers and expressions • Equations and inequalities • Functions
Space and figures	• Knowing figures • Measurements • Figures and their transformations • Figures and their positions	• Knowing figures • Measurements • Figures and their transformations • Figures and their positions	• Knowing figures • Figures and their transformations • Figures and their coordinates • Figures and proofs
Statistics and probability	• Statistical data activities for beginners • Phenomenon of uncertainty	• Simple statistical data processing • Possibility	• Statistics • Probability
Practical and integrated applications	• Practical activity	• Integrated applications	• Thematic studies

Ministry of Education of the People's Republic of China (2001)

In the first and second stages of the *2001 Standard*, the content was expanded to include figures and their transformations, and figures and their positions, which strengthened the understanding of solid figures. For geometry in the third stage, figures and their transformations, and figures and their coordinates were added, breaking though the original Euclidean geometric proof system and adopting first experimental geometry and then geometric proof. Visible geometry and local axiomatic methods were adopted to explore the characteristics of figure. There was a great change in traditional geometry in the *2001 Standard*. The requirement for formal proofs was decreased, and the use of transformations and coordinates to understand real space and solve geometry problems increased. Attention was paid to the practical meaning of quantity and quantity of units, and the emphasis was on selecting the proper tools for the measuring process, developing estimation skills and the ability to apply mathematics to real life, and geometric applications in the real world. There was a reduction in emphasis on formal reasoning skills on the basis of stressing the real background of space and figure knowledge, the requirements for complex and difficult geometric proofs were removed, and the scope of formal proof was limited to the triangle and quadrilateral. Also, the propositions need for a proof were listed so as to let the students experience the significance and process of the logic and master the basic methods of proof. The *2001 Standard* called for self-exploration to understand gradually the relationships of shape, size and mutual position, preliminary understanding of the characteristics and properties of some special graphics, and using measurement, computing, actual operations, figure transformations, algebraic and simple reasoning to solve simple geometric problems. This process was intended to develop spatial sense, visual geometry, figure design, and reasoning ability. Therefore, the *2001 Standard* greatly strengthened and promoted the teaching content of geometry.

In the first two stages of statistics and probability, the topics of phenomenon of uncertainty, understanding possibility, median, and mode were added. In the third stage, probability was added. All statistics content in three stages required students to experience the process of data handling, which included collecting, sorting, analyzing, and predicting or reasoning.

The area of practical and integrated applications was also added to the *2001 Standard*. The purpose was to allow students to experience the relationships among mathematics, life, society, and other subjects and the inner connection in mathematics, and to develop innovation skills, practical abilities, and positive attitudes toward mathematics.

To illustrate the flexibility and selectivity of the mathematics curriculum, the sequence and formation of the content were not specified, which means that the teaching materials could be arranged flexibly.

The curriculum implementation recommendations included recommendations for teaching, evaluation, and developing teaching materials and curriculum resources. These recommendations were concerned mostly with teaching and learning approaches, the presentation of knowledge in the teaching materials, and academic evaluation methods. As well, typical cases were described to facilitate the users (teachers, textbook writers, education administrators) to understand the mathematics curriculum standard exactly and not water it down.

8.1.3 Writing the Textbook for the Mathematics Curriculum Standard for Compulsory Education

In the process of developing and publishing the *2001 Standard*, a number of publishing houses applied to write experimental textbooks for Grades 7–9. By the end of 2006, nine sets of these textbooks had been approved by the National Examination Committee for Primary and Secondary Schools: *PEP Edition Mathematics* (Grades 7–9) (edited by Q. Lin, and published in succession since 2004); *BNU Edition Mathematics* (Grades 7–9) (edited by F. Ma, and published in succession since 2001); *SHT Edition Mathematics* (Grades 7–9) (edited by X. Zhang and Wu (2005), and published in succession since 2005; *ECNU Edition Mathematics* (Grades 7–9) (edited by J. Wang, and published in succession since 2001); *JST Edition Mathematics* (Grades 7–9) (edited by Y. Yang and Dong, and published in succession since 2004); *HBE Edition Mathematics* (Grades 7–9) (edited by J. Yang, and published in succession since 2003); *QD Edition Mathematics* (Grades 7–9) (edited by T. Zhan, and published in succession since 2006); *ZJE Edition Mathematics* (Grades 7–9) (edited by L. Fan, and published in succession since 2005); *HNE Edition Mathematics* (Grades 7–9) (edited by S. Yan and Qiu, and published in succession since 2003).

The above nine sets of experimental mathematics textbooks for primary schools were examined by the National Examination Committee for Primary and Secondary Schools according to the *Interim Measures of Textbook Writing for Primary and Secondary Schools*, and each one was thought to have unique distinguishing features.

For example, the *Experimental Textbook for Mathematics Curriculum Standards for Compulsory Education* (*Grades 7–9*), published by the People's Education Press, had the characteristics that the structure and knowledge system were arranged properly and the concept introduction and content presentation were natural. It had a scientific basis and inherited the advantages of traditional teaching materials; it supplied interesting resources and materials and included a number of mathematical activities and project learning to promote inquiring and cooperative communication, initiative and enthusiasm for learning, and awareness of the relationship between presentation and mathematics substance. It was based on problem solving, and the combination of basic knowledge, skills, and application. As well, it focused on the integration of mathematics and information technology (Ministry of Education Office for the Examination and Approval of Basic Education Textbooks, 2006).

The Beijing Normal University Press textbook was realistic, and interesting material was selected from nature, society, and other subjects that embodied the formulation and application of mathematics knowledge. Concepts and theorems were elicited from specific problems, the teaching content was diversified, and the topics were taught through the process of *think*, take *action, have a discussion* (Office for the Examination and Approval of the Basic Education Textbooks of Ministry of Education, 2006).

The Shanghai Scientific and Technical Publishers' textbook had the following characteristics. It maintained the advantages of the traditional materials but the content was arranged in such a way as to be adaptable to the students' cognition, and with a focus presenting contexts and problems designed to promote mathematical thinking and initiative. The activities encouraged independent learning, cooperation, observation, communication, and exploring space. Examples were related to the countryside and the application of Information Technology so as to foster the development of rural students (Office for the Examination and Approval of the Basic Education Textbooks of Ministry of Education, 2006).

The East China Normal University Press version presented diversified, rich teaching content, sequenced to reflect the cognitive levels of junior middle school students. The curriculum guided students to participate in the learning process and used self-study reading materials that considered the aspects of problem, analysis, conclusion, thinking, and exploring. The summaries of each chapter included knowledge structure graphs to help students build knowledge systems (Office for the Examination and Approval of the Basic Education Textbooks of Ministry of Education, 2006).

The Jiangsu Science and Technology Press textbook was concerned with the combination of teaching content presentation and mathematics essence. The teaching was based on the students' previous cognitive understanding. It concentrated especially on important concepts and rules and created some useful paradigms, as well as providing students with independent study opportunities presented as Mathematics Activity, Mathematics Laboratory, and project learning. The emphasis was on fostering the application of mathematical knowledge and the application of information technology. The use of computer software was introduced in many places, combined with the course content (Office for the Examination and Approval of the Basic Education Textbooks of Ministry of Education, 2006).

The Hebei Education Publishing House textbook emphasized the connection of mathematics and reality and presented material related to life and the application of mathematics, under the headings of Observation, Thinking, Talking, and Doing. The project-based activities were related to students' practical abilities. A great effort was made to promote situational mathematics and mathematical inquiry, which helped to stimulate students' enthusiasm about learning mathematics (Ministry of Education Office for the Examination and Approval of the Basic Education Textbooks, 2006).

The Qingdao Publishing House textbook reflected an organizational model of problem situation—model building—interpretation—application and development. It included a number of lively curriculum resources, presented in the format of scenario introduction, problem background, and practical application, which was tied to the students' lives and had practical applications. Efforts were made to inspire students to learn mathematics, under the headings of Communication and Discovery, Experiment and Inquiry, Wide-angle Lenses, and Pleasure Park, which focused on guiding students to think and take an interest in mathematics. Information technology was combined with the mathematical content (Office for the Examination and Approval of the Basic Education Textbooks of Ministry of Education, 2006).

The Zhejiang Education Publishing House textbook had the following characteristics. It emphasized the application of modern mathematics curriculum theory and had an improved writing style that was helpful for concept introduction, content organization, and giving examples. It advocated research learning methods and emphasized provision for students to engage in activity and autonomous learning. The topics were arranged under the headings of Inquiring Activity, Doing, and Designing Problems. The curriculum resources were abundant, which increased the interest of the textbook, and its presentation form was diversified, which helped to stimulate students' interest in mathematics learning. The problems emphasized the cultural value of mathematics and information technology integrated with the mathematics teaching material (Ministry of Education Office for the Examination and Approval of the Basic Education Textbooks, 2006).

The Hunan Education Publishing House textbook emphasized combining the imparting of knowledge and the development of mathematical thinking. Each topic was approached under the headings of Observation, Abstraction, and Analysis. It paid attention to both the science of mathematics and its elementary levels, and the logic of the content presentation and the structure of the system were clear. Student participation was encouraged, through the headings Work Your Brain, Try to

Speak, Try to Do, and Have a Try. Again, the emphasis was on the cultural value of mathematics, including historical material, and mathematics with science and technology and human society (Office for the Examination and Approval of the Basic Education Textbooks of Ministry of Education, 2006).

The above nine sets of textbooks were put into use from autumn of 2001 and used successively in the experimental areas.

8.2 Developing and Testing New Curriculum for Senior High School

In April 2000, the Ministry of Education established the Mathematics Curriculum Standards Development Group for Senior High Schools. Professor S. Yan from Beijing Normal University, Professor D. Zhang from East China Normal University, and Professor S. Wang from Capital Normal University were appointed as group leaders. The members of the group included practicing teachers, teaching-research staff, mathematics education researchers, mathematicians, publishing house personnel, and examination researchers.

The group initiated several research projects: an international comparative study for senior high school mathematics curriculum development, a mathematics curriculum content study, a status investigation and analysis, an investigation of the effects of contemporary social, economic, cultural, technological development and mathematics progress, and an investigation of the psychological development of senior high school students and its impact on the mathematics curriculum, and mathematics curriculum evaluation.

8.2.1 Development of the Mathematics Curriculum Standards for Senior High Schools (Experimental Version)

It took three years for the experimental version to go through the stages of basic research, standard drafting, asking for advice, modifying the first draft, and forming an experimental version. The outcome was issued by the Ministry of Education with the name *Mathematics Curriculum Standards for Senior High Schools* (*2003 Standard*).

The first stage, the drafting process, was based on basic research and topic research. During the process, many forums were organized in the cities of Beijing, Shanghai, Guangzhou, and Changchun so as to obtain advice from education, mathematics, and technology circles. The new content was tested in more than forty senior high schools in Beijing, Xinjiang, Guangzhou, and Changchun, and advice was adopted from many experts and experiment results.

The second stage was the formation of a draft for seeking advice. Again many forums were organized in Beijing, Nanjing, and Chongqing, with participants including mathematicians, mathematics education researchers, teaching-research staff, and middle school mathematics teachers.

The third stage was the formation of a draft for examination. Once again a series of forums was conducted. Comments were received from six academics from the Chinese Academy of Sciences (Zhang Gongqing, Wang Yuan, Wang Zikun, Ma Zhiming, Cui Junzhi, and Shi Zhongci) and 12 from educators and researchers nationwide.

The fourth stage was the formation of an experimental draft. The Ministry of Education organized experts to review the draft for examination, and then the standard developing group studied the comments from experts, adopted the advice, and modified the draft for examination.

8.2.2 Content of the Mathematics Curriculum Standards (Experimental Version) for Senior High School

The *2003 Standard* included four parts: Foreword, Curriculum Objectives, Content Standards, and Curriculum Implementation Recommendations. The structure was consistent with the Mathematics Curriculum Standards for Full-time Obligatory Education (experiment version). Details are as follows:

Foreword

The Foreword described the nature, rationale, and Design Considerations. It positioned the mathematics curriculum for senior high schools as one of the main curricula after compulsory education, which included the basic content of mathematics and was a basic curriculum to enhance the quality of citizens.

The rationale consisted of ten goals:

- Build a common foundation and provide a developmental platform.
- Provide diverse courses and address personality differences.
- Cater for learning styles and develop initiative and exploration.
- Improve mathematical thinking ability.
- Develop awareness of mathematics applications.
- Recognize the "double basics" (basic knowledge and basic skills).
- Embody the cultural value of mathematics.
- Integrate information technology and the mathematics curriculum.
- Establish a rational and scientific evaluation system.

The curriculum design section described the curriculum framework for senior high schools, proposed guidelines for students to select courses, and listed the main action verbs in the *Standards*. It was more selective than the earlier of senior high school mathematics curriculum. It was no longer guided by subjects like geometry and algebra, but was more comprehensive. Modules were divided into required and elective courses. The required course consisted of five modules, and the elective course was split into four series of thematic modules.

All students in senior high schools had to study the required course. It was divided into five modules covering the basic content that students should master before graduation and was also the basis for the elective courses. In addition, it was the base for the entrance examinations for Arts and Physics. There was one basic module, and other four modules were determined by student or school selection.

The elective courses were divided into two categories, limited and optional. The limited elective courses consisted of two series for students with different development directions. Series 1 was for those who wanted to develop in the direction of the humanities or social sciences. Series 2 was for those who wanted to study science or economics. These elective courses were necessary and fundamental for most students in senior high schools.

The optional elective courses also consisted of two series and were intended for students who wanted to broaden or improve their mathematics quality further, depending on their chosen orientations, according to the following suggestions.

- Students would meet the requirement to graduate from senior high school if they could complete 10 credits of the required course.
- On the basis of 10 credits for the required course, students who wanted to develop in the direction of the humanities or social sciences had two choices: one was learning electives 1-1 and 1-2 in Series 1, for which they could get 4 credits, then they could select two thematic modules from Series 3 of the elective course for two credits, giving a total of 16 credits. The other was for those who wanted to obtain a higher mathematics qualification, who could select four thematic modules from Series 4 of the elective course and gain four credits in addition to the aforementioned 16 credits, giving a total of 20 credits.
- Students who wanted to develop in the science direction (including economics) had two choices on the basis of 10 credits per required course. One was selecting elective 2-1, elective 2-2, and elective 2-3 in Series 2 of the elective course and get six credits, and then selecting two thematic modules from series 3 to get two credits and two from series 4 to get two credits, giving a total of 20 credits. Students who wanted to obtain higher qualifications could select four thematic modules from Series 4 of the elective course for four credits in addition to the original 16 credits, giving a total of 24 credits.

These combinations had a certain flexibility. Students could apply to transfer between combinations and could get the corresponding credit transfer by completing a test.

The *2003 Standards* divided the action verbs by levels of knowledge and skills, processes and methods, and affect and attitudes.

Curriculum Objectives

The Curriculum Objectives were divided into overall and concrete objectives.

The overall objective, based on nine years' compulsory mathematics education, was to improve students' mathematics literacy as future citizens so as to meet the needs for personal development and social progress.

Six concrete objectives were as follows:

- Acquire the basic knowledge and skills and comprehend the basic mathematics concepts and the essence of mathematics, understand the background and applications of concepts experience mathematical thinking, mathematical discovery, and creativity according to different forms of autonomous learning and inquiry activity.
- Improve basic abilities in spatial imagination, abstract generalization, inference and proof, operations and solutions, and data processing.
- Improve the abilities of proposal, analysis, and solving problems mathematically (including practical problems) as well as mathematical expression and communication, and acquire mathematical knowledge independently.
- Develop skills of application and innovation, and to think about and evaluate some mathematics models in the real world.
- Stimulate interest and establish confidence in learning mathematics develop scientific attitudes and a spirit of "never saying die".
- Develop a mathematical perspective and recognize the value of mathematics in science, application, and culture, and form critical thinking habits. Exhibit rational thinking and experience the esthetic significance of mathematics, leading to a dialectical and historical materialism world outlook.

These six objectives could be divided into three dimensions: knowledge and skills, processes and methods, and affect and attitudes, as proposed by the curriculum reform.

Content Standards

The Content Standards described the contents, requirements, and cases for reference for the five modules and four series.

The specific content for the five modules of the required course was as follows:

- Mathematics 1: set, function concept, and fundamental elementary functionI (exponential, logarithmic, and power functions)
- Mathematics 2: preliminary solid geometry, preliminary analytic geometry
- Mathematics 3: preliminary algorithm, statistics, probability
- Mathematics 4: fundamental elementary functionII (trigonometric function), plane vector, triangular identity transformation
- Mathematics 5: solutions for triangles, sequences, and inequalities.

In this content, the algorithm topic was newly added, and vectors, statistics, and probability were strengthened; the others were traditional content from the past although there were some changes in their objectives, keynotes, and modes of handling. On one hand, the content was considered necessary for students to further understand the relationship between the numbers of changes in the real world and grasp spatial relationships, and data collecting and processing to analyze the way things are going and solve some practical problems. On the other hand, the content was a necessary foundation for all students in senior high schools, either as a step into society, to enter into vocational or technical education, or continue to university studies.

Compared with previous mathematics curricula for senior high schools, there was more emphasis on background knowledge generation and development, and real-world application.

There were two types of elective course, limited and optional. The contents of these courses are listed below.

Limited Elective Course

Series 1 and Series 2 consisted of two and three modules, respectively, with specific content as follows:

- Series 1 course consisted of two modules:
  - Elective Course 1-1: common logic language, conic section and equation, derivatives and their applications.
  - Elective Course 1-2: statistics cases, inference and proof, numeral system expansion, complex number introduction, block diagrams.
- Series 2 course consisted of three modules:
  - Elective Course 2-1: common logic language, conic sections and equations, spatial vectors, and solid geometry.
  - Elective Course 2-2: derivatives and their applications, inference and proof, numeral system expansion, complex number introduction.
  - Elective Course 2-3: counting principles, statistics cases, and probability.

In Series 1 and Series 2 of the Limited Elective Course, some content overlapped, for example, as common logic language, numeral system expansion, and complex number introduction. Some content was similar in name but different in content requirements, as was the case for the example with conic sections and equations, derivatives and their applications, statistics cases, and inference and proof. Also some content arranged in different series, such as block diagrams, only occurred in Series 1 and spatial vectors and solid geometry, counting principles and probability only occurred in Series 2. These differences were because Series 1 was for students who wanted to develop in of the humanities and social sciences, and Series 2 was prepared for those targeting the sciences and economics.

- Optional Elective Course,
  - Series 3 and Series 4 for Optional Elective Course were compiled, respectively, of six and 10 themes.
  - Series 3 consisted of six themes:

Theme 1: selected readings for history of mathematics
Theme 2: information security and ciphers
Theme 3: spherical geometry
Theme 4: symmetry and groups
Theme 5: Euler's formula and closed surface sorting
Theme 6: trisection of an angle and expansion of number fields.

- Series 4 was made up of 10 themes:

Theme 1: selected readings for geometric proof
Theme 2: matrices and transformations
Theme 3: sequences and differences
Theme 4: coordinates and parametric equations
Theme 5: selected readings for inequality
Theme 6: elementary number theory
Theme 7: optimization and primary experimental design
Theme 8: overall planning methods and graph theory
Theme 9: risk and decision making
Theme 10: switch circuits and Boolean algebra.

The Series 3 themes mainly used plain language to explain the content and basic ideas so as to broaden students' horizons and encourage them to recognize the value and beauty of mathematics. The Series 3 Elective Course was evaluated by a combination of qualitative and quantitative methods, but was not considered for the college entrance examination. In Series 4, the content was introduced in a simple way, with students needing to calculate, prove, and handle some problems using mathematics knowledge. After completing Series 4, they needed to write a report on their learning and solve some simple problems using what they had learned. Some colleges could also choose several themes from series 4 for the college entrance according to the needs of recruitment professionals.

Implementation Recommendations

The *Implementation Recommendations for the 2003 Standards* included teaching recommendations, evaluation recommendations, and recommendations for developing teaching materials.

The teaching recommendations were student-oriented and guided students to make study plans, to lay a good foundation and develop their abilities; to pay attention to the relationships in mathematics and improve their understanding of it in its entirety, to emphasize the connections between mathematics knowledge and practice and develop their application skills, and to focus on the cultural value of mathematics and the formation of a scientific outlook, as well as to improve the approaches to teaching and learning to encourage active learning and to use modern information technology properly to improve the teaching quality.

The evaluation recommendations emphasized the processes of assessing students' mathematics learning, evaluating their basic knowledge and skills correctly, and implementing diverse evaluation strategies to promote student development.

The recommendations for developing teaching materials advocated the diversification of textbook writing, and making appropriate adjustments for the stipulated teaching content of each module to address the unique styles and features appropriate to different materials.

The structure of the *2003 Standards* was similar to that of the *2001 Standards*. It expounded 10 aspects of the mathematics curriculum in senior high schools. The structure adopted was "Required Course + Limited Elective Course + Optional Elective Courses." The specific objectives included basic knowledge and basic skills, basic abilities, expanded abilities, application and innovative thinking, and interest and attitudes. The Implementation Recommendations advocated student-oriented learning and guiding students to select appropriate curricula and make study plans. Attention was paid to the evaluation of the learning process and students' abilities, with a focus on implementing diversified evaluation methods to promote students' development. It advocated that diverse textbooks and materials could have their own styles and features (Ministry of Education of the People's Republic of China, 2003).

8.3 Writing of Experimental Textbooks for Senior High School

In June 2003, the writing of experimental textbooks for senior high schools started. Some publishing houses organized mathematicians and mathematics education experts to do this writing.

By 2006, there were six sets of experimental textbooks approved by the National Examination Committee for Teaching Materials for Elementary and Secondary Schools. These were the PEP *M*athematics *T*extbooks A Edition (S. Liu editor in Chief, published in increments since 2004); PEP Mathematics Textbooks B Edition (C. Gao editor in Chief, published in succession from 2004); the Beijing Normal University Edition Mathematics Textbooks (S. Yan and S. Wang editors in Chief, published in increments from 2004); the Jiangsu Education Edition Mathematics Textbooks (Z. Shan editor in Chief, published in increments from 2004); the Hubei Education Edition Mathematics Textbooks (M. Qi editor in Chief, published from 2005); and the Hunan Education Edition Mathematics Textbooks (J. Zhang and S. Li editors in Chief, published from 2005).

The above six sets of experimental textbooks for senior high schools were examined according to the *Interim Measures of Textbook Writing for Primary and Secondary Schools*, and each one had its own unique characteristics.

For example, the textbook edited by S. Liu was published by the People's Education Press. It reflected fundamentality, times, typicality, and acceptability, and it carried forward the advantages of traditional teaching materials. It included many thinking problems and side notes to inspire student thinking actively and prompts that could deepen their understanding of the content. It contained a lot of pictures close to life and utilized many headings to reflect the relationship between the teaching and the outside world, to broaden the students' horizons and give them experience of the applications and cultural value of mathematics. A lot of effort was made to develop students' hands-on abilities and application skills, as well as their use of information technology (Office for the Examination and Approval of the Basic Education Textbooks of Ministry of Education, 2006).

The textbook edited by C. Gao was also published by the People's Education Press. Its focus was on knowledge development and the promotion of mathematical exploration and thinking. The advantages of traditional teaching materials were retained and narrated briefly, and attention was paid to links with junior high school courses so as to facilitate a smooth transition from junior to senior high school. Algorithmic thought was incorporated into the relevant sections. Moreover, there was a focus on integrating information technology and the mathematics curriculum, with some use of the free software "Scilab" developed by a Sino-French cooperation. In particular, the courseware for the solid geometry chapter had rich graphics and also provided visual graphics that could be changed. The textbooks encouraged students to use modern information technology to help themselves understand concept and form spatial concepts (Office for the Examination and Approval of the Basic Education Textbooks of Ministry of Education, 2006).

The textbook edited by S. Yan and S. Wang was published by the Beijing Normal University Publishing Group. It focused on the fundamentality of the mathematics curriculum and retained the advantages of traditional teaching materials, while some traditional content was improved and developed under the new historical conditions. The introduction of laws of deriving, examples of configuration, selection of exercises, and allocation of teaching references was beneficial to reinforce "double-base" training and implement classroom teaching. This book advocated research learning methods and carefully designed learning situations and paid attention to integrating some important mathematical thoughts (e.g., function, symbolic–graphic combinations) to establish the internal relationships between different sections, which ensured the integrity of the course. Some innovations, including the presentation of teaching materials, and vivid curriculum resources were used to stimulate students' interest in learning, expand their mathematical views, and improve their mathematics culture (Office for the Examination and Approval of the Basic Education Textbooks of Ministry of Education, 2006).

Z. Shan's textbook was published by the Jiangsu Education Publishing House. This set of textbooks had "a shallow entrance and deep meaning," which means the content was fundamental, interesting, and hierarchical. It focused on the links between teaching materials and mathematical knowledge, as well as strengthening the links with other disciplines and practical life. The presentation of the content was brief, and the degree of difficulty was appropriate, which left relatively broad space for teachers. Information technology was integrated extensively, including the use of Excel software and drawing functions (Office for the Examination and Approval of the Basic Education Textbooks of Ministry of Education, 2006).

The textbooks edited by M. Qi were published by the Hubei Education Publishing House. This set of textbooks was particularly strong in arranging the content structure and knowledge systems and providing accurate teaching content. The teaching resources focused on application and the culture of mathematics, as well as some creative cases (such as talking from the mathematical model of SARS and the safety probability of the Shenzhou 5 launch), which were relevant to the students' ages. It emphasized the process of forming knowledge and advocated research-based learning and some project learning to develop students' mathematical thinking abilities. The headings included "Reading and Discussion" and "Thinking and Practice" were helpful for students to broaden their horizons and develop autonomous learning. Great importance was attached to the integration of teaching content and the information technology, with the heading "Information Technology Links". This set of textbooks was particularly readable (Office for the Examination and Approval of the Basic Education Textbooks of Ministry of Education, 2006).

The contribution by J. Zhang and S. Li was published by the Hunan Education Publishing House. The structure of content arrangement and presentation for this set of textbook was novel, the introduction and application of the situation was clear and smooth, and the curriculum development was extensive and deep. The language usages and exercise were creative and encouraged active learning. It made many beneficial attempts to guide students to understand the nature and purpose of the mathematics, and to combine the mathematical content, culture, and value, as well as integrating modern information technology (Office for the Examination and Approval of the Basic Education Textbooks of Ministry of Education, 2006).

Each of the above six versions of mathematics textbooks for senior high schools had unique characteristics, but all of them presented the curriculum content by a mixed system, and the use of headings and content representation had improved greatly since earlier versions. These six sets of textbooks were put into use incrementally since the fall of 2004.

8.4 Revising Mathematics Curriculum for Nine-Year Compulsory Education Curriculum

During March 2005, many NPC (National People's Conference) deputies and CPPCC (Chinese People's Political Consultative Conference) members put forward joint proposals to halt the experimental work on the mathematics curriculum standards. Their appeal was on the basis that the mathematics curriculum standards destroyed the mathematics system for thousands of years, which lead to teachers being unable to teach well, students being unable to learn well, and the mathematics teaching quality falling seriously (Jiang, 2005). In response to these opinions, the Ministry of Education re-established the Mathematics Curriculum Standards Revision Group, who set about the revised work of the mathematics curriculum for compulsory education. The head of this group was Professor N. Shi, who was the headmaster of the Northeast Normal University. The group members were six professors of mathematics, five professors of mathematics education, a teacher-researcher of mathematics, two mathematics teachers, a total of 14 people. The group conducted a nationwide investigation of the implementation of the 2001 Standards and analyzed new progress in international mathematics education reforms. These led to suggestions for revision, after several discussions and consultation with the community, and eventually to the *Mathematics Curriculum Standard for Compulsory Education* (*Revised Version*). In December 2011, the Ministry of Education officially issued the *Mathematics Curriculum Standards for Compulsory Education (2011 Edition) (2001 Standards).*

8.4.1 The Revision Process for the Mathematics Curriculum Standard for Compulsory Education (Revised Version)

8.4.1.1 Survey of Implementation Status for *2001 Standards*

When the revision work started in June 2005, a survey was developed for teachers from 12 provinces, Tianjin, Guangdong, Jilin, Ningxia, Shanxi, Shandong, Chongqing, Jiangsu, Hebei, Beijing, Zhejiang, and Hainan, and a field survey was carried out in 12 experimental areas of six provinces, Haikou in Hainan, Shaoguan in Guangdong, Qingdao in Shandong, Xianyang in Shanxi, Lingwu in Ningxia, and Wuxi in Jiangsu. The revision group visited classes in middle and primary schools, organized forums, and communicated with teachers from middle and primary schools to acquaint themselves with the experimental situation and make suggestions for the revisions.

The survey results showed that the concept of the new mathematics curriculum was approved by more and more teachers and that the curriculum reform had strengthened the connections of curriculum content and modern society, that the new ways of learning improved students' self-studying abilities, and that the teachers focused more on the process of learning. The new curriculum promoted the teachers' professional development and research on teaching. The formation of teaching materials was new and the topic was rich, all of which stimulated students' interest in learning and their exploration and application of mathematics. At the same time, a number of issues were reflected by survey. For example, the objectives of the *2001 Standards* were not clear enough and not easy for teachers to operate. The spiraling presentation of the curriculum content had not been handled well; for example, some content appeared twice with a long interval in between, and some content was repeated frequently. The new way of teaching caused polarization of students in the primary grades, and the new evaluation methods were difficult to operate. The results of this investigation had important reference value for the curriculum standard revision.

8.4.1.2 A Further Investigation of New Developments in International Mathematics Education Reform

When the *2001 Standards* were first begun, a comprehensive investigation and analysis were conducted to understand the situation of mathematics education reform at home and abroad. It was a time of many new developments in the international mathematics curriculum reform; the USA, Germany, and other countries had conducted new research in mathematics education, and Hong Kong and Taiwan also had new mathematics curriculum developments. The revision group carried out a comparative study of the mathematics curriculum reforms for these countries and regions.

8.4.1.3 Seminars to Discuss and Revise the 2001 *Standards*

During the revising process, 12 seminars were held, which included nine full discussions and three partial members' discussions.

On May 16, 2005, the revision group held its first conference with the Ministry of Education. J. Zhou, the Minister of Education, met group members, and Vice-Minister X. Chen gave important instructions on the significance of the curriculum standards revision.

From July 2005 to April 2010, eleven meetings were held in Jilin, Chongqing, Changchun, Beijing, Nanjing, Ningbo, and other places. Discussions focused on the basic thought and principles of revision, specifically for the Foreword, Basic Rationale and Design Considerations, Curriculum Objectives, Content Standards, and Implementation Recommendations. The first draft of the revised Standards was developed and distributed for feedback. The revised version was completed on April 25, 2010.

8.4.1.4 Soliciting Opinions

During the process of modifying the curriculum standards, many activities were organized to seek advice. The main events were as follows:

- In June 2006, when the first draft of *the Standards (Revised Version)* was completed, members of the revision group asked for advice from more than 30 experts, scholars, and practicing teachers nationwide.
- On September 8, 2006, Professor N. Shi invited academics from the Chinese Academy of Sciences and mathematicians to give advice about *the Standards (Revised Version)*. B. Jiang, D. Li, Z. Wu, Z. Hou, and Z. Bai attended the discussion and also X. Chen (the Vice-Minister of Education) attended the forum.
- At the beginning of 2007, Professor N. Shi introduced the modifications at a Chinese Spring Festival tea party to ask mathematics experts for advice.

- In July, 2007, Professor N. Shi discussed the *2001 Standards* with some members to solicit their opinions for the revised version.
- In July 2007, the Ministry of Education for Basic Education distributed the draft to teaching-research offices of ten provinces and ten national and provincial experimental areas, and forty experts, to seek their advice.
- In addition, various forums were organized to seek advice from Professors W. Xiang and D. Zhang, and some mathematicians, mathematics educators, and educators in primary and secondary schools.

8.4.2 Changes to the Revised Mathematics Curriculum Standard for Compulsory Education

8.4.2.1 Changes to the Structural System

From the *2001 Standards* to the *2011 Standards,* the following changes were made to the structure.

- The Foreword was rewritten.
- In the Content Standards, the four knowledge domains for the three stages of schooling were unified so as to highlight the comprehensive application of knowledge. The previous name of space and figure was changed to graphics and geometry.
- The Implementation Recommendations were integrated for the three stages of schooling, the recommendations for teaching, evaluation, and textbook compiling were unified, and the development and utilization of curriculum resources were increased.
- The action verbs and cases were added to the appendices in a uniform way.

The *2011 Standards* further clarified and unified the four parts for the three schooling stages, namely the objectives and content of *Number and Algebra, Graphics and Geometry, Statistics and Probability,* and *Comprehension and Practice.* Core words about students' mathematical qualities were elaborated in detail, for example, number sense, symbol sense, spatial concepts, geometric sense, the concept of data analysis, operational capability, reasoning ability, and mathematical modeling. All of these were prepared for teachers to understand and grasp the core idea of the curriculum content.

8.4.2.2 Changes in the Basic Rationale and Curriculum Objectives

In the revised standards, the expressions of the function of mathematics education and the meaning of mathematics were modified, and the Basic Rationale of the *2001 Standards* was adjusted so as to make the expression more exact and easy to understand and implement. It elaborated the nature and objectives of the mathematics curriculum: "The mathematics curriculum for compulsory education is for all students and meets the needs of all students' personality development; as a result, everyone can get a good mathematics education, and different persons can have different types and levels of development in mathematics."

The overall design of the Curriculum Objectives retained the structures of the overall objective and objectives for the schooling stages. The revised standards clearly stated *Four Basics*: "Through mathematics learning in the compulsory education stage, students can obtain the necessary basic knowledge, basic skills, basic ideas, and basic activity experiences in mathematics which are necessary for social life and further developing." Basic knowledge and basic skills are the "double-basic" tradition of Chinese mathematics education, and basic ideas and basic activity experiences are important symbols of students' mathematical literacy. In the revised standards, "Solving Problems" was modified to "Problem Solving", to highlight students' awareness of problems and the development of comprehensive problem-solving abilities, especially the ability to discover and raise problems in specific situations, and improve the ability to analyze and solve problems (Shi, Ma, & Liu, 2012).

8.4.2.3 Changes in Content Standards

In the *2011 Standards,* the curriculum content was divided into the four aspects of number and algebra, figures and geometry, statistics and probability, and comprehension and practical.

In the content structure, there were no changes made to number and algebra. In the first stage, the content consisted of knowing numbers, number operations, common quantities, and exploring patterns. In the second stage, it was knowing numbers, number operations, expressions and equations, direct and inverse proportion, and exploring patterns. In the third stage, it was number and form, equations and inequalities, and functions (Shi et al., 2012).

There was no change to the first and second stages of figure and geometry but in the third stage, the previous four parts were reduced to three. Figures and their transformations and figures and their coordinates remained the same, while knowing figures, and figures and proofs were combined to become figures and their nature. The *2011 Standard* adopted the approach of proving geometry combined with experimental geometry.

For statistics and probability, a considerable adjustment was made to the content structure, making it more hierarchical. The contents of the first stage were reduced, focusing mainly on classification, simple data collecting, and sorting. The second stage was divided into two parts, statistical processes of simple data and the possibility of random phenomenon. The third stage was divided into sampling and data analysis, and event probability. The adjustments were made mainly because the first-stage content had been found to be too hard for the lower-grade students, which had necessitated repetition in the second and third stages. After adjustment, the three stages were clearly distinct, and the difficulty increased gradually (Shi et al., 2012).

The comprehension and practical content was modified greatly to emphasize the kinds of problems and learning activities students could participate independently. The objective was to help students to accumulate experience in doing mathematics activities and to develop their applied skills and innovation (Shi et al., 2012).

8.4.2.4 Changes to Implementation Recommendations

In the *2011 Standards,* changes were made to the Implementation Recommendations centralized to avoid unnecessary repetition and enhance the operability. To promote better understanding of rationale and the teaching processes and methods, some targeted cases were added. The Basic Rationale of the *2011 Standards* was consistent with the 2001 version, except for some adjustments to expressions. The new version included ten core words to describe students' mathematical quality in detail and clearly put forward the requirements of "Four Basics" (basic knowledge, basic skills, basic thinking methods, and basic experiences in activities) in the curriculum objective, which clarified the relationship of the process objectives and results objectives, and the development of innovation and ability. In the Implementation Recommendations, there was a clear description of the teacher's role and the relationship between teachers and students (Ministry of Education of the People's Republic of China, 2011). These changes made the requirements of the *Standards* more exact and in line with the teaching practice.

The textbooks were revised to align with the changes made in the revised *2011 Standards*, and these were approved in the fall of 2012 by the National Textbook Approval Committee for Primary and Secondary Schools.

All in all, the mathematics curriculum reform at the beginning of the twenty-first century established the concept of people-oriented mathematics. During the process, a developmental paradigm emerged (monographic study → comprehensive study → first draft → modify the draft → form the final version) as well as a curriculum paradigm (nature of curriculum/basic rationale → curriculum objective → curriculum content (content standard) → implementation recommendations). The curriculum structure for compulsory education is in the form of "learning areas + schooling stages", and the senior high school uses a modular form with a framework of "required course + elective course", thus forming a comprehensive and selective course structure. The curriculum content reflects the times and is fundamental, universal, developmental, and selective. The teaching materials have been developed in a pattern of one standard with multiple textbooks. All of these features characterize our country's mathematics curriculum system.

References

Jiang, B. (2005). What has the mathematics class lost under the new curriculum standards? *Guang Ming Daily (education weekly)*, 3. Retrieved from http://xueshu.baidu.com/s?wd=paperuri%3A%28034991746c176c6013445039b5e5780c%29&filter=sc_long_sign&tn=SE_xueshusource_2kduw22v&sc_vurl=http%3A%2F%2Fwww.cqvip.com%2FMain%2FDetail.aspx%3Fid%3D15255246&ie=utf-8&sc_us=12345630984252604110.

Ministry of Education Office for the Examination and Approval of Basic Education Textbooks. (2006). *An overview of experimental textbook for compulsory education curriculum standards*. Beijing: People's Education Press.

Ministry of Education of the People's Republic of China. (2003). 普通高中数学课程标准 *(实验)* [Ordinary high school mathematics curriculum standards (experimental)]. Beijing: People's Education Press.

Ministry of Education of the People's Republic of China. (2001). *Mathematics Curriculum Standards for Full-time Compulsory Education* (Experimental Draft). Beijing: Beijing Normal University Press.

Shi, N., Ma, Y., & Liu, X. (2012). 义务教育数学课程标准修订过程与主要内容 [The revision process and main contents of mathematics curriculum standards for compulsory education]. *Curriculum, Teaching Material and Method, 3,* 50–56.

Sun, X. (2007). Overview of recent developments in mathematics in China. *Mathematics Bulletin, 6,* 14–19.

Chapter 9
Features and Characteristics of Chinese New Century Mathematics Textbooks

Haidong Li and Na Li

Abstract Unified curriculum standards (or teaching and learning syllabus) provide uniform guidelines from all teaching and learning activities in different grades levels across China under a centralized education system. Without exception, the textbooks using in schools are also designed in accordance with the beliefs advocated by these official documents. This chapter reports the common features and characteristics (such as the emphasis on the knowledge learning and its applications) of several series school textbooks were and are used in the twenty-first century with the major beliefs presented in the standards or syllabus. It aims to show the essences kept in the New Century mathematics textbooks which reflects the major reforms happened on the textbooks development.

9.1 Background

The Chinese education system is centralized, because it uses unified curriculum standards (previously called the "teaching and learning syllabus") across the nation and uniform guidelines for all teaching and learning activities in different grade levels, and serves as a direct channel for major educational reforms (Li, Zhang, & Ma, 2014; Liu & Li, 2010; Tang, Peng, Cheng, Kuang, & Song, 2013).

In China, mathematics textbooks are developed according to the national curriculum standards, namely the *Full-time Compulsory Education Mathematics Curriculum Standard* (*Experimental Version*) (Ministry of Education, 2001) and the *Senior Secondary School Curriculum Standard* (*Experimental Version*) (Ministry of Education, 2003). The two standards reflect the two phases of Chinese fundamental education. A nine-year compulsory education system, which includes

H. Li (✉)
People's Education Press, Beijing, China
e-mail: lhd@pep.com.cn

N. Li
Faculty of Mathematics and Statistics, Central China Normal University, Wuhan, China

Y. Cao and F.K.S. Leung (eds.), *The 21st Century Mathematics Education in China*, New Frontiers of Educational Research,
https://doi.org/10.1007/978-3-662-55781-5_9

six-year primary education and three-year junior secondary education, is followed by a three-year senior secondary education.

Usually, different versions of textbooks have been developed under the same curriculum standard; this is referred to as "one standard, many versions." These versions have common characteristics and features, but also have their own unique characteristics and features.

Compared with the previous textbooks, the current versions have changed greatly in knowledge, content organization, and presentation. The nature of these changes will be addressed first in this chapter. As well, a set of typical textbooks, developed by the People's Education Press and utilized in junior secondary schools, will be introduced as an example to illustrate the characteristics and features of Chinese mathematics textbooks.

9.2 Common Features and Characteristics

The Curriculum Reform for the New Century has led to many new ideas for developing textbooks, as stated in the curriculum standards issued by the Ministry of Education. Under the unified guidance of the curriculum standards, there are common features and characteristics in these versions of textbooks, as described below.

9.2.1 Emphasis on the Relationship Between Knowledge Learning and Its Applications

New Century textbooks emphasize the relationship between mathematical knowledge learning in the classroom and its application in the real world. To introduce the new knowledge, practical problems occurring in students' lives serve as the teaching materials, to help students easily experience and find mathematical knowledge abstracted from life. This is not only shown in the single lessons but also in the whole chapters. The textbooks create chances for students to solve problems through model constructing. These factors can contribute to students developing a deep understanding of mathematical knowledge and its applications, and mathematical modeling as well as their abilities to apply mathematics.

For example, to introduce the concept of Function to senior secondary school students, all the textbook versions present a number of examples for students to try to solve before making a generalization for function definition. Version A, published by the People's Education Press, presents three examples: the relationship between the height and time of a shell launching, the relationship between time and the area of the ozone hole in the atmosphere, and the relationship between Engel's coefficient and time (Liu, 2007, pp. 15–16). Version B, also published by the

People's Education Press, presents five examples: the relationship between the distance and time of free fall; the relationship between people's curiosity and their ages; the relationship between the height and growth period of corn; the relationship between GDP and time; and the relationship between current and resistance under constant voltage (Gao, 2006, pp. 31–32). The version issued by the Beijing Normal University Publishing Group generalizes the concept of Function from the example of the relationship between mileage and time on the national highway (Yan & Wang, 2006, pp. 25–26). The version issued by the Jiangsu Education Publisher also presents three examples: the relationship between the population and year; the relationship between height and time of free fall; and the relationship between the temperature and the time in one day (Shan, 2008, p. 21). All of these examples are selected in view of their relationships with students' life experiences, or with real applications and science. In these contexts, it is indicated that mathematics has an intensive relationship with the development of society and social life. This makes students feel that mathematics learning is worthwhile.

Moreover, mathematical knowledge is shown alongside its applications in actual life, which is usually the major content of one chapter. Taking the chapter of *Liner Equation with One Unknown* as an example, the previous textbooks usually presented in the style of "the concept-solution—application," while the new textbooks have improved upon this style. The new style is to use equations to analyze and solve actual problems, with the actual problems being described throughout the chapter. An equation can be regarded as a kind of mathematical model. In one such example, the concept and solution of an equation are introduced and discussed to promote the process of using liner equations with one unknown to solve practical problems. The structure of this chapter (Lin, 2007, pp. 78–114) can be seen in Fig. 9.1.

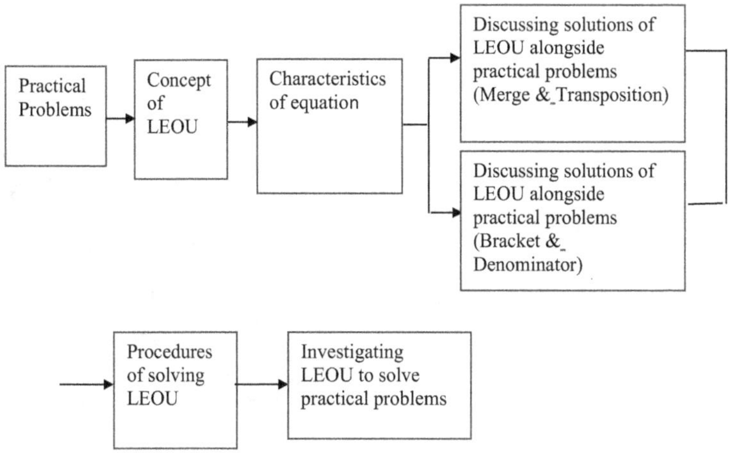

Fig. 9.1 Content structure of chapter Liner Equation with One Unknown (LEOU)

This approach provides much more opportunity and time for students to learn mathematics in practical contexts, and enhances their ability to construct mathematical models and to analyze and solve problems by using equations.

To understand this style better, the topic of *Linear Inequality in One Unknown*, issued by the Jiangsu Scientific Publisher, will be given as another example (Yang & Dong, 2006, pp. 6–20). The first section of this chapter is titled Inequality in Life. To introduce the concept of inequality, several practical problems are provided that are familiar to students in their lives. Students can generalize the concept of inequality from these examples with the guidance of the teacher. When the solutions for linear inequality in one unknown are discussed, a problem relating to a young tree's growth height is discussed, which provides an actual context. The next section, *Solve Problems via Linear Inequality in One Unknown*, focuses on solving problems. Several practical problems are provided to demonstrate the problem-solving process by means of establishing proper inequalities. Practical problems can be found from the beginning to the end of this chapter, which reflects the importance of mathematical knowledge in actual life.

9.2.2 Emphasis on Knowledge Development, Heuristics and Investigation

One of the beliefs underpinning the New Century education reform is that education should be student-centered. Thus, lots of mathematical tasks are designed to help students experience the knowledge acquirement and application through observing, experimenting, conjecturing, reasoning, discussing, and reflection activities.

When designing mathematical tasks, the New Century textbooks demonstrate knowledge in the sequence of "knowledge background—knowledge establishment—description of knowledge relationship." This process can attract students'

Fig. 9.2 Counting, ordering, and producing 1, 2, 3, …

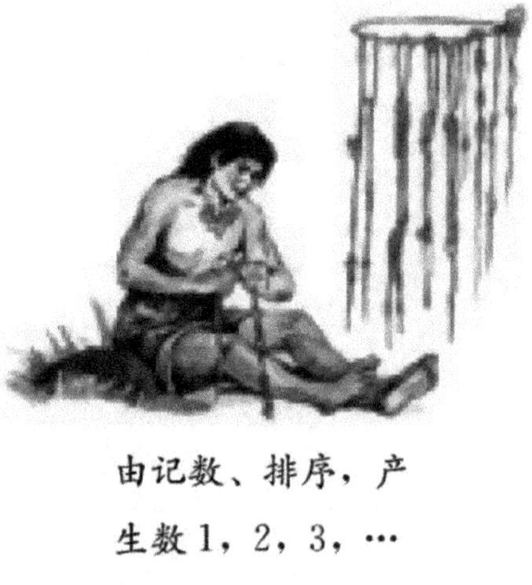

Fig. 9.3 Describing “Null” “Vacancy,” and then 0

Fig. 9.4 Dividing, measuring, and then fraction $\frac{1}{2}, \frac{1}{3}, \ldots$

learning interest and help them to understand the mathematical knowledge and methods and form the good habits of mathematical thinking and application. For example, the textbook, published by the People’s Education Press, presents three figures (see the following Figs. 9.2, 9.3, and 9.4) before introducing negative numbers (Lin, 2007, p. 2).

The above three figures provide backgrounds for introducing positive integers, zero, and positive fraction origins. Based on these, students can generalize that the development and processing of numbers only occurred throughout history as a response to people’s needs. Negative number learning is also treated in a similar way. The whole process reflects number development and relationships among different numbers.

Furthermore, the New Century textbooks encourage active participation in mathematics activities, which enlivens the learning and provokes students’ interest in mathematics. Many investigation tasks have been designed for textbooks, for instance the topic of *Making an Eye Chart* in the Beijing Normal University textbooks (Ma, 2007, pp. 170–173), which uses the following activities:

- Measuring the height and width of “E” and investigating the relationships between Es of different sizes;
- Copying the Es corresponding to the respective vision levels of 0.1, 0.2, 0.3, 0.5, and 1.0, moving in horizontally and investigating their relationship with the knowledge of similarity;
- Making eye charts for 3 and 8 m vision testing distances, respectively, based on the Vision Table of 5 m testing distance;
- Investigating the conversion relationship among all the “Es” on eye charts that need different testing distances.

Students are familiar with eye charts in their daily lives, but they rarely notice the mathematics knowledge contained in the chart design. Given such a context, they can investigate the principle of designing the chart and find the ratio between the size of the E and the testing distance. By this mathematical activity, students calculate the height and width of the E based on their knowledge of similar triangles. Then, they can improve their ability to analyze practical problems by using mathematical knowledge and enhance the integration of theory with practice.

9.2.3 Improving the Content Presentation to Inspire Students' Interest in Mathematics

Regarding to this characteristics, two aspects are considered. One is on the presentation style which means in an appearance way. The other is on the content related to its nature. The two are explained in the following.

9.2.3.1 Improving the Content Presentation Style to Lead Students' Thinking and Investigating

The previous textbook versions usually gave mathematical definitions, concepts, and theorems directly, then explained them in detail, and finally introduced how to use them to solve problems. In the New Century textbooks, content is presented mostly as the sequence of **Problem Background—Problem Analysis—Abstracting Solution—Extending Application**. In this process, mathematical thinking can be shown through various activities, as illustrated below.

- Observing: Making conjectures through observing (i.e., introduction).
- Thinking: Drawing conclusions from thinking and enhancing the comprehension of conclusions through reflection (i.e., deduction and extension).
- Investigating: Finding conclusions through hands-on activities (often followed by the conclusions).

- Discussion: Finding conclusions through communication with other students, which is appropriate for deepening the recognition of content, the extension of the conclusion, and the relationship between different contents.
- Generalizing: Making generalizations about some mathematical features through observing, thinking, investigating, and discussion, which is appropriate for the generalizing of conclusions and the guidance of mathematical thinking.

For example, the knowledge of function monotonicity in Version A of the People's Education Press and the Beijing Normal University version is dealt with as follows. The reasons for investigating features of function are given first (Liu, 2007, pp. 27–29). Then, three function images are shown, and students are asked to find the variation laws of functions through observing (one of the laws is Monotonicity). The following requires the students to observe the images of $f(x) = x$ and $f(x) = x^2$ and articulate their monotonicity in natural language. One question "How to use analytical expression of $f(x) = x^2$ to describe the monotonicity of this function?" is posed to help students describe the function monotonicity in mathematical language.

The Beijing Normal University Publisher version (Yan & Wang, 2006, pp. 40–41) starts with observing expressions and images of $y = x + 1$ and $f(x) = x^2$ to find the laws of the function value y changing with the independent variable x and articulate it in natural language. Then, an activity named *Thinking and Communication* is set up to make students describe how general function changes based on the images of this function, and respond to the question how to describe the variation law of the function value in mathematical language.

It can be seen that the concept of monotonicity is not given directly and upfront in the new textbooks. Instead, they analyze some specific functional examples and pose inquiry questions to guide students to describe monotonicity, not only in natural language but also in mathematical language. This can help students to have a deep understanding of mathematical knowledge.

Another example is the conditions of congruent triangles; the textbook of the People's Education Press provides the following eight investigation problems (Lin, 2008, pp. 6–13):

- Investigation 1: If two triangles satisfy one or two of the conditions that three sides and three angles are correspondingly equal, must they be congruent?
- Investigation 2: If two triangles satisfy the condition that three sides are correspondingly equal, must they be congruent?
- Investigation 3: If two triangles satisfy the condition that two sides and their included angle are correspondingly equal, must they be congruent?
- Investigation 4: If two triangles satisfy the condition that two sides and one opposite angle are correspondingly equal, must they be congruent?
- Investigation 5: If two triangles satisfy the condition that two angles and their included sides are correspondingly equal, must they be congruent?
- Investigation 6: If two triangles satisfy the condition that two angles and one opposite side are correspondingly equal, must they be congruent?

- Investigation 7: If two triangles satisfy the condition that three angles are correspondingly equal, must they be congruent?
- Investigation 8: If two triangles satisfy the condition that the hypotenuse and one right-angle side are correspondingly equal, must they be congruent?

Through the above investigation activities, students are guided to explore the conditions of congruent triangles and master the methods, procedures, and presentation of the proof from conducting experiments to logical reasoning. As for the presentation, the new textbooks strengthen the investigation activities rather than listing conclusions directly, which reflects the focus on the students' learning process.

9.2.3.2 Enhancing Readability and Affinity

Compared with the previous ones, much attention has been paid to the readability of the New Century textbooks. Photos, figures, tables, words, and mathematical symbols are integrated to help students understand the mathematical knowledge and stimulate their interest in learning mathematics.

In terms of the external form, the textbook enlivens the presentation form by posing interesting questions, which guide students to ponder and comprehend the learning content. And there are not only black-and-white textbooks, but also colored and bicolored versions to be chosen. Along with improved readability, the inclusion of illustrations and pictures adds interest. For instance, in addition to the necessary illustrations of the mathematical content, all the new textbooks show a lot of pictures of actual items. Some of them, related to mathematical content, support the teaching, while some are purely decorative illustrations to make the layout lively and stimulate students' interest.

In terms of the content, the new textbooks present mathematical knowledge in ways that students like rather than in totally didactic ways. For example, a reading and thinking column, presented as a dialogue between numbers and the letter X, is presented in the chapter titled Addition and Subtraction of Integral Expressions in the People's Education Press's textbook (Lin, 2007, pp. 61–62), inspired by the form of a science sketch. Through the controversial dialogue, the meaning of using letters to represent numbers can be reflected. Students can realize the impact of progression from arithmetic to algebra in mathematics history. Textbook exercises are also presented in age-appropriate ways. The following exercise from the People's Education Press's textbook is an example (Lin, 2008, p. 81):

> On his way abroad, mathematician Loo-Keng Hua saw the passenger next to him reading an intellectual problem in a magazine. The problem was to find the cube root of 59819. Hua blurted out that it was 39. The people around him were all astonished and wanted to know how to get the answer so quickly.
>
> How do you think Hua calculated it so quickly and accurately? Try to solve the following questions:
>
> 1. Since $10^3 = 1000$ and $100^3 = 1000000$, can you know how many digits of $\sqrt[3]{59319}$?

2. If you know the single digit of 59819 is 9, can you know the single digit of $\sqrt[3]{59319}$?
3. If the last three digits 819 of 59819 were struck out of get 59, and $3^3 = 27$, $4^3 = 64$ can know the tens digit of $\sqrt[3]{59319}$?

The exercise is from the content of "cube root" and is known as an "extensional exploration" activity. It is also an estimation problem. A simple mathematics problem is integrated into a story, which makes the problem more interesting. This way of introducing a problem can attract more students' interest and help them to learn mathematics pleasantly instead of the old boring ways of doing exercises.

9.2.4 Providing Mathematics Context Knowledge to Embody Mathematical Cultural Value

It is one of the requirements that mathematical culture should be valued in the New Century textbooks. Most textbooks have provided a lot of reading materials about mathematicians' lives, mathematics development, and interaction between mathematics and the development of human society, which can help students to experience mathematicians' academic spirits and scientific attitudes, enable them to understand the history of the development of mathematics, and enable them to appreciate the scientific value, application value, and cultural value of mathematics. All of these can help students to broaden their horizons to understand the value of mathematics and to enhance students' cultural literacy.

In the section on *Number and Algebra*, junior secondary textbooks introduce the history of algebra and algebra language. The history of positive and negative numbers, the production and evolution of some important mathematical symbols, reading materials related to equations and their solutions, such as the *Nine Chapters in the Mathematical Art* and *Qin Jiushao's Method*, and the origin, development, and evolution of function concept, are all presented in these textbooks. In the section on *Space and Shape*, *Euclid's Geometry* is introduced to help students sense the value brought about by a deduced geometry system for mathematics development and human civilization. The introduction of several classical proofs of the *Pythagorean Theorem*, such as the Euclidean method and Zhao Shuang's method, and some important issues related to the *Pythagorean Theorem*, is provided to make students feel the flexible, graceful, and delicate features of mathematical proof and the rich cultural connotation of the *Pythagorean Theorem*. A brief introduction to the history of π is shown to help students to have an overview of methods, numerical value, formulae, and features involved in π and its value in modern life (e.g., computing the value of π accurately has been one of the best methods for evaluating the performance of a computer). Introducing cyclotomy in the context of Ancient Greece and China, with appropriate content, can give students a sense of approximation thinking and mathematical connotations in different cultures. For mathematical appreciation, the *Golden Section* and *Seven Bridges of Königsberg* can be introduced to students to encourage mathematical thinking and appreciate the aesthetic value of mathematical propositions and methods. Included in the *Statistics and Probability* section are the origins of probability theory, coin toss experiments, Buffon needle problems, and geometric probability, which enable students to have a brief understanding of the random phenomenon and contribute to their further study.

Here are a few selected examples from junior secondary textbooks published by the People's Education Press. These examples are offered as elective content: denoting the tolerant error in machining by positive and negative numbers; Chinese as the first to use negative numbers; the dialogue between the number 1 and the letter X; the history of the equation; the origin of geometry; the measurement of length; representing a location by latitude and longitude; why to prove; modern and ancient representations and solutions to liner equations; why $\sqrt{2}$ is not a rational number; Yanghui (Pascal)'s triangle; proofs of the *Pythagorean Theorem*; the *Helen—Qin Jiushao formula*; the *Golden Section*; π; the origin of probability theory; probability and winning; wonderful fractal graphics; ancient trigonometric tables; the generation and application of view.

In addition to providing specialized reading materials on mathematical history and culture, mathematical culture is also incorporated into the introductions to chapters, marginalia, and project learning. For example, a poem is presented in the chapter on *Exponential and Logarithmic Functions* in the first senior textbook from Hunan Education Press. The essence of this chapter is embodied in the poem. Infusing mathematical knowledge with some literary flavor can help students to enjoy learning mathematics in a delightful atmosphere. One example of such a poem is (Zhang, 2005, p. 71):

> Moring mists hinder transportation, nuclear mushroom clouds cover the sky.
> Fossil years calculate clever, text searching faster than wind.
> Index and logarithm reflect each other, cube and square look symmetrical.
> Do not affect the interpretation of the world, three kinds of function make outstanding contributions.

In the objective world, there is a lot of monotonically increasing and decreasing phenomena in number. Exponential, logarithmic, and power functions are three basic function models that describe the increasing or decreasing process. The poem not only shows mathematical essence, but also connects mathematics and culture. This can help to improve students' cultural literacy.

9.2.5 *Stress the Integration of Information Technology and Mathematics Curriculum to Improve the Effectiveness of Mathematics Teaching and Learning*

IT is a powerful cognitive tool which should be utilized. This is another requirement of the New Century curriculum. Thus, New Century textbooks have used the power of information technology to help students understand the nature of mathematics. In textbooks, learning how to use a scientific calculator is compulsory. Students can do complex calculations, check calculations, and attempt some exploration activities with the help of a calculator. In addition to the use of calculators, textbooks integrate information technology into content as appropriate, which makes calculators and computers new learning tools. By means of computer software, students can observe the process of formation and change of learning objects. The software, with conversion and measurement functions, can help students to summarize constant positions and quantity relationships in graphic motions. Textbooks integrated with information technology can help students to find the nature of mathematics. There are several ways to integrate information technology into the mathematics curriculum.

One way is to provide examples embedding information technology, such as setting up specialized activities that involve the application of information technology, exercises to be completed via a computer, and mathematical experiments to introduce information technology in detail or allow students to do mathematical experiments. Second, indicating information technology applications in appropriate places, such as, "This can also use a calculator or computer to …" "IT is recommended to use …" can prompt information technology use. The *Application of Information Technology*, in the junior middle school textbooks published by the People's Education Press, incorporates data calculations using spreadsheets, exploring the positional relationship between two straight lines, drawing to find a rule, drawing statistical charts with the help of a computer exploring the nature of axial symmetry, drawing function images through computers, exploring features of inverse functions, doing statistics with computers, exploring the features of rotation, exploring the features of quadratic functions, and exploring homothetic features.

With computers and calculators appearing in classrooms, students have opportunities to see the formation and change of learning objects. They are also given chances to observe, experiment, conjecture, and find mathematical facts. Computers and calculators have some advantages that other learning tools do not. Students can do some activities that were difficult to complete in the past. For example, in the senior high school textbook issued by the Jiangsu Education Press (Shan, 2008, pp. 85–87), the topic of "data fitting," which requires the use of information technology, is presented in the section on function application. In this topic, the method of drawing function images with EXCEL is introduced first. The values of independent variables are generated by the "arithmetic trend" function, and the corresponding values are obtained via the EXCEL function of relative referencing through dragging. Then, three problems are presented to be solved: one on population modeling, one on the modeling of car brakes, and one on the law of celestial movement. Using the given data tables, students are required first to draw a scatter plot in EXCEL, then select an appropriate fitting function, and finally forecast further populations, analyze a car's speed, and interpret Kepler's third law.

9.2.6 Summary

Overall, the new textbooks stress mathematics learning through application to students' lives, attention to their developing, heuristic, and inquiring abilities, and attracting their mathematics learning interest and appreciation of culture, with attention also given to the application of information technology.

9.3 A Specific Example: The Textbooks for Grades 7–9 by People's Education Press (Lin, 2012–2014)

To show more characteristics of Chinese mathematics textbooks, we take the latest version of the People's Education Press textbooks as an example. Before the year 2000, most of the textbooks were issued by this publisher. Even after 2000, more than 60% of the junior middle school textbooks and nearly 80% of the senior middle school ones were provided by this press. Due to its popularity, we selected this version as the example. These textbooks will be described from the perspectives of specific content, the structures of each lesson and each chapter and each volume, and the unique characteristics and features.

9.3.1 Specific Content

The textbooks follow the *Compulsory Education Mathematics Curriculum Standards* (*2011 Edition*). All the content required by the standards for Grade 6 to Grade 9 has been covered in these textbooks. The four areas *Number and Algebra*, *Graphics and Geometry*, *Statistics and Probability*, and *Comprehensive and Practical Activities* are presented in a mixed way. The following table shows the details of this content.

Grade 7 (Semester 1)

Chapter 1 Rational Numbers 1.1 Positive and Negative Numbers 1.2 Rational Numbers 1.3 Addition and Subtraction of Rational Numbers 1.4 Multiplication and Division of Rational Numbers 1.5 Power of Rational Numbers	**Chapter 2 Addition and Subtraction of Integral Expression** 2.1 Integral Expressions 2.2 Addition and Subtraction of Integral Expressions

(continued)

(continued)

Chapter 3 **Linear Equations with One Unknown** 3.1 From Arithmetic Formula to Equation 3.2 Solving a Linear Equation with One Unknown (1)—Transpose and Merge 3.3 Solving a Linear Equation with One Unknown (II)—Remove Parentheses and Denominator 3.4 Practical Problems and Linear Equations with One Unknown	**Preliminary of Geometric Figure** 4.1 Geometrical Figures 4.2 Line, Ray, and Segment 4.3 Angle 4.4 Topic Learning: Learn How to Make a Rectangular Packing Box

Grade 7 (Semester 2)

Chapter 5 Intersecting and Parallel Lines 5.1 Intersecting Lines 5.2 Parallel Lines and Determination 5.3 Properties of Parallel Lines 5.4 Translation	**Chapter 6 Real Numbers** 13.1 Square Roots 13.2 Cube Roots 13.3 Real Numbers
Chapter 7 Rectangular Coordinate System 7.1 Rectangular Coordinate System 7.2 Simple Application of Coordinate Method	**Chapter 8 System of Linear Equations in Two Unknowns** 8.1 System of Linear Equations in Two Unknowns 8.2 Elimination—Solve Systems of Linear Equations in Two Unknowns 8.3 Practical Problems and System of Linear Equations in Two Unknowns 8.4 Solutions to Systems of Linear Equations in Three Unknowns
Chapter 9 Inequality and System of Inequalities 9.1 Inequality 9.2 Linear Inequality with One Unknown 9.3 System of Linear Inequalities with One Unknown	**Chapter 10 Data Collection, Processing, and Presentation** 10.1 Statistical Survey 10.2 Histogram 10.3 Topic Learning: Water Saving from the Perspective of Data

Grade 8 (Semester 1)

Chapter 11 Triangle 11.1 Segments Related to Triangle 11.2 Angels Related to Triangle 11.3 Polygons and the Sums of their Interior Angles	**Chapter 12 Congruent Triangles** 12.1 Congruent Triangles 12.2 Determinations of Congruent Triangles 12.3 Features of Angular Bisectors
Chapter 13 Axisymmetric 13.1 Asymmetry 13.2 Draw Axisymmetric Figures 13.3 Isosceles Triangle 13.4 Topic Learning: The Shortest Path Problem	**Chapter 14 Multiplication of Integral Expressions and Factorization** 14.1 Multiplication of Integral Expressions 14.2 Multiplication Formula 14.3 Factorization

(continued)

(continued)

Chapter 15 Fractional Expression 15.1 Fractional Expression 15.2 The Operation of Fractional Expressions 15.3 Fractional Equation	

Grade 8 (Semester 2)

Chapter 16 Quadratic Radical 16.1 Quadratic Radical 16.2 Multiplication and Division of Quadratic Radical 16.3 Addition and Subtraction of Quadratic Radical	**Chapter 17 Pythagoras Theorem** 17.1 Pythagoras Theorem 17.2 Inverse Theorem of Pythagoras Theorem
Chapter 18 Parallelogram 18.1 Parallelogram 18.2 Special Parallelograms	**Chapter 19 Linear Function** 19.1 Function 19.2 Linear Function 19.3 Topic Learning: Plan Selection
Chapter 20 Data Analysis 20.1 Central Tendency of Data 20.2 Volatility of Data 20.3 Topic Learning: Data of Physical Health Test	

Grade 9 (Semester 1)

Chapter 21 Quadratic Equations with One Unknown 21.1 Quadratic Equations with One Unknown 21.2 Solving Quadratic Equations with One Unknown 21.3 Practical Problem and Quadratic Equations with One Unknown	**Chapter 22 Quadratic Functions** 22.1 The Images and Features of Quadratic Functions 2.2 Studying Quadratic Equations with One Unknown from the Perspective of Function 22.3 Practical Problems and Quadratic Functions
Chapter 23 Rotation 23.1 Rotation of Graphics 23.2 Centrosymmetric 23.3 Topic Learning: Graphics Design	**Chapter 24 Circle** 24.1 Properties of Circle 24.2 Positional Relationship between Dot and Circle, and Line and Circle 24.3 Regular Polygon and Circle 24.4 Arc Length and Sectorial Area

(continued)

(continued)

Chapter 25 Preliminary Probability 25.1 Random Events and Probability 25.2 Finding Probability by Enumeration Method 25.3 Finding Probability by Frequency Estimation	

Grade 9 (Semester 2)

Chapter 26 Inverse Proportional Function 26.1 Inverse Proportional Function 26.2 Practical Problems and Inverse Proportional Function	**Chapter 27 Similarity** 27.1 Similarity of Graphics 27.2 Similar Triangles 27.3 Homothetic
Chapter 28 Trigonometric Function of Acute Angle 28.1 Trigonometric Function of Acute Angle 28.2 Solving Right Angled Triangle and Its Application	**Chapter 29 Projection and View** 29.1 Projection 29.2 Three-View 29.3 Topic Learning: Learn How to Make a Stereo Model

9.3.2 Structures of Each Lesson, Each Chapter, and Each Volume

In each lesson, the structure is determined by the content, which is basically composed of the following parts (see Fig. 9.5). The background for the content of this lesson is provided first. Then, the problem is analyzed through mathematical activities, such as thinking, exploration, and generalization activities, to explain the knowledge underpinning the lesson. After this, a summary of previous content, illustrated examples, and exercises are prepared. At the end of this lesson, the summary is given, along with possible extensions for the lesson and other review or consolidation tasks. Selective content may also be added.

To provide a more detailed description of these textbooks, the structures of each chapter and each lesson are shown in Fig. 9.6.

This figure shows the sequence arrangement of each chapter. First, the name, a picture indicating the topic, and an introduction are presented at the beginning of each chapter. Then, sections on specific content are organized in a sequential way. Each section (e.g., one or two lessons) usually includes exercises for students to do. Sometimes, topics relevant to this chapter are recommended. At the end of each chapter, summary, review, or consolidated tasks are provided.

Each volume of these textbooks also has its own structures. The following introduce the characteristics of each volume of the textbooks.

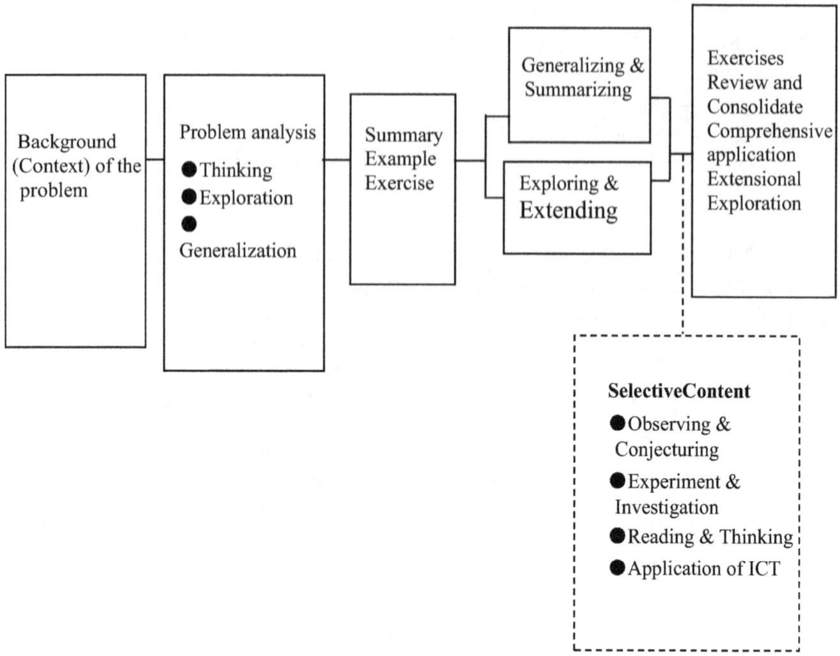

Fig. 9.5 Structure of each lesson

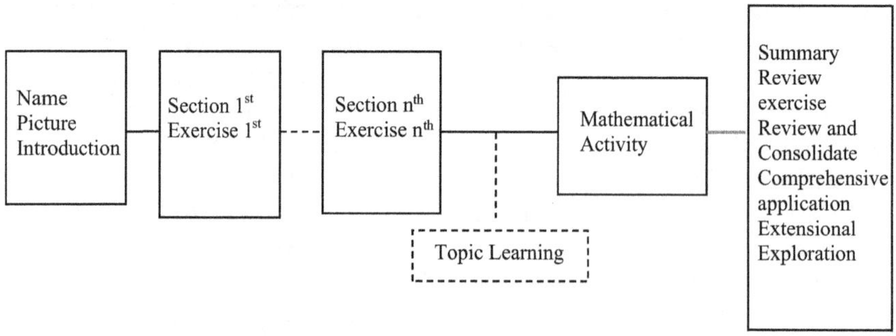

Fig. 9.6 Structure process of each chapter

- Introduction of each volume

The introduction is presented at the beginning of each volume of junior middle school textbooks. In addition to introducing the main content, the teaching columns and learning methods are also presented. The introductions are presented in a friendly way, which shows respect for students and views them as being at the center of the learning.

- The first page of each chapter

The name, theme map, and introduction to each chapter are shown on the first page. The style is usually lively, with illustrations and pictures, to introduce what will be learnt in the chapter. In addition to the name of the chapter, one or several physical pictures related to the chapter's topics are used as the theme map. The introduction is not only presented in a narrative way but also in a problem-based way. First, the real situation or mathematical context is provided, then the main content, learning methods, and mathematical thinking methods required in the chapter.

- Activities in each section

Each section presents the content in the sequence of **Problem Background—Problem Analysis—Abstracting Solution—Extending** Application. In the extension section, some kinds of thinking, investigating, or generalizing activities are included to motivate students to show their thinking processes and lead to generalizations. As well as these activities, Tips, and Clouds are provided to help students understand the content. Relevant content background is included in the Tips, and problems which can help students understand the topic are shown in the Clouds.

- Exercises

There are three kinds of exercises in the new textbooks: those used in classroom teaching, those appearing at the end of each section, and those for review. The first of these are used by students and the teacher during classroom teaching, to help students consolidate what they have learnt. The exercises at the end of each section are for use in classroom teaching or as homework. These exercises can not only help students to consolidate what they have learnt in the section, but also contribute to knowledge application. The exercises arranged at the end of a chapter are mainly for review purposes. These exercises are not simply grouped by difficulty levels but according to their teaching functions of consolidation, comprehensive application, and extended exploration. The first of these reviews addresses the knowledge learnt in a section or a chapter. Comprehensive application exercises call for the application of this knowledge to solve practical and mathematical problems. Extended exploration activities are optional and can be used to extend the knowledge addressed in the section or chapter.

- Selective content

Selective contents include expansion of the body of knowledge, the background of the body content and its applications, the history of mathematics, the introduction of mathematical thinking methods, and the application of information technology.

Based on the characteristics of the selective content, the materials are categorized under the headings of *Reading and Thinking*, *Observation and Conjecture*, *Experiment and Exploration*, and *Application of Information Technology*.

- Mathematical activity

There are some comprehensive, practical, and open-ended mathematical activities in each chapter, which require the comprehensive use of mathematical knowledge. These are optional for either teaching or reviewing the chapter.

- Summary of the chapter

Each chapter has a summary of the knowledge which can be used to help students to review what they have learned. There are two parts to a summary. One is the *Structure of Knowledge* in the chapter, which presents the main knowledge points, development context, and linkages. The other is *Review and Reflection*, which summarizes the main content and the thinking methods. Problems which can motivate students to think in depth are proposed as key or difficult knowledge points, which can help students have a better understanding of the core contents of the chapter and the relevant mathematical thinking methods.

9.3.3 Characteristics and Features

The mathematics knowledge in the textbooks has the characteristics of being universal, basic, and developmental. The textbooks follow the development of technology, consider the needs of a developing society, and focus on the students' long-term development. To ensure students develop fully in basic mathematics knowledge, basic skills, basic thinking, and basic activity experience, ample learning resources are provided in the textbooks. All of these purposes are requirements of the *Mathematics Curriculum Standard of Compulsory Education* (*2011 Edition*) (Ministry of Education, 2012). In addition to the five common characteristics listed in the above section, there are other characteristics:

Constructing the system in line with mathematical logic and students' psychological status, and presenting core concepts and thinking methods spirally

In the previous mathematics curriculum standards, mathematical content was classified into algebra and geometry. However, in the 2011 version, there are four areas: "*Number and Algebra*, *Graphics and Geometry*, *Statistics and Probability*, and *Comprehensive and Practical Activities* (Ministry of Education, 2012)". In the new textbooks, algebra and geometry textbooks are not presented separately. Instead, the four kinds of contents are arranged in a unified way.

The order of different content, coordination, and interconnections between different content, subjects related to mathematics, and students' cognitive understanding are all considered in constructing the textbook system. This system is in line with mathematical logic and students' psychological status in order to facilitate students' understanding of mathematics and mathematical ability.

The relationships between different content are emphasized in the textbooks, i.e., between *Number and Algebra*, *Graphs and Geometry*, and *Statistics and Probability*, showing the entirety of mathematics. The process of solving problems by drawing on knowledge from different content areas helps to give students a comprehensive view of mathematics. For example, in order to reflect the intrinsic link between *Numbers and Figures*, a *Rectangular Coordinate System* is arranged in advance (Chapter 6, 2nd Semester, Grade 7). Since the coordinate tool can fully reflect the number-shape combination thinking, teaching the knowledge of coordinate systems is beneficial for analyzing translation transformation, knowing the essential characteristics of symmetry transformation, handling some graphics issues, and deepening students' understanding of function and system of linear equations in two variables, etc.

Textbooks not only provide the content required by the curriculum standards, but also focus on students' age and cognitive characteristics. Core mathematical concepts and mathematical thinking methods are laid out step-by-step. Reasonable and effective learning approaches are designed and suggestions are provided for teachers. For example, *Function* is the core learning content for *Number and Algebra*, but it is also one of the most difficult concepts in junior secondary education. Considering the difficulty of this topic, the content is scattered and appears in a spiraling way. In accordance with the number sequence, *Equation* and *Function* are presented alternately. Thus, the sequence is: *Linear Equation (System)*, *Linear Function*, *Quadratic Equation*, and then *Quadratic Function*. There are three chapters on Function. *Linear Function* (including *Concept of Function*, *Proportional Function*, and *Linear Function*) is presented in the second volume of Grade 8. *Quadratic Function* is presented in the first volume of Grade 9 and *Inverse Proportional Function* in the second volume of Grade 9. This sequencing can help to overcome the difficulties caused by knowledge presented in a linear way, and to deepen the students' understanding of equations and functions in stages.

In another instance, developing the ability of logical thinking is the key issue in promoting students' mathematical literacy. In view of this, textbook activities are designed carefully to develop logical thinking processes. These activities are based on different levels: giving brief reasons, giving reasons, simple reasoning, and symbolic reasoning. For example, proof appears in Chapter 5 (*Intersecting Lines and Parallel Lines*) in the second semester of Grade 7, but only at the level of completing critical steps and giving reasons. The requirements are increased gradually in subsequent activities. Logical reasoning ability can be developed naturally across topic areas rather than be limited to the areas of *Graphics and Geometry*.

Emphasizing thinking and enhancing learning methods to help students have a deep understanding of core content

The value of thinking in core mathematics is emphasized in the textbooks, and students are guided to use relevant learning methods. Most mathematics knowledge is presented by posing problems. Through problem thinking and problem-solving, students can experience the process of generalizing mathematical concepts and abstracting mathematical principles. Hence, students can experience the mathematical processes used by "real" mathematicians.

For example, algebraic operations, such as addition and multiplication, are used in analyzing correlations between different quantities. In the operation, the universality of operation law can help us analyze the correlations of known and unknown quantities effectively. In solving actual problems, algebraic tools are used to express the quantity (or expression) of actual contexts, or reflect their relationships (equation and function) and changing processes (function). Through this, actual problems are solved using algebraic tools. In the *Number and Algebra* section, extension of number systems, expansion of expressions, enrichment of equations, and the introduction of variables and functions are presented successively. This is a progression from simple to complex, from concrete to abstract, from constant to continuous variables, which reflects the basic method of studying algebra-generalization methods. In the process, the function of rational number is given full attention in relation to real numbers, operations of integral expressions, fractional expressions, and quadratic radicals. Mathematical thinking and generalization are enhanced. Rules of calculation and laws of operation are compared. In the summary, the research method of extension from number to expression is elaborated.

In the area of *Graphics and Geometry*, the basic thinking, content, and methods for studying geometric problems are shown in the textbooks. Taking the topic of the parallelogram as an example, the research perspective from to specializing to generalizing is shown. That is, specialized parallelograms—rectangles, diamonds, and squares—can be obtained from the specialized sides and angles of parallelogram. Then, their properties can be investigated from the location and quantity relationships among their constituent elements (sides, angles, and diagonals). The methods of determination can be studied from the reciprocal relationship between the properties and determination of these figures.

In the case of *Statistics*, the core is data analysis and inference. In the textbooks, the focus is very much on finding rules through investigating data about actual problems. Students gather information through data collection, organization, description, and analysis. They learn to choose the appropriate method based on the problem backgrounds and experience the randomness of data. This can develop student's ideas about data analysis and statistical thinking.

The role of mathematics in other subjects also receives attention in the textbooks: arithmetic to algebra, experimental geometry to argument geometry, constant to variable, and certain mathematics to random mathematics. The role of basic knowledge and methods is emphasized in accomplishing the transition.

This establishes foundations for further learning and mathematics application. For example, the strategy of delimitating unknowns is used in solving the system of linear equations with two variables. The strategy of delimiting an unknown and decreasing the index of an unknown can help students catch the nature of thinking in solving equation problems.

Encourage basic mathematical thinking

As has been established earlier in this chapter, the development of basic mathematical thinking is the most important aim of the school mathematics curriculum. Mathematical thinking can also be used to assess students' mathematical literacy. It is the foundation of mathematics production and development, the basis of investigations, and the essence of the mathematics curriculum. The actual or theoretical background of each concept or theory is emphasized in its introduction. This can help students know the natural development of knowledge. Through problem-solving, students experience the process of generalizing concepts and abstracting mathematical principles. This can help them to understand the research methods of mathematics.

The role of the chapter introduction and summary

A good introduction plays an important role in stimulating students' learning interest, enhancing the teaching of mathematical thinking, and developing the ability to identify and pose questions. The role of the textbook chapter introduction is to pave the way for the exploration from practical problems to mathematical problems and to introduce the content in a natural way.

The summary of each chapter focuses on generalizations arising from the mathematical thinking methods. This can help students to review what they have learnt through the chapter and help the teachers to improve their teaching of mathematical thinking. There are two kinds of summary in each chapter. One is an overview of the knowledge, linkages, and interrelationships covered in the chapter and the other is for review of the types of mathematical thinking they have learnt.

9.4 Brief Summary

In summary, New Century textbooks not only keep good traditional aspects but also show new visions. Traditionally, Chinese textbooks emphasized presenting content in a systematic and rigorous way and stressed students' calculation skills, logical thinking, and imaginary ability. These are all evident in the New Century textbooks but, in addition, the New Century textbooks use content related to students' lives, highlight the knowledge, development, and application, and focus on students' ability to self-explore and collaborate with others in learning.

References

Gao, C. (Ed.). (2006). 普通高中课程标准实验教科书(必修)数学第一册(B版) *[Experimental mathematics textbooks (compulsory, B Version) of general high school under the curriculum standard]* (2nd ed., Vol. 1). Beijing: People's Education Press.

Li, Y., Zhang, J., & Ma, T. (2014). School mathematics textbook design and development practices in China. In Y. Li & G. Lappan (Eds.), *Mathematics curriculum in school education* (pp. 305–326). Dordrecht: Springer.

Lin, Q. (Ed.). (2007). 义务教育课程标准实验教科书数学七年级上册 *[Experimental mathematics textbooks (Grade 7) under curriculum standard for compulsory education]* (3rd ed., Vol. 1). Beijing: People's Education Press.

Lin, Q. (Ed.). (2008). 义务教育课程标准实验教科书数学八年级上册 *[Experimental mathematics textbooks (Grade 8) under curriculum standard for compulsory education]* (2nd ed., Vol. 1). Beijing: People's Education Press.

Lin, Q. (Ed.). (2012–2014). 义务教育课程标准实验教科书数学七年级上册-九年级下册 *[Experimental mathematics textbooks (Grade 7–9) under curriculum standard for compulsory education]* (1st ed.). Beijing: People's Education Press.

Liu, J., & Li, Y. (2010). Mathematics curriculum reform in the chinese mainland: Changes and challenges. In F. K. S. Leung & Y. Li (Eds.), *Reforms and issues in school mathematics in east Asia: Sharing and understanding mathematics education policies and practices* (pp. 9–31). Netherlands: Sense Publishers.

Liu, S. (Ed.). (2007). 普通高中课程标准实验教科书(必修)数学第一册(A版) *[Experimental mathematics textbooks (compulsory, A Version) of general high school under the curriculum standard]* (2nd ed., Vol. 1). Beijing: People's Education Press.

Ma, F. (Ed.). (2007). 义务教育课程标准实验教科书数学七年级上册 *[Experimental mathematics textbooks (Grade 7) under curriculum standard for compulsory education]* (5 ed., Vol. 1). Beijing: Beijing Normal University Publishing Group.

Ministry of Education, P. R. C. (2001). 义务教育数学课程标准(实验稿) *[Full-time compulsory education mathematics curriculum standard (Experimental version)]*. Beijing: Beijing Normal University Publishing Group.

Ministry of Education, P. R. C. (2003). 普通高中数学课程标准(实验) *[Senior secondary school curriculum standard (Experimental version)]*. Beijing: People's Education Press.

Ministry of Education, P. R. C. (2012). 义务教育数学课程标准 (2011 版) Mathematics Curriculum Standard for Compulsory Education (2011 version). Beijing: Beijing Normal University Publishing Group.

Shan, Z. (Ed.). (2008). 普通高中课程标准实验教科书(必修)数学第一册 *[Experimental mathematics textbooks (compulsory) of general high school under the curriculum standard]* (2nd ed., Vol. 1). Jiangsu: Jiangsu Education Publisher.

Tang, H., Peng, A., Cheng, B., Kuang, K., & Song, N. (2013). Characteristics of "two basics" teaching in secondary mathematics classrooms in China. In Y. Li & R. Huang (Eds.), *How chinese teach mathematics and improve teaching*. New York: Routledge.

Yan, S., & Wang, S. (Eds.). (2006). 普通高中课程标准实验教科书(必修)数学第一册 *[Experimental mathematics textbooks (compulsory, A Version) of general high school under the curriculum standard]* (3rd ed., Vol. 1). Beijing: Beijing Normal University Publishing Group.

Yang, Y., & Dong, L. (Eds.). (2006). 义务教育课程标准实验教科书数学八年级下册 *[Experimental mathematics textbooks (Grade 8) under curriculum standard for compulsory education]* (2nd ed., Vol. 1). Nanjing: Jiangsu Scientific Publisher.

Zhang, J. (Ed.). (2005). 普通高中课程标准实验教科书(必修)数学第一册 *[Experimental mathematics textbooks (compulsory) of general high school under the curriculum standard]* (1st ed., Vol. 1). Changsha: Hunan Education Press.

Part III
Mathematical Classroom Instruction

Chapter 10
A Study of Mathematics Classroom Teaching in China: Looking at Lesson Structure, Teaching and Learning Behavior

Yiming Cao, Lianchun Dong and Xinlian Li

Abstract This chapter reports a video-based analysis that examined lesson structure, teaching behavior, and learning behavior in Chinese mathematics classrooms. In particular, mathematics lesson structure refers to classroom activities and the sequences and procedures by which teachers and students complete them. To analyze teaching behavior, two typical teacher actions were selected: teachers' heuristic explanations and teacher questioning. Learning behavior included students' participation, their questioning, and their presentation.

As the key component of education, classroom teaching connects teachers, students, and the curriculum and thereby determines whether the implementation of curriculum reform is successful or not. In the Chinese language, the broader meaning of Jiaoxue (教学) is the teacher's teaching and students' learning, whereas the narrower meaning covers only the teacher's teaching, namely the activities used to stimulate, sustain and enhance students' learning (Yao, 2004). Regardless of which meaning is considered, there is no doubt that the key issues in classroom teaching include lesson structure, the teacher's teaching behavior, and the students' learning behavior. Therefore, the improvement of classroom teaching should begin by considering these three issues.

Y. Cao (✉) · X. Li
Beijing Normal University, Beijing, China
e-mail: caoym@bnu.edu.cn

L. Dong
Minzu University of China, Beijing, China

Y. Cao and F.K.S. Leung (eds.), *The 21st Century Mathematics Education in China*, New Frontiers of Educational Research,
https://doi.org/10.1007/978-3-662-55781-5_10

10.1 Structure of Mathematics Lessons

10.1.1 Introduction

As an agent bridging teaching theories and teaching practices, a teaching model can be regarded as the application of teaching theories in practice on the one hand, and a systematic and synthetic summarization of teaching practices on the other. Thus, there is a very close connection between a teaching model and the practitioners (Cao, 2007). Generally speaking, a teaching model is developed under the umbrella of teaching theories, and thereby guided, controlled, and restricted by theoretical principles and paradigms. Teaching practices provide a base from which the development of a teaching model can occur, but a teaching model is different from simply presenting some individual teachers' experiences or reflections, or demonstrating teaching methods. Instead, it is the sublimation of teaching methods, emphasizing the roles of theoretical principles and paradigms interwoven in the development of the teaching model per se. Therefore, a teaching model can work as a bridge interacting theories and practices and can be used to guide teaching practices and to support the introduction and development of new theories.

There is a variety of perspectives on the concept of the teaching model. Joyce, Weil, and Calhoun (2002) argued that teaching models discuss, in a systematic way, educational goals, teaching strategies, curriculum design and textbooks, and the relationships among social and psychological theories so that teachers can have a variety of choices in implementing their teaching. Some US researchers also pointed out that teaching models are a sequence of procedures oriented toward learning results (Wu, 1998). In the community of Chinese educational researchers, the teaching model is perceived as a theory about lesson designing and organizing which is developed in instructional practices.

Based on the above-mentioned perspectives and the characteristics of mathematics teaching, the study of teaching models for mathematics lessons ought to focus on two aspects: One is the teaching model in general, and the other is the features that exist exclusively in mathematics lessons. Thus, two types of teaching model should be explored for mathematics lessons. The mathematics-specific teaching model can reflect some typical features of the lessons. The subject-general teaching model is shared across all school subjects. Regardless of which type it is, a teaching model should have a relatively reliable structure. For a mathematics-specific teaching model, this structure refers to both the components (e.g., teaching and learning activities) constituting mathematics instruction and the relationships among these components.

In this study, mathematics lesson structure refers to classroom activities and the sequences and procedures by which teachers and students complete them. These activities, sequences, and procedures are all indispensable.

This study set out to answer the following research questions:

- Are there any recurrent classroom activities or procedures in one teacher's instructional practices?
- Are there any reliable lesson structures or repeating sequences evident in a series of ten lessons taught by one teacher? If yes, are these reliable lesson structures or repeating sequences also observable in other teachers' classrooms? If no, what variations can be observed in other teachers' classrooms?

The study focused on two year 7 mathematics teachers' classrooms in Shanghai; it should be noted that these should not be seen as representatives of mathematics classrooms all over mainland China.

10.1.2 Research Design

Participants

The participants were drawn from a database for a larger research project, the LPS Project, which was conducted by the International Center for Classroom Research in the University of Melbourne, Australia. The LPS Project collected data from four sites in China, Hong Kong SAR, Macao SAR, Shanghai, and Suzhou. There were three teachers in the Shanghai database, and a sequence of 15 consecutive lessons was video recorded. The teaching topics for the three teachers were the same: *Linear equations with two unknowns* and *Systems of two simultaneous linear equations*. Teacher SH1 and teacher SH2 were quite similar in their teaching paces and designs, whereas teacher SH3 was different (each teacher has a unique label, such as SH1, in the LPS study and this study retained the same labels). Therefore, teachers SH1 and SH3 were selected for this study, and a detailed analysis was conducted for the first 10 lessons of the 15 consecutive lessons.

Research instruments

Thirteen codes developed by Japanese researcher Yoshinori Shimizu were adopted in this study to analyze the videos of the two mathematics teachers. *StudioCode*, software for video analysis, was used to facilitate the analysis in this study.

Coding systems

The thirteen codes were used to analyze the lesson structure. An earlier version of this coding system was developed in the TIMSS video study to analyze mathematics lessons, and the naming of the codes was based on the identification of key events in the teaching. Based on the coding systems developed by the TIMSS researchers, Shimizu made further revisions and eventually the system had 13 codes. The details of these 13 codes are shown below (Shimizu, 2002).

- Reviewing the Previous Lesson (RP)
- Checking Homework (CH)
- Presenting the Topic (PT)
- Formulating the Problem for the Day (FP)
- Presenting the Problem for the Day (PP)
- Working on Sub-problem (WS)
- Working on the Problem Individually or in Groups (WP)
- Presentation by Students (PS)
- Discussing Solution Methods (DS)
- Practicing (P)
- Highlighting and Summarizing the Main Point (HS)
- Assigning Homework (AH)
- Announcement of the Next Topic (AN).

There were three steps in applying these codes to analyze the participating teachers in this study. The first step was to watch the three teachers' videotapes and look for classroom behaviors that matched the descriptions of the codes. During this process, ambiguous episodes for which it was difficult to find one code were discussed by all researchers to reach agreement. In the second step, the 4th and 5th lessons by teachers SH1 and SH3 were coded independently by two researchers, and the identified ambiguous episodes were discussed to achieve an agreement. In the first two steps, the codes and descriptions were refined, and in the third step, the refined coding systems were used, with the assistance of *StudioCode*, to code all the ten lessons by the two participating teachers.

10.1.3 Findings

Students' acquisition of new knowledge in whole-class teaching.

Whole-class teaching was common in both classrooms which is not surprising as, it can be argued, this is the core of Chinese mathematics classrooms. For all the lessons taught by the two teachers, no episodes matched the code "Announcement of the Next Topic," and there was only one episode in teacher SH3's lessons that matched the code "Checking Homework," although, this latter did work as a base on which the new knowledge could be introduced. Although the two teachers did not mention the knowledge to be learnt in the next lesson and certainly did not ask the students to preview the knowledge, they used "Reviewing the Previous Lesson" (SH1: 9/10; SH3: 6/10) to make connections between old and new knowledge.

Intensive lessons with strong emphasis on problem solving

For both teachers SH1 and SH3, "Presenting the Problems for the Day" was observed in every lesson. In teacher SH1's class, this code was observed at least once and at most four times every lesson, with an average of 2.8 times per lesson. In teacher SH3's class, it was observed at least once and at most seven times every lesson, with an average of 2.9 times. Generally, in a Chinese mathematics lesson, there is a clear mathematics problem or task which the whole lesson is designed to analyze and solve.

Four models for teaching problem solving were observed in the two teachers' classes.

Model one: "Presenting the Problems for the Day → Discussing Solution Methods" ("PP → DS").

This model was observed very frequently in the two teachers' lessons. Teacher SH1 used it in every lesson, while it was observed in seven out of the ten lessons by teacher SH3. Teacher SH1 used this model more frequently than teacher SH3. In teacher SH1's lessons 2, 3, 4, 6, 7, 8, and 9, all mathematics tasks were taught by following the "PP → DS" model, whereas in lessons 5 and 10, 70% of the tasks followed this model. About 50% of all mathematics tasks in lesson 1 were taught using the model. In teacher SH3's case, all mathematics tasks in lessons 2,4, and 8 followed the "PP → DS" model. Nearly 67% of the mathematics tasks in lesson 5 were taught through this model, and the proportions were, respectively, 57, 50, and 25% in lessons 3, 1, and 7. The "PP → DS" model was not observed in lessons 6, 9, or 10.

In mainland China, the contents covered in a mathematics lesson are usually very intensive and thereby mathematics teachers choose to pose tasks and then solve them by organizing whole-class discussion. The "PP → DS" model is efficient in satisfying the demands of teaching intensive content. Furthermore, in the process of classroom discussion, teachers can regulate the teaching pace so that their students can acquire knowledge in a short period of time. The disadvantage of the "PP → DS" model is the constraint on students' thinking, because it requires them to follow the teacher's pace and participate in the environment created by the teacher. If students fail to follow the teachers' pace, they might miss opportunities to learn, and if some students' thinking is faster than the teacher's, the mathematics lessons might become boring and cause these students to have negative emotions. Nevertheless, due to the big class size, it is impossible for the teachers to cater to every single student's demands. In this regard, the "PP → DS" model is relatively suitable for Chinese mathematics lessons.

In this study, some sub-models were identified for the "PP → DS" model:

- "PP → DS → PP → DS," "Presenting the Problems for the Day → Discussing Solution Methods → Presenting the Problems for the Day → Discussing Solution Methods": In this sub-model, after presenting a mathematics task, the teacher discusses the possible solutions for the task. Then, the teacher presents another task and organizes the discussion again.
- "PP → DS → P," "Presenting the Problems for the Day → Discussing Solution Methods → Practicing": In this sub-model, after presenting a

mathematics task, the teacher discusses the possible solutions for the task. Then, the teacher asks the whole class to do some exercises to consolidate the learned knowledge.

There were two types of "PP → DS → P" identified in this study: "PP → DS → P → HS," "Presenting the Problems for the Day → Discussing Solution Methods → Practicing → Highlighting and Summarizing the Main Point," and "PP → DS → P → DS," "Presenting the Problems for the Day → Discussing Solution Methods → Practicing → Discussing Solution Methods." In other words, after students finished the practicing, the teacher had two options. One option was to summarize the key points in the problem solving. Alternatively, the teacher could organize classroom discussion about how to solve the practice tasks.

- "PP → DS → HS," "Presenting the Problems for the Day → Discussing Solution Methods → Practicing → Highlighting and Summarizing the Main Point." In the "PP → DS → HS," after presenting a mathematics task, the teacher discusses the possible solutions for the task and then summarizes the key points about problem solving.

  In summary, the "PP → DS" model is quite popular and conventional in Chinese mathematics lessons and was observed in the classrooms of both teachers SH1 and SH3. However, this model was observed in higher frequencies in teacher SH1's classroom than in teacher SH3's classroom. In this regard, the lessons taught by teacher SH1 were more traditional than those taught by teacher SH3.

Model two: "Presenting the Problems for the Day → Working on the Problem Individually or in Groups" ("PP → WP")

This model was observed in the classrooms of both teachers SH1 and SH3. It was used infrequently by teacher SH1 (1/10) but frequently by teacher SH3 (7/10). Teacher SH3 taught all mathematics tasks using the "PP → WP" model in lesson 6. In lesson 7, the proportion of mathematics tasks taught via the "PP → WP" model was 75%. The percentages were 50, 50, 43, and 33% in lessons 1, 9, 3, and 5, respectively.

Due to teachers' various teaching styles, the "PP → WP" model can be used in quite different ways. The main difference between the "PP → WP" model and the "PP → DS" model lies in the different roles of the teachers. In the "PP → DS" model, the teacher plays a leading role, and the students are lead to participate in the discussions. In contrast, in the "PP → WP" model the teacher plays a guiding role to create an environment for students to explore and solve the mathematics tasks. In the "PP → WP" model, the teacher does not control the students' thinking, and students have sufficient opportunities to think about and discuss mathematics. However, if the teacher's control of students' learning is loose, some might get off the track during the discussion. Thus, the teacher needs to keep communicating with students to track the progress of students' discussions. This is, however, impossible in big classes, as in China. It is because of this that the "PP → WP" model was observed less frequently in the two participating teachers' classes. Nevertheless, the "PP → WP" model is accepted by more and more mathematics teachers who are making attempts to integrate it into their everyday practices.

In this study, some sub-models were identified for the "PP → WP" model.

- "PP → WP → PS," "Presenting the Problems for the Day → Working on the Problem Individually or in Groups → Presentation by Students": In this sub-model, the teacher asks students to share their thinking in public after students' group work.
- "PP → WP → DS," "Presenting the Problems for the Day → Working on the Problem Individually or in Groups → Discussing Solution Methods": In this sub-model, the teacher organizes classroom discussions after students' group work.

In summary, the "PP → WP" model provides students with more opportunities to learn and discuss, and thereby are accepted increasingly by mathematics teachers in mainland China.

Model three: "Presenting the Problems for the Day → Presentation by Students" ("PP → PS")

This model was observed in the two teachers' lessons, but with lower frequencies. By examining the observed episodes using the "PP → PS" model, it could be seen that these episodes all occurred in lesson ten for both teachers, which was a tutorial lesson in both cases. Although this model was also observed in teacher SH3's lesson 9, it was only in the last several minutes of the lesson. Thus, in this study, it appears that the "PP → PS" model was usually used when the students had a strong basis of knowledge and therefore had mature thinking and were able to elaborate on the teachers' questions.

In this study, some sub-models were identified for the "PP → PS" model.

- "PP → PS → HS," "Presenting the Problems for the Day → Presentation by Students → Highlighting and Summarizing the Main Point": In this sub-model, the teacher gives some summarizations after students' presentations of their work and thinking.
- "PP → PS → DS," referring to "Presenting the Problems for the Day → Presentation by Students → Discussing Solution Methods." In this sub-model, the teacher organizes classroom discussions after students' presentations of their work and thinking.

Model four: "Presenting the Problems for the Day → Working on Sub-problem → Discussing Solution Methods → Working on Sub-problem → Discussing Solution Methods" ("PP → PS").

This model was only observed in teacher SH1's lesson 1. This implies that the observed mathematics lessons focused mainly on problem solving, with less consideration given to the connections between problems.

Generally, mathematics lessons in China focus mainly on problem solving and the teachers usually use various strategies to teach it.

Stable structure of mathematics lessons.

By examining the sequences of ten lessons taught by teachers SH1 and SH3, it could be seen that both teachers had stable lesson structures. Both started the lesson with a review (SH1: 9/10, SH3: 6/10). After the teaching of problem solving, both made summaries (SH1: 9/10, SH3: 10/10) and assigned homework (SH1: 10/10, SH3: 10/10).

As mentioned previously, mathematics lessons in China usually cover a large amount of mathematics knowledge. Therefore, teachers do not usually have time to consider issues beyond the current lesson goals. However, in order to maintain coherence among lessons, the teachers usually review the prior knowledge at the beginning of each one. The review section can help students to construct an understanding of the connections within mathematics knowledge, and also facilitates the teaching of the current lesson, especially when it has strong connections with previous knowledge. Furthermore, because a large amount of mathematics knowledge has to be covered in one lesson, students tend to feel too overwhelmed to find the connections within the mathematics knowledge. Therefore, it is essential for the teacher to give some summaries. Assigning homework is also a traditional component of the mathematics lesson. In order to consolidate the large amount of mathematics knowledge learnt in one lesson, it is necessary for students to do some exercises. Since there is a lack of time to do them in mathematics lessons, the consolidating exercises are usually assigned to students as homework.

Of course, the lesson structure is linked to the teaching guide books. Each mathematics teacher has a teaching guide book in which some detailed suggestions are given. These suggestions include lesson components such as "Review and Introduction," "Summarization and Highlighting," and "Assigning Homework." Many mathematics teachers accept these suggestions and develop some habits of implementing them. This is one typical feature of mathematics lessons in mainland China.

Introducing mathematics problems as a straightforward presentation of problems rather than following "Presenting the Topic → Formulating the Problem for the Day → Presenting the Problems for the Day."

In the observed sequence of lessons, two of those taught by teacher SH1 and three taught by teacher SH3 used the model of "Presenting the Topic → Formulating the Problem for the Day → Presenting the Problems for the Day." In contrast, straightforward presentations of mathematics problems were observed in each lesson of the sequence. This could imply that the two teachers preferred to present the mathematics problems in a straightforward way.

The reason behind this preference has something to do with the large amount of knowledge covered in each lesson. In interviews with the teachers, they commented on this, saying "If we followed the 'Presenting the Topic → Formulating the Problem for the Day → Presenting the Problems for the Day' model, it would take a much longer time and thereby the lesson goals could not be completed." They also believed that "the presenting of a situation usually brings some distractions to the students and causes them to find it difficult to engage in the lesson." Therefore, when selecting the situation, teachers tend to be very careful. These situations cannot take up much time, and nor can they distract students.

10.2 Teaching Actions

10.2.1 Teacher Heuristic Teaching

In the Chinese language, *Qifa* (启发) refers to a situation where pupils obtain a deep and clear understanding of some knowledge with assistance and guidance from other people. In modern Chinese, *Qi* (启) usually refers to teachers' actions of creating environments in which students have strong motivation and positive attitudes toward the appreciation, imagination, reasoning, and memorization of knowledge. *Fa* (发) usually means students' actions of developing strong motivation to learn and to carry out learning activities. *Qifa* (启发) (heuristics teaching) can be regarded as a pedagogical principle which can be interwoven into the whole process of teaching of every subject. It can also be seen as a teaching method or strategy.

In mathematics teaching, *Qifa* (启发) has been applied substantially and developed. In the book *How to Solve It*, Polya discussed a dictionary-style set of heuristics. He believed that modern heuristics should include the whole process of solving a problem and that teachers should provide students with opportunities for independent explorations. By selecting mathematics tasks appropriately, teachers should allow students with medium mathematics capabilities to access the independent exploration of mathematics. In this way, students can learn how to think mathematically and acquire critical and creative thinking abilities. Freudenthal (1972) believed that teachers need to motivate students to learn and experience mathematics. If we teach it as a product, then the only opportunity for students' activities relies on its application, which is mathematics problems. In mathematics problems, there is no real mathematics, but only the imitation of it. The incorrect impressions left in students' minds when they leave school can be seen as a result of inappropriate teaching.

With the ongoing educational reform, there have been some variations in China regarding the *Qifa* (启发) teaching. The "Study-Ask-Practice-Think-Act" approach, proposed by Confucius, was developed by Zhu Xi to "Study-Ask-Think-Criticize-Act" and then by Cai Yuanpei to "Self Study-Guide-Synthesize-Instinct-Beauty." Similarly, in Western society, heuristic teaching has shifted from "Question-Ask-Reflect-Guide-Conclude," as introduced by Socrates, to Plato's "Conversation-Debate-Think-Truth" and then to Dewey's "Situation Creation-Question Formulation-Hypothesis proposal-Reasoning-Application and Test."

In Western society, a shift has occurred from traditional heuristic teaching to exploration. In this regard, Polya did some pioneering work by introducing "heuristics." In the 1930s to 1940s, he talked about understanding the process of problem solving, especially the mental activities essential to it and then focused on the exploration method, eventually reviving the heuristic teaching in plain and modern ways (Polya, 2014).

In modern China, a famous mathematics educator, Fu Zhongsun (傅种孙), focused on thinking about mathematics in philosophical ways. He proposed that "the key aspect is not to know how but to know the process of knowing how," which emphasized the processes and ways of mathematical reasoning.

In the 1980s, a group of mathematics educators and researchers in China started to investigate mathematics methodologies such as Polya's ideas. They also put these ideas into educational reforms in textbooks and teaching, and introduced a series of academic conferences about PM and MM. After the educational experiments, MM came to be regarded as modern *Qifa* teaching.

Nowadays, *Qifa* teaching has a variety of styles, such as Question-Ask, Activity, Cognitive, and Non-cognitive. The meaning of *Qifa* has also been broadened from simple to sophisticated. In this study, the nature and meaning of *Qifa* teaching was explored so as to present the practices of mathematics teaching in China.

10.2.2 Research Design

Participants

From the LPS Project China database (including Hong Kong SAR, Macao SAR, Shanghai and Suzhou), two representative teachers, SH1 and SH2 (as mentioned above, these are the unique labels assigned in the LPS study and retained for this study), were selected, and out of the sequence of 15 lessons, lessons 6–10 were selected for further analysis.

Research instrument

StudioCode, a software package developed collaboratively by ICCR at The University of Melbourne, and a software company for video analysis, was used to facilitate the analysis in this study.

Code development:

In the TIMSS study, two methods were used to code classroom conversations. One coding method was about utterances, and the other one was based on this but analyzed the conversation in further detail. In the second method, a sequence of Elicitations-Responses was added to depict teacher–student conversations. Although it involved research about conversation and *Qifa*, the conclusions were about the time spent on teacher talk and student talk. By looking at the LPS and TIMSS research designs, we developed a simplified two-dimensional coding system to examine the form and function of *Qifa* teaching in Chinese mathematics classrooms.

The codes for the form of *Qifa* teaching were as follows:

- Answering questions: individual answers, whole-class answers, follow-up question, teacher–student co-answers, teacher self-answers
- Problem solving: student seat work, student questioning, student discussion, individual discussion, group discussion
- Teacher explanations: whole-class explanation, individual guidance

The codes for the function of *Qifa* teaching were as follows:

- Bridging: connection, sub-tasks, problem situations
- Understanding: student reading tasks, teacher reading tasks, student explanations, teacher explanations, demonstrations
- Guidance: analyzed relationships, detailed methods, general rules, reflective questioning
- Evaluation: hints, comments, praise, explanations, and comments.

This coding system was used to analyze the teachers' questioning and explanations, thus illustrating the characteristics of Chinese mathematics classes. In particular, it was used to analyze how the Chinese teachers used *Qifa* teaching to teach and explore mathematics and to carry out teacher–student interactions.

10.2.3 Findings

10.2.3.1 With Teachers' Guidance, Chinese Mathematics Classrooms Are Highly Unified and Designed

By looking at the coding results, it could be seen that teacher SH1 spent 86.49% of lesson time on explanations, which included explanations to the whole class (83.87%). The time spent on students' seat work was 15.27%. The whole process of teaching was within the teacher's design and plan. There were no students asking questions. About 1.448% of the time was spent on students' discussion.

The time used by teacher SH2 on explanations was shorter than that spent by teacher SH2. The proportion was 69.97%. About 46.17% of the time was used for explanations to the whole class. The time for students' seat work accounted for 47.17% and, as a result, about 94% of the teaching was within the teacher's design and plan. It was observed once that a student asked a question. The proportion of time for students' discussion was 3.43%.

The emphasis on teacher's systematic explanations, basic knowledge, and basic skills is a typical feature of Chinese mathematics teaching. Ausubel argued that the main task for schools is to deliver clear, stable, and systematic subject knowledge to students by stating that in most classrooms, especially for older students, students have meaningful accepted learning (Ausubel, 1963). The priority of schools is to develop systematic knowledge by implementing meaningful learning so that students can have desirable cognitive frameworks. This argument agrees with some Chinese teachers' practices and is generally accepted in most Chinese schools.

Mathematics classes in China are neither teacher-centered not student-centered. Instead, they are teacher-guided but student-centered, or both teacher-centered and student-centered.

There has been a lengthy debate about student–teacher relationships. In the 1980s, a teacher-guided but student-centered relationship was proposed but, by the end of twentieth century, the "both teacher-centered and student-centered" relationship was accepted increasingly by the Chinese research community. These relationships are different from teacher-centeredness or student-centeredness in western research communities. In Chinese classes, although most of the time is within the teachers' control, a large amount is spent on teacher–student interactions and students' seat work.

For teacher SH1, the proportion of time used in activities such as individual questioning, whole-class questioning, follow-up questioning, teacher–student co-answering, student seat work, student discussions, individual discussions, and classroom discussions was 42.41%. This represented one half of the time devoted to teacher explanation (86.94%). In other words, teacher explanation is actually an activity in which both the teacher and the students participate. Further, students' independence can also be reflected in the time for student seat work. About 15.27% of the time was used for this. In summary, mathematics teaching has a strong feature of teacher-guided and student-centered learning.

For teacher SH2, the proportion of time used in activities such as individual questioning, whole-class questioning, follow-up questioning, teacher–student co-answering, student questioning, student discussion, individual discussion, and group discussion was 68.15%. This proportion was nearly the same as the proportion of teacher explanation (69.97%). In other words, teacher explanation is actually teacher facilitation of discussions. In addition, 47.17% of the time, on average, was spent on student seat work. This reflects that teacher SH2's teaching focused on problem-solving activity participated in by both the teacher and the students. Allowing students to explore mathematics independently reflects the pedagogical principles of "teaching for not teaching" and "teaching how to fish," based on the Chinese proverb which means "Give a hungry man a fish and you'll feed him for a day. Teach him how to fish and he'll feed himself for a lifetime." This is student-oriented, rather than teacher-centered or student-centered.

10.2.3.2 Chinese Mathematics Teachers Value *Qifa* and Guidance Through Teacher–Student Conversations

A typical feature of Chinese mathematics lessons is that teacher's explanations follow the sequence of "bridging, understanding, hints, and evaluation," in which the teacher and students investigate mathematics together. For teacher SH1, the proportion of time used in activities, such as individual questioning, whole-class questioning, follow-up questioning, teacher–student co-answering, student seat work, student discussion, individual discussion, and classroom discussion, was 42.41%. For teacher SH2, the proportion of time used in activities, such as individual questioning, whole-class questioning, follow-up questioning, teacher–student co-answering, student questioning, student discussion, individual discussion, and group discussion, was 68.15%. A further coding of the functions of the above activities was conducted.

Teacher SH1 helped students to make connections between new and old knowledge by reviewing old knowledge (5.40%). By setting up sub-tasks (6.48%), the teacher guided students to explore new mathematics knowledge.

Teacher SH2 also helped students to make connections between new and old knowledge by reviewing old knowledge (1.1240%). By setting up sub-tasks (18.358%), setting up problem situations (0.26%) and allowing students' independent explorations (17%), the teacher guided students to explore new mathematics knowledge.

Teachers SH1 and SH2 both presented mathematics tasks and helped their students to understand the tasks (9.824, 3.736%), through: student reading tasks (1.606, 2.424%), teacher reading tasks (6.18, 1.458%), student explanations (1.148, 0.342%), teacher explanations (1.984, 2.486%), and demonstrations (0, 1.524%).

When teaching problem solving, teacher SH1 and SH2 were not simply telling their students the procedures, but guiding them via questions, conversations, and individual guidance (113.8 times, 89.8 times). Teacher guiding included analyzing the relationships, detailed methods, general rules, reflective questions, and comments.

10.2.3.3 Teachers Design Tasks Carefully, Give Feedback in Time, and Provide Personalized Facilitation and Guidance

Teacher SH1 and his students worked on the problem solving together, and he used "question-answer" methods to facilitate and guide the students' understanding. On average, the frequency of teacher–student conversations was 110.4 per lesson. For teacher SH2, this number was 65 per lesson. Teacher SH2 preferred to giving individual guidance when the students were doing seat work. The average number of incidents of individual guidance was 42.8 per lesson, out of which 24.8 were given in detailed ways. In contrast, for teacher SH1, the average number was 14.8 per lesson, out of which 3.4 were given in detailed ways. There was a significant difference between the two teachers, in that teacher SH1 spent much more time on "problem solving together" and on whole-class guidance, whereas teacher SH2 allowed students to explore the mathematics on their own and then gave individual guidance. For individual guidance, teachers SH1 and SH2 both used two approaches. One way was simply to stop at the student's seat and check the progress, without clear guidance. Sometimes the teacher might nod his/her head or say "Well done" to indicate agreement. The other way was giving detailed guidance but not telling the answers.

10.2.3.4 There Is no Interaction and Collaboration Between Students. There Is a Lack of Active Student Learning and Exploration

In China, *Qifa* teaching and teacher–student communication are both organized by the teacher. Students' communications with one another are barely observed. In the observed lessons, for teacher SH1, teacher-led discussion accounted for 1.448% of the lesson time. The longest teacher-led discussion was 1 min and 35 s and the shortest was 8 s. For teacher SH2, the teacher-led discussion accounted for 3.43% of the lesson time. The longest teacher-led discussion was 3 min and 8 s and the shortest was 3 s. In an interview, the teacher said it would be very difficult to handle the classroom teaching if the students were allowed to communicate freely, since it would interfere significantly with the sequence of teaching. In the ten lessons analyzed, student questioning was only observed once and, on this occasion, the student just corrected the teacher's mistakes made when writing the procedures on the blackboard.

There was only one occasion when teacher SH2 asked students whether they had any questions before the end of a lesson. No students asked any questions, except that one student said it was too easy to ask any questions. It could be seen that these students lacked awareness about asking questions.

10.3 Teacher Questioning

10.3.1 Introduction

There has been a long history of research about effective questioning. In Eastern society, this can be dated back to Confucius, who said "不愤不启,不悱不发。举一隅不以三隅反,则不复也" ("In my method of teaching, I always wait for my student to make an effort himself to find his way through a difficulty, before I show him the way myself. I also make him find his own illustrations before I give him one of my own. When I have pointed out the bearing of a subject in one direction and found that my student cannot himself see its bearings into other directions, I do not then repeat my lesson."). In Western society, this approach can be dated back to Socrates' question–answer methods used to obtain the contradictions in students' thinking and, thereby, to facilitate their understanding by inquiry (Cao, 2007; Cao, Li, & Clarke, 2011).

Nowadays, questioning is still one of the most frequently used strategies in classrooms. Japanese researcher Fuji Yoshihiro argued that this is vital to teaching. It plays an important role in lessons and is a significant form of teacher–student interaction. Questioning can facilitate the development of students' reasoning and thinking, and it is also an essential method for accomplishing lesson goals (MOE, 2012).

10.3.2 Research Design

Five school districts in developed mainland Chinese cities were selected; these cities are spread across the eastern, southern, western, and northern regions of mainland China. In each selected school district, some key and non-key schools were selected and 5–7 teachers teaching year 7 were chosen in each selected school. In some schools, there were fewer than five year 7 teachers. In these cases, all the year 7 teachers were selected. In total, 132 teachers in 42 schools were selected, and 218 mathematics lessons were videotaped.

Out of the 218 lessons, a further selection was made according to experts' reviews and coding analyses. The detailed procedures were as follows. First, based on the lesson evaluation rubrics which were obtained as the results of the "MIST-China" project, all the lessons were evaluated and ranked by their scores. The 30 lessons with the highest scores were selected and assigned to four experienced teacher coaches and expert teachers, who watched the videos and re-ranked all 30 lessons. By combining the results of the lesson evaluation rubrics and the four experts' reviews, 15 lessons were selected eventually. These lessons were coded by using Instruction Quality Assessment (IQA), and the 5 lessons with the highest scores were finally chosen for analysis in this study.

The software NVivo was used to analyze the teacher strategies used in the "Initiation-Response-Follow-up" sequences and the time proportions used. The coding system is shown in Tables 10.1 and 10.2 (Dong, Seah, & Clarke, 2014).

10.3.3 Findings

10.3.3.1 The Most Frequently Used Strategies Were Asking for Products/Results and Repeating Students' Answers, and These also Took up the Longest Times

Clearly not every category could be observed in each lesson. However, the categories that could be observed in each lesson were supplementing, seeking confirmation, clarification, checking, cueing, explanation, elaboration/justification, repeating, products/results, and strategies. Out of these commonly used categories, the most frequently used were those asking for products/results and repeating students' answers; these also took the longest times.

Asking for a product/result is an initiation question, and these were usually observed at the beginnings of questioning sequences. Sometimes, it just involved checking the homework with students, which took a very short time. Sometimes, this technique was used to request students' solutions to problems, and it was usually followed by some follow-up questions asking for further details and elaborations. Repeat questions were usually observed after students' responses and were usually followed by questions asking for elaborations or cueing questions.

Table 10.1 Codes for initiation questions

Category	Abbreviation	Descriptions	Explanation/examples
Understanding checking	Check	Questions used to check whether students can follow or agree with teacher's teaching or other students' opinions	It only requires yes/no answers (such as "Do you agree?")
Review	Review	Questions used to elicit previously learnt concepts, propositions or formulae, procedures, or some facts from students	"What is the text description of this formula we learnt in the last lesson?"
Information extraction	Information	Questions requiring students to identify and select information from text descriptions, graphs, tables and diagrams so as to elicit students' interpretations of these texts, graphs, tables, diagrams	"What is the first one?"
Exemplification	Example	Questions requiring students to provide examples	Could you give me an example?
Product/result	Result	Questions used to elicit results of mathematical operation, or the final answer of the problem solving from students	The answers are usually nouns. The questions usually start with How many … How much …
Strategy/procedure	Strategy	Questions used to elicit the procedures or strategies of problem solving	The questions usually start with What … How …
Explanation	Explain	Questions requiring students to provide explanations	The questions usually start with Why …

10.3.3.2 Review, Information Extraction, and Exemplification Questions Took More Time

Although there were fewer episodes categorized for review, information extraction, and exemplification, the identified episodes usually took longer. In other words, the questioning sequences, including these three question types, were usually longer compared with other questioning sequences. Normally, when these three question types were observed, the topic under discussion was not a specific question and thereby required more time. Moreover, these three question types were used more frequently in the discussion of sophisticated tasks, which also required more time.

Table 10.2 Codes for follow-up questions

Category	Abbreviation	Descriptions	Explanation/examples
Seeking confirmation	Confirm	Questions used to confirm understanding of students' intended meanings in their responses	"Any right-angle triangle?"
Clarification	Clarify	Questions requiring students to show more details about their answers or solutions	"How did you get this 16?"
Repeat	Repeat	After the students have responded to the teacher's question, the teacher repeats or rephrases this question	
Cueing	Cue	Questions used to direct students to focus on key elements or aspects of the situation in order to enable problem solving, especially when the students fail to make progress or when students give the incorrect responses	"What is the problem asking you?"
Elaboration/justification	Justify	Questions asking for additional information under the circumstances, when the students' previous responses fail to provide sufficient mathematical information to fully answer teacher's questions or to justify their answers	"Why did you choose this method to solve this problem?"
Supplement	Add	Questions used to request for supplement	"Did anyone do this problem in a different way?"

10.3.3.3 Most Time Spent on Asking for Review

Most of the teachers spent time reviewing the knowledge learnt in the previous lesson, including theorems, formulae, and representations, in the first ten minutes of the lesson. Teachers also asked review questions during the discussion of mathematics tasks. In this way, they indicated that they valued the connection between new and old knowledge.

10.3.3.4 The Main Initiation Question Types Were Asking for Products/Results, Understanding Checks, and Asking for Explanations

The most frequently used initiation question types were asking for products/results and understanding/checking, followed by asking for explanations. These question types were observed in all lessons but one.

10.3.3.5 The Main Follow-up Question Types Were Asking for Product/Result, Understanding Check and Asking for Explanation

The most frequently used initiation question types were repeat, cueing, and supplementing, which were followed by seeking confirmation.

10.4 Students' Behavior in Mathematics Classrooms

10.4.1 Students' Engagement

With the implementation of the curriculum reform, it has been increasingly accepted that the students should be the active participants in mathematics lessons. In the *School Curriculum Standards (2011 Version)*, it is emphasized that teaching should be an activity in which teacher and students participate and interact with one another proactively so as to develop together. The *Standards* also mentioned that "students should learn actively and the teacher is the learning organizer, guide and co-operator," and "mathematics teaching activities ought to arouse students' interest, and stimulate their motivation and mathematical thinking" (MOE, 2012). Therefore, teachers should pay attention to students' participation and facilitate their engagement in mathematics learning. Classrooms are the places where students' mathematics literacy can be developed, and the process of this development can be determined by the students' participation in the classroom. Therefore, it was significant to examine students' participation in this study.

The word "participation/involvement/engagement" comes from the field of social sciences, for example, management and organizational behavior sciences (Zeng, 2001). It describes a situation when an individual joins in a group activity. In the field of education, it emphasizes the equity when teachers interact with students, and thus it can be regarded as a relationship.

Although many researchers have investigated student engagement, there is still a lack of agreement about what student engagement refers to. Most researchers have maintained that it is a compound concept meaning students' devotion and efforts used to improve learning, understanding, and knowledge/skills acquisition in learning activities. It is a compound including cognitive, emotional, and behavioral engagement (Kong, 2003). In this study, student engagement refers to the engagement in the teaching and learning; it refers mainly to students' behavior, but it can also reflect, to some extent, their cognitive and emotional engagement.

10.4.2 Research Design

Eight lessons were collected from Beijing and Chengdu and analyzed in this study. The data analysis was conducted using NVivio.

The previous studies divided students' engagement into three dimensions, namely cognitive engagement, emotional engagement, and behavioral engagement. When analyzing lesson videotapes, it is difficult to quantify students' cognitive and emotional engagement, but their behavioral engagement can reflect their cognitive and emotional engagement. For example, if students have high emotional engagement, they will also participate actively in learning activities, which in turn reflects their emotional engagement. Therefore, the focus of this study was students' behavioral engagement. On the basis of the previous studies by Kong (2003) and Si (2011), student engagement was analyzed from nine aspects: students' answers, group learning, students' exercises, students' questions, students' reading, students' explanations, students' hands raised, students' demonstrations on the blackboard and others.

10.4.3 Findings

10.4.3.1 The Main Aspect of Students' Engagement Is Answering Questions

In the eight lessons analyzed from Beijing and Chengdu, the frequencies of students giving answers were higher than any other aspects of student engagement. The time taken up by students' answers accounted for more than 10% of the lesson time. In other words, teacher–student interactions were mainly questions and answers. In contrast, the least frequent occurrence was student questioning, which was only observed six times in all the eight lessons.

10.4.3.2 Students' Answers Were Mainly Choral Responses and There Were Few Individual Responses

In all eight lessons, the students' answers were mainly choral responses and there were barely any individual responses. The cognitive levels of individual responses were much higher than those of the choral responses. The choral responses were mainly mechanical or reciting from memory. Cao and He (2009) found that teacher–whole class interaction was the main type of teacher–student interaction and that question–answer was the main type of teacher–student interaction. The implication of this finding is that the teacher's questions in Chinese mathematics lessons are mainly for the whole class rather than an individual student. It could also be seen from this analysis that teacher questions for the whole class tended to be simple and did not require sophisticated reasoning activities or higher level student engagement. With this whole-class teacher questioning, it was difficult for the teacher to monitor every student's progress since, in the choral responses, some individual responses were overlooked or not heard.

10.4.3.3 The Cognitive Levels of Students' Answers Were not High

In the analyzed videos, 652 students' answers were observed; out of these 367 were lower level answers, such as mechanical or reciting from memory, and 285 were higher level answers, indicating some degree of comprehension, analysis or creativity. On average, the cognitive levels of students' answers were not high.

10.4.3.4 Positive Correlation Between Teachers' Questions and Students' Answers

In the analyzed videos, 652 students' answers were observed. By using SPSS, teacher questions and student answers were analyzed as two variables. The results showed that the cognitive levels of teacher questions were correlated positively to the cognitive levels of students' answers.

10.4.3.5 Difference Between Students' Engagement in Beijing and Chengdu

By comparing the data in Beijing and Chengdu, it was found that there were some significant differences in the following aspects.

First, student exercises and students' answers took the longest time in the Beijing lessons.

In the Beijing lessons, student exercises altogether took 45 minutes, the time equivalent to the duration of one lesson, and the total time spent on student answers was also about 45 minutes on average. In contrast, in the Chengdu lessons, student exercises altogether took 50% of the time of one lesson on average. The frequencies of these student engagements were also different. Altogether 16 student exercises were observed in the Beijing lessons, whereas there was only one in all the Chengdu lessons.

Second, student explanations and group learning took the longest time in the Chengdu lessons.

In the Chengdu lessons, the average amount of time spent on student explanations was equivalent to the time of one lesson, and the same was the case for group learning. In contrast, in the Beijing lessons, student explanations altogether took 30% of the time of one lesson on average, but there was no group learning.

In summary, in the two cities' lessons, student exercises, student explanations, and group learning were different. In the Chengdu lessons, various student behavioral engagements accounted for 66% of the lesson time, whereas the number was 42% for the Beijing lessons. Through teacher interviews, it was established that the teaching styles in the two cities' lessons were different. In the Beijing lessons, traditional lecturing was dominant, whereas in the Chengdu lessons, the dominant teaching style was teacher-guided.

10.5 Student Questioning

10.5.1 Introduction

How to develop students' questioning capabilities has been an important topic in educational research. Polya (2014) believed that the beginning of problem solving is asking yourself questions, and thus questioning was investigated as one strategy for solving problems. In recent years, many researchers have conducted investigations about questioning, but the questions were mainly asked by the teachers rather than the students. The studies of student questions have mainly been about the value of students asking questions, the factors influencing their questioning and the strategies that can improve students' questioning (Ren, 2007).

In the *School Curriculum Standards (2011 Version)*, it is elaborated that "students are expected to ask, understand and solve mathematics questions by using the learnt mathematics knowledge" (MOE, 2012). If students are encouraged to ask questions, this can help them with solving problems and the development of critical and creative thinking, including the use of text, graphs, and tables to describe mathematical situations (Ren, 2007).

In Chinese mathematics lessons, mathematics problems are usually identified, proposed, integrated, and solved by the teachers. For example, Li (2014) found that only 13.8% of primary school students and 5.7% and 2.9% of junior and senior high school students, respectively, asked questions when they became confused in class. Based on this, it could be argued that Chinese students are weak at asking questions; hence, it is very urgent to examine how to improve their capabilities in this regard.

10.5.2 Research Design

Twenty junior secondary level mathematics lessons were selected for this study and a checklist for student questioning was developed by drawing on the teacher-questioning checklist. The checklist for student questioning is as shown in Table 10.3. This checklist was used when observing the 20 selected lessons. Whenever a student question was identified, the checklist was filled in by the researchers.

10.5.3 Findings

Based on the results of this study, it appears that student questioning is not popular in Chinese mathematics lessons. Only 21 student questions were identified in all 20 lessons. In each of these cases, the questioner was an individual student rather than a representative of a group. After one student or one group shared their explanations, other students would not ask questions straightaway even if they might have had some questions in mind. Instead, they usually waited for a signal from the teacher, who would determine who could ask questions and when. There were barely any questions asked during the communication between students or groups. Very few opportunities were provided to the students to ask questions. The identified student questions were mostly low in cognitive level. All of these features reflected that, in Chinese mathematics lessons, students tend to rely on teachers to solve problems. There is a need to improve students' capabilities in asking questions in mathematics lessons.

The factors influencing students' capabilities to ask questions are discussed below.

- Teaching methods: Most teachers used lecturing methods, which affected the time for the students to think and ask questions. In the context of the education reform, some new teaching methods have been introduced into mathematics lessons, such as DJP (Dao-Jiang-Ping which means Guide-Lecture-Discussion in English) and group learning. These new teaching methods have great potential to encourage students to ask questions.

Table 10.3 Checklist for student questioning

Lesson topic:		Lesson length:		Session No.	
Teacher:		Year level:		Teaching style:	
Questioner	Target	Time	Content	Question level	Teacher feedback
.					

- The time for asking questions: The questioners were all individual students in this study. There were only two episodes in which the teachers asked students whether they had any questions. This implies that the teachers had strong control of the mathematics lessons.
- Teachers' attitudes and feedback to student questions: Most of the teachers responded positively to the students' questions. On most occasions, the teachers told the answers to the students directly.
- Teachers' evaluation: The way the teachers evaluated student questions influenced the students' motivations to ask them. In this study, most teachers' comments on student questions were very simple, and very few teachers praised student questions or encouraged the students to ask more.

10.6 Students' Presentations in Mathematics Classrooms

10.6.1 Introduction

Some results (Cao, 2009) regarding teacher-student interaction occurred across countries from TIMSS 1999: In the eighth grade level mathematics lessons, students had sufficient opportunities to answer questions in simple ways (1–4 or 5–9 words in each answer) but fewer opportunities to answer questions in detail (more than 10 words in each answer). In other words, the teaching styles were mainly teacher lecturing, and students were passive learners. This situation was very typical in Hong Kong lessons.

Some researchers (Cao, 2009) investigated mathematics lessons taught by one teacher in Suzhou and two in Shanghai and found that the proportion of teachers' utterances with 1–4 words was relatively low whereas the proportion of teachers' utterances with more than 25 words was relatively high. For the two Shanghai teachers, the proportion of teachers' utterances with 1–4 words was less than 10% but the proportion with more than 25 words was more than 50%. In contrast, the length of students' utterances was 1–4 words on average, and the proportion of students' utterances with more than 10 words was relatively low. Less than one-third of the students' utterances were longer than 25 words. This implies that the interactions between the teacher and the students were very superficial and shallow. It also means that the teachers' explanations tended to be in-depth, and long explanations were favored by most Chinese teachers in mathematics lessons.

In the process of data collection for the LPS project, through lesson observations and interviews with teachers and students, it was found that students' public writing on the blackboard and students' public presentations were significant in shifting to the students into the role of active learners. During the above activities, the students could share their own thinking, reasoning, and strategies and thereby they were more engaged in mathematics lessons.

10.6.2 Research Design

The data for this study were collected from the LPS project. A sequence of 12 consecutive year 7 mathematics lessons was videotaped, taught by a Beijing teacher (BJ1), and a sequence of 15 consecutive year 7 lessons taught by a Shanghai teacher (SH2). *StudioCode* was used to code and analyze these lessons.

By observing the lessons in this study, two types of students' public behavior were identified: their public writing on the blackboard and their public presentations. Students' public writing on the blackboard refers to their actions of working on the teacher's assigned tasks on the blackboard. In most cases, the students do not talk, but just write down their solutions. Students' public presentation refers to students' actions of sharing their thinking about mathematics concepts, or about the strategies for solving problems. These two types of students' public behavior were examined in this study.

- Codes of students' public writing on the blackboard

The codes that were used to analyze the students' public writing on the blackboard are shown in Table 10.4.

- Codes for students' public presentations

The codes used in the analysis of students' publications are shown in Table 10.5.

10.6.3 Findings

The features of students' public writing on the blackboard are listed below:

- Regarding the types of students' public writing or public presentations, most of what they wrote or presented in public was assigned by the teacher. Sometimes the teacher encouraged the students to compete with their public writing or presentations. Sometimes students' public presentations were assisted by using a visual presenter.
- The teachers preferred to choose more than one student to present or work on one task, and then the different opinions or ideas were compared and discussed by the whole class.
- Both the observed teachers paid attention to their students' roles as active learners. The students were the center of the teaching activities. The average time used for students' public behavior was 5.27 minutes for teacher BJ1, which accounted for 13.18% of the lesson time. For teacher SH2, the average time used for students' public behavior was 12.43 minutes, representing 31.08% of the lesson time.
- The main purpose of the students' public behavior was to review and consolidate the old knowledge, and the second purpose was to guide students' thinking and engage them in classroom activities.

Table 10.4 Codes of students' public writing on the blackboard

Codes		Explanations
Types of public writing (BL)	Assigned (BL1)	Teacher assigns students to do public writing
	Competition (BL2)	Students run competition for public writing
	Supplement (BL3)	Students supplement the previous students' public writing
	ICT (BL4)	Students' public writing is assisted with ICT equipment
Forms of public writing (BS)	One students one task (BS1)	One student writes his or her solutions about one task
	Many students one task (BS2)	Many students write their solutions about one task
	Many students many tasks (BS3)	Many students write their solutions about many tasks
Tasks for public writing (BT)	Standard (BT1)	Standard mathematics tasks such as the following: Given AB = CD, ∠1 = ∠2, which of the following could be used to judge ΔABD ≌ ΔCDB () A. SAS B. ASA C. AAS D. SSS 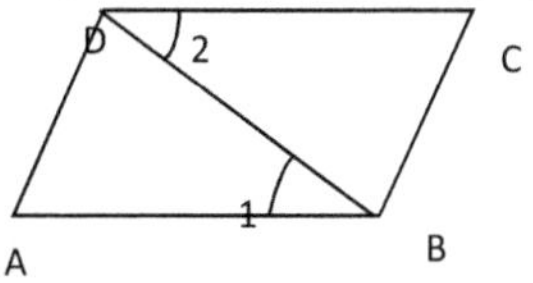
	Drilling (BT2)	Some challenging tasks such as the following: In the same context with the task above, given AB = CD,∠1 = ∠2. Prove that ΔABD ≌ ΔCDB.
	Exploration (BT3)	Some tasks requiring students to explore and find the solutions, such as the following: What do we need to know so as to prove that ΔABD ≌ ΔCDB? 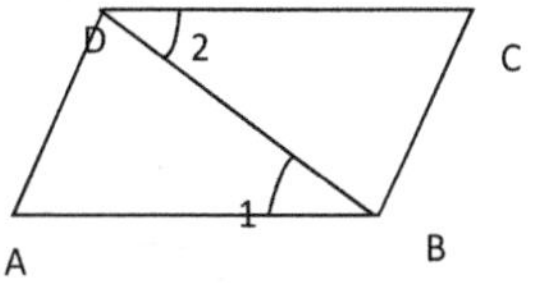
	Problem (BT4)	Very open tasks such as the following: Given AB = CD,∠1 = ∠2, what conclusions can you obtain? 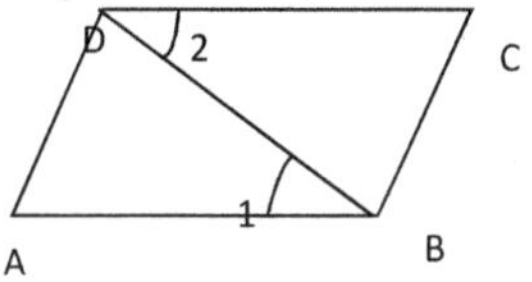

(continued)

Table 10.4 (continued)

Codes		Explanations
Purposes of public writing (BM)	Check (BM1)	Check the old knowledge before teaching new topics
	Introduction (BM2)	Introduce the new topics
	Consolidation (BM3)	Consolidate what has been learned
Comments on public writing (BJ)	Teacher comments (BJP)	Teacher gives comments on public writing
	Student comments (BSP)	Students give comments on public writing
	Co-commenting (BJSP)	Teacher and students give comments on public writing
Teacher's evaluation on public writing (BP)	Positive (BP1)	Teacher gives positive evaluation of students' public writing
	Negative (BP2)	Teacher gives negative evaluation of students' public writing
	Interruption (BP3)	Teacher stops students' public writing before it is finished
	Supplement (BP4)	Teacher supplements students' public writing
	Correction (BP5)	Teacher corrects the mistakes in students' public writing
	Comments (BP6)	Teacher gives comments after students finish public writing
	Stimulation (BP7)	Teacher gives some guidance and hints so that students can make progress in public writing

- Teacher SH2's teaching style was to promote fluency in the students' problem solving, whereas teacher BJ1 was trying to create an open environment for students' acquisition of problem-solving skills.
- Teacher BJ1 paid more attention to the students' performances in public. If the presenting student failed to present correctly or completely, some other students were selected to give assistance. These actions allowed more students to get engaged in the lesson. Teacher BJ1 placed more emphasis on guiding and stimulating the students' learning. The time spent on guiding and stimulating accounted for 56.54% of the time spent on teacher talk. Teacher BJ1 catered to differences in the students' mathematics capabilities and adapted his guidance to different students' demands. In contrast, teacher SH2 gave a large number of comments after students' public writing or presentation so as to provide summaries to the whole class. The time used for these comments took up 68.72% of the time spent on teacher talk. By giving summaries and comments, the students were able to obtain a clearer system of mathematics knowledge and the strategies for problem solving.Special acknowledgments to Siyang Che, Hongyun Zou, and Yulin Zhuang for their agreement to sharing some of their research results on teacher questioning, student engagement, and student questioning.

Table 10.5 Codes of students' public presentation

Codes		Explanations
Types of presentation (JL)	Assigned (JL1)	Teacher assigns students to do public presentation
	Competition (JL2)	Students run competition for public presentation
	Supplement (JL3)	Students supplement the previous students' public presentation
Forms of presentation (JS)	One student (JS1)	One student presents one task
	Many students (JS2)	Many students present one task
Contents of presentation (JN)	Presenting strategies' outlines (JN1)	Students briefly present the outline of problem solving strategies
	Presenting process and details (JN2)	Students present the process and details of problem solving
Ways of presentation JF)	Verbal presentation (JF1)	The presentation is made verbally
	Both verbal and written (JF2)	Students make presentations verbally while writing down some key words in blackboard
Ways of commenting (JJ)	Teacher comments (JJP)	Teacher gives comments on public presentation
	Students comments (JSP)	Students give comments on public presentations
	Co-commenting (JJSP)	Teacher and students give comments on public presentation
Teacher evaluation (JP)	Positive (JP1)	Teacher gives positive evaluation of students' public presentation
	Negative (JP2)	Teacher gives negative evaluation of students' public presentation
	Interruption (JP3)	Teacher stops students' public writing before it is finished
	Supplement (JP4)	Teacher supplements students' public presentation
	Correction (JP5)	Teacher corrects the mistakes in students' public presentation
	Comments (JP6)	Teacher gives comments after students finish public presentation
	Stimulation (JP7)	Teacher gives some guidance and hints so that students can make progress in public presentation
	Substitute (JP8)	Teacher takes the role of presenter when students become stuck in the process of presentation
	No response (JP9)	Teacher gives no responses to students' presentation but moves to the next stage of teaching

References

Ausubel, D. (1963). *The psychology of meaningful verbal learning*. New York: Grune & Stratton.

Cao, Y. (2007). 中国数学课堂教学模式及其发展研究 [A study of mathematics teaching models and their development in China]. Beijing: Beijing Normal University Publisher.

Cao, Y. (2009). 数学课堂教学实证系列研究 [A study of mathematics classroom teaching]. Nan'ning: Guangxi Education Publisher.

Cao, Y., & He, C. (2009). 初中数学课堂师生互动行为主体类型研究——基于LPS项目课堂录像资料 [A study of teacher-student interaction types in junior level mathematics lessons-based on th LPS database]. *Mathematics Education Journal, 18*(5), 38–41.

Cao, Y., Li, J., & Clarke, D. (2011). 数学课堂中启发式教学行为分析——基于两位数学教师的课堂教学录像研究 [A study of *Qifa* teaching in mathematics lessons-based on the videotapes of two mathematics teachers' teaching]. *China ICT Education, 10,* 100–102.

Dong, L., Seah, W., & Clarke, D. (2014). Teacher questioning in mathematics classes in China and Australia: A case study. In J. Vincent, G. FitzSimons, & J. Steile (Eds.), *The proceedings of the MAV 51st annual conference*, La Trobe University, Melbourne, 4–5 December. Bundoora, VIC: The Mathematics Association of Victoria.

Freudenthal, H. (1972). *Mathematics as an educational task.* Berlin: Springer.

Joyce, B., Weil, M., & Calhoun, E. (2002). *Models of teaching* (8th ed.). UK: Pearson.

Kong, Q. (2003). 数学教学过程中的学生参与 [Student engagement in mathematics teaching]. Shanghai: East China Normal University Publisher.

Li, P. (2014). 小学语文中高学段学生课堂提问现状调查研究 [A survey of student questioning in upper primary level Chinese lessons] (Unpublished Doctoral Thesis), Jiangxi Normal University.

MOE. (2012). 全日制义务教育数学课程标准 *(2011版)* [School mathematics curriculum standards (2011 version)]. Beijing, P.R. China: Beijing Normal University Publisher.

Polya, G. (2014). *How to solve it: A new aspect of mathematical method.* USA: Princeton university press.

Ren, B. (2007). 数学问题提出的教育价 [The educational value of asking mathematics questions]. *Career, 12,* 50–51.

Shimizu, Y. (2002). Discrepancies in perceptions of lesson structure between the teacher and the students in the mathematics classroom. In *Interactive symposium, "International perspectives on mathematics classrooms", at the annual meeting of the American Educational Research Association, New Orleans, in 1st to 5th April, 2002.*

Si, H. (2011). 基于视频案例下代数课堂教学中学生参与度研究 [Research on Student Engagement Based on Video Case of the Algebra Classroom]. Master Thesis, Hangzhou Normal University.

Wu, L. (1998). 教学的原理、模式和活动 [Principles, models and activities for teaching]. Nan'ning: Guangxi Education Publisher.

Yao, L. (2004). 有效教学 [Effective teaching] (Unpublished Doctoral Thesis), East China Normal University.

Zeng, Q. (2001). 学生的参与及其发展价值 [Student engagement and its developmental values]. *Teaching School Subjects, 1,* 4–7.

Chapter 11
Task Design in Mathematics Classrooms

Zhenhong Shao

Abstract In various international tests and competitions, such as the International Mathematics Olympiad (IMO) and the Program for International Student Assessment (PISA) in which Shanghai students participated, Chinese students' outstanding performances have attracted a worldwide interest in basic Chinese education. In particular, Western mathematics education researchers have expressed interest in the characteristics of Chinese mathematics teaching; also a lot of studies have been done by Chinese scholars, as reported here.

11.1 Review of Domestic Literature on Mathematical Tasks

In various international tests and competitions, such as the International Mathematics Olympiad (IMO) and the Program for International Student Assessment (PISA) in which Shanghai students participated, Chinese students' outstanding performances have attracted a worldwide interest in basic Chinese education. In particular, Western mathematics education researchers have expressed interest in the characteristics of Chinese mathematics teaching; also a lot of studies have been done by Chinese scholars, as reported here.

One example is the study of mathematical problems. Chinese mathematics education researchers began to study this field from the mid-1980s. Much of the early research work focused on the empirical descriptions of "problem-solving," "solving problems," "applying mathematics," and research on the meaning, classification, and psychological mechanisms of mathematical problems, as well as work on the mathematics curriculum, evaluation, and how to teach mathematics in the context of mathematical problem-solving reported in the late twentieth century (Qi, 2007).

Z. Shao (✉)
Beijing Jingshan School, Beijing, China
e-mail: shaobnu@163.com

Y. Cao and F.K.S. Leung (eds.), *The 21st Century Mathematics Education in China*, New Frontiers of Educational Research,
https://doi.org/10.1007/978-3-662-55781-5_11

The "Chinese learners' paradox" led to a series of studies of Chinese mathematics classrooms, especially of "variant Mathematics Teaching." Chinese scholars have distinguished and analyzed conceptual and procedural variance in mathematics teaching from an experimental perspective. For specific topics, they have also analyzed what can be changed, how to implement this change, and what the degree of change should be in the variant mathematics teaching model. As well, Chinese scholars have described variations in mathematics teaching as a special way of teaching which can promote effective learning from the theoretical perspective (Bao & Huang, 2003).

From 2001, active mathematics experiences have been considered as the goal of the mathematics curriculum, as stated clearly in the *Mathematics Curriculum Standards in Full-time Compulsory Education (Pilot Manuscript).* In 2011, the *Compulsory Mathematics Curriculum Standards (2011 Version)* described the concept of the active mathematic experience as one of four important basics of mathematics teaching and learning (basic knowledge, basic skills, basic mathematical activity, and basic mathematical thinking). After this, the use of mathematics activities has been studied widely in Chinese mathematics education (Tu & Ning, 2002). The concept of mathematics activities, how to design them, and what we should pay attention to in the design, as well as their value, function, and effects, have all been studied by Chinese scholars. The implementation of mathematics teaching and teaching strategies to expose students to active experiences and learning activities have also been investigated (Guo & Shi, 2012).

Related to mathematical problems, variation and activity-based mathematics have been the main foci of the research on mathematical tasks.

A mathematical task is defined as a classroom activity, the purpose of which is to focus students' attention on a particular mathematical idea (Stein, Grover, & Henningsen, 1996). Mathematical task research was first introduced into China by Gu (2001), who translated the book *The Mathematics Instructional Task* written by Stein, Smith, Henningsen, and Silver (2000, pp. 462–520) into Chinese in 2000. Some other Chinese scholars used the terms "instructional task (jiaoxue renwu)" or "mathematical task (shuxue renwu)" in their papers and illustrated the concept and basic structure of mathematical tasks in Chinese mathematics classrooms. Based on these studies, some other scholars did some case studies to examine the cognitive levels of mathematical tasks in the Chinese classroom and gave some advice about how to design and teach them (Yang, 2012; Yuan & Lu, 2006). A Chinese scholar Shao (2014, 2015) used quantitative and qualitative methods to compare the characteristics of mathematics tasks used in lessons in China and the USA by coding 15 high-quality videotaped lessons.

Nevertheless, there have been very few studies on the topics of "mathematical tasks," "instructional tasks," or "mathematical instructional tasks" in China. Most of the studies have involved classification of the cognitive levels of mathematics tasks using the system proposed by Stein. Thus there is great value in conducting research on mathematical tasks, especially those of high cognitive levels, by case study.

11.2 Task Design

Mathematical tasks can convey messages about what mathematics is; also, the ways in which they are implemented can potentially influence students to do mathematics in different ways (Doyle, 1988). A good task can enable students to expand their knowledge and to be more motivated. The selection of appropriate tasks is an important factor in effective teaching (Doyle, 1983, 1988; Stein, Grover, & Henningsen, 1996). The concepts, roles, and classifications of mathematics tasks have been studied internationally, but there has been little domestic research on task design. This study analyzed the characteristics of tasks used in Chinese mathematics classes from video recordings of high-quality lessons chosen from normal classrooms. Specifically, the focus was on student involvement in complex tasks. The tasks were analyzed from three aspects, (1) using "real-life" background knowledge; (2) the task presentation formats; and (3) their cognitive levels.

The samples were chosen from the database of the China-MIST program (*A Comparative Study of Chinese and American Regional and School Levels of Mathematics Instruction,* funded by the US National Science Foundation). In 2011, 218 junior high school mathematics classrooms were videoed from five districts, representing a range of big cities in China; 15 high-quality videotaped lessons were chosen and analyzed quantitatively and qualitatively by mathematics education experts.

11.2.1 "Real-Life" Context in Task Design

According to the *Compulsory Mathematics Curriculum Standards (2011 Version)* (Ministry of Education, 2011), mathematics content that reflects students' real lives can help them to understand, think, and explore better. In relation to the organization of curriculum content, the new curriculum standards have suggested emphases on process, dealing properly with the relationship between process and results; intuition and its relationship to the abstract; and direct experience and its relationship with indirect experience. The presentation of course content should be hierarchical and diverse.

The TIMSS 1999 (Hiebert, 2003, pp. 10–30) investigated mathematics questions and real-life contexts based on the analysis of videotaped Grade 8 lessons from seven countries. Questions in the study were divided into two categories, "pure mathematical problems," that is those described only by abstract mathematical languages and those linked to real-life contexts. The study found that the proportion of questions linked to real-life contexts ranged from 9 to 42%, with an average of 22%.

Domestic scholar Cao (2006, 2007) also studied the use of real-life contexts in mathematics teaching. He found that about 11% of mathematics problems were related to real life, and that most teachers were generally positive about using such problems. In reality, it was found that some teachers would add some real-life situations, but these were not necessarily relevant to the question, which impacted upon the achievement of the teaching goal.

The famous Netherlands mathematics educator Freudenthal stressed the idea of mathematics as a human activity and advocated that it should be connected to reality, stay close to children and be relevant to society, in order to be of human value. According to him, education should give students the "guided" opportunity to "reinvent" mathematics by doing it. This means that, in mathematics education, the focal point should not be on mathematics as a closed system, but on the activity, on the process of mathematization.

The above reviews have focused on two types of mathematics problems, pure ones and those linked with real life. Discrepancies have been found in the quality of real-life problems. For example, some only pay lip service to an article or a real-life event, while others set the problem in a real-life scenario, which can be solved by using mathematical knowledge. In order to analyze the characteristics of life-related mathematics tasks, we have classified them into three types (Table 11.1).

We categorized the mathematical tasks in the 15 lessons into "pure mathematical questions," "practical situation problems," and "solving practical problems" (see Fig. 11.1; Table 11.2).

11.2.1.1 Breakdown of Questions by Type

There were 129 mathematical tasks in the 15 videotaped lessons, meaning at least five mathematical tasks per class. The highest frequency was a lesson in which 17 tasks were solved; the average was 8.7 tasks per lesson.

11.2.1.2 Mathematical Tasks Described by Mathematical Language

Of the 129 mathematical tasks, 84.4% (109 tasks) were categorized as "pure mathematical questions," which were described by "abstract mathematical language." Twelve tasks (9.4%) were "practical situation problems," which were

Table 11.1 Categories of mathematical tasks connected to real life

Category	Explanation of the category	Example
Pure mathematical questions	Question described in mathematical language	One side of an isosceles triangle is 10 cm and the other side is 5 cm, what is its perimeter?
Practical situation problem	Question only mentions an article, or a real-life event	Explore the definition and nature of the isosceles triangle by using origami
Solving practical problems	Using the mathematical knowledge to solve a real-life question	You want to know about our classmates' interests in news, sports, movies, entertainment, and drama. How can we find out about their five favorite television programs?

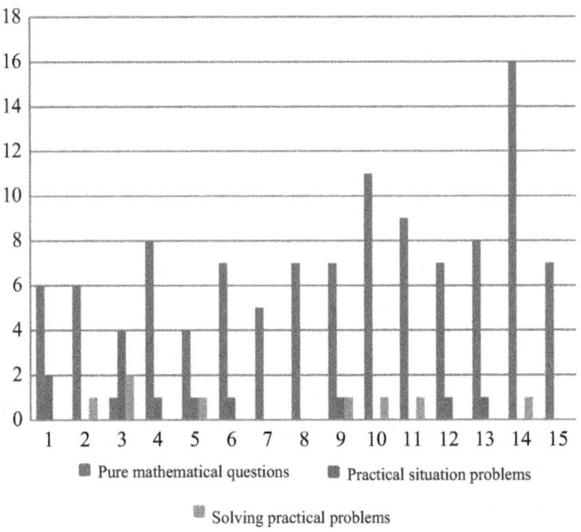

Fig. 11.1 Comparison of different categories of "real-life" contexts in mathematical tasks

Table 11.2 Comparison of different categories of "real life" contexts

Tasks context	China	
	Quantity	Percentage (%)
Pure mathematical questions	109	84.4
Practical situation problems	12	9.4
Solving practical problems	8	6.2
Total quantity	129	

linked to real life, but only by mentioning an article or a real-life event. Eight tasks (6.2%) involved "solving practical problems," talking about real-life tasks, which could be solved by using mathematical knowledge. It has already been mentioned earlier in this chapter that the TIMSS 1999 (Hiebert, 2003) of Grade 8 classes in seven countries found 9–42% of mathematics questions to be related to real life, with an average of 22%. Based on this result, the proportion of real-life tasks designed in Chinese classrooms is lower than the international average.

11.2.1.3 Solving Practical Problems to Introduce New Knowledge

Of all the 129 tasks in the 15 videotaped lessons, eight were categorized as "solving real-life problems," but only five of these were designed to introduce a new topic or new knowledge. In these cases, the tasks were introduced at the beginning of the class. When students attempted to complete them, they realized that they would need some new knowledge to be able to solve them. Then the new knowledge for this lesson was introduced. For example, the first task in one Chinese class was to solve whether 1000 digital photos could be put onto a U disk; the content of this lesson was division with the same base power.

11.2.2 Forms of Mathematical Tasks

From the observation of the mathematics classroom videos, we found that mathematical tasks can had different functions at different positions during a lesson. At the beginning of the class, some teachers reviewed the old knowledge by asking students questions about the knowledge learned in previous lessons and some teachers reviewed this through practice exercises. These can be referred to as "reviewing old knowledge" tasks. Some teachers arranged one or two mathematics problems for their students to think about by themselves, so that they could be ready for the new learning very quickly at the beginning of the class. These can be called "warm up" tasks. The most important activity is the learning of new knowledge. Generally in the videos, new concepts, theorems, and rules were introduced; they were illustrated by one or more examples, then teachers would provide several questions for students to practice. In summary, the different ways in which tasks were used in the mathematics classroom were classified as "review tasks," "warm up tasks," "learning tasks," and "practice tasks."

This categorization is shown in Fig. 11.2 and Table 11.3.

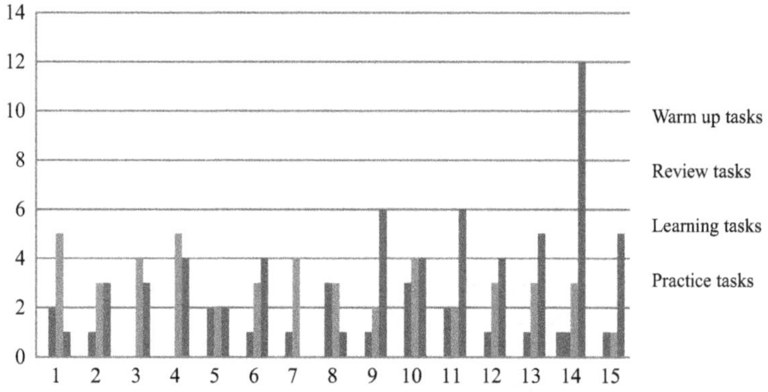

Fig. 11.2 Forms of mathematical tasks

Table 11.3 Presentation forms of the mathematical tasks

Classification	China	
	Quantity	Percentage (%)
Warm up tasks	1	0.8
Review tasks	20	15.5
Learning tasks	47	37.2
Practice tasks	61	46.5

11.2.2.1 Types of Tasks

From Fig. 11.2, we can see that practice tasks were the most frequent in the Chinese mathematics classrooms. 46.5% of the tasks involved practice, and the average number was four per lesson. This was followed by "learning tasks" (37.2%), "review tasks" (15.5%), and then "warm up tasks" (0.8%). Reviewing old knowledge is a common way to start a lesson in China, so tasks of this nature constituted 16.4% of the total. Only one warm up task was found in the 15 videotaped lessons. According to the classification, "learning tasks" consisted of two parts, examples and activities for learning new concepts, theorems, and rules.

11.2.2.2 Acquisition and Consolidation of New Knowledge

Both learning tasks and practice tasks can lead to the acquisition and consolidation of new knowledge. With 46.5% of the tasks being "practice tasks" and 37.2% being "learning tasks," 83.7% of the total tasks were related in some way to the acquisition or consolidation of new knowledge. This suggests a high level of acquisition and consolidation of new knowledge.

11.2.3 Cognitive Levels of Mathematical Tasks

Careful choice and presentation of mathematical tasks is vital to success in mathematics teaching (Doyle, 1988; Hiebert & Wearne, 1993; Stein, Grover, & Henningsen, 1996). Doyle (1983, 1988) suggested that what students learn depends largely on the kinds of tasks chosen by the teacher. Henningsen and Stein (1997) explained that the choice of tasks can limit or broaden students' views of the subject matter in which they are engaged.

Not all mathematics tasks can provide the same opportunities for learning. Some have the potential to engage students in complex forms of thinking and reasoning, while others focus on memorization or the use of rules or procedures. Not all tasks used by a teacher should place higher-level cognitive demands on students. Different goals are better served by different kinds of task. For example, if the goal is to increase students' fluency in retrieving basic facts, formulae, and rules, then a task that focuses on memorization may be appropriate. However, students need opportunities, on a regular basis, to engage with tasks that lead to deeper understandings of mathematical concepts, processes, and relationships (Stein, Smith, Henningsen, & Silver, 2000, pp. 462–520). Instructional materials should provide the opportunities for students not only to access core knowledge and understanding of the main mathematical concepts, formulae, and theorems, but also to propose and solve problems.

Stein et al. (2000) classified mathematical tasks according to four levels of cognitive demand that make different contributions to learning. Memorization and procedures without connections are lower-level tasks. Tasks categorized as procedures with connections and doing mathematics are classified as higher-level cognitive demand. Here, we will describe the characteristics of mathematical tasks at each cognitive demand level (Table 11.4).

According to Stein's classification of cognitive level demands, we classified and counted all the tasks in the mathematics lesson videos. The data for the learning tasks and practice tasks, as well as the cognitive demand level, are shown in Table 11.5; Figs. 11.3, and 11.4.

11.2.3.1 Proportions of Tasks

From Figs. 11.3 and 11.4 and Table 11.5, we can see very clearly that almost half of the mathematical tasks used in the mathematics classes were at level 2, that is "procedure without connection" tasks. This was followed by "procedure with connection," level 3. The next was level 4, the tasks that involved "doing mathematics." The least represented were those which did not involve mathematical activities.

11.2.3.2 Cognitive Demand Levels of "Review Tasks"

We analyzed and classified all of the review tasks in the videotaped sample lessons according to their levels of cognitive demand (see Fig. 11.5). From Fig. 11.5, we can see that all the review tasks had lower cognitive demand levels, the majority being level 1, memorization tasks, and level 2, procedure without connection. There was only one review task at cognitive demand level 4.

11.2.3.3 Cognitive Demand Levels of "Practice Tasks"

We also analyzed and classified all of the practice tasks from the videotaped sample lessons according to cognitive level demand (Fig. 11.6). It can be seen that 45 of the 61 practice tasks were at level 2, procedure without connections. The next was level 3, procedure with connections. The cognitive demand level 4, doing mathematics, tasks were not represented well. There was only one practice task at level 4 in all the videotaped lessons.

11.2.3.4 Learning Tasks Are Relatively of High Cognitive Demand

As well as analyzing the practice tasks, we analyzed the learning tasks (Fig. 11.7) Most of these were levels 3 and 4, with the highest proportion of learning tasks being at level 4. There were also some tasks at level 2, "procedures without connections," but we did not find any "memorization" learning tasks.

Table 11.4 Characteristics of mathematical tasks at each of the four levels of cognitive demand (Stein et al., 2000)

Lower-level demands	Higher-level demands
Memorization tasks • Involve either reproducing previously learned facts, rules, formulae, or definitions OR committing facts, rules, formulae, or definitions to memory • Cannot be solved using procedures because a procedure does not exist or because the time frame in which the task is being completed is too short to use a procedure • Are not ambiguous—such tasks involve exact reproduction of previously seen material, and what is to be reproduced is clearly and directly stated • Have no connection to the concepts or meaning that underlie the facts, rules, formulae, or definitions being learned or reproduced	Procedures with connections tasks • Focus students' attention on the use of procedures for the purpose of developing deeper levels of understanding of mathematical concepts and ideas • Suggest pathways to follow (explicitly or implicitly) that are broad general procedures that have close connections to underlying conceptual ideas as opposed to narrow algorithms that are opaque with respect to underlying concepts • Usually are represented in multiple ways (e.g., visual diagrams, manipulatives, symbols, problem situations). Making connections among multiple representations help to develop meaning • Require some degree of cognitive effect. Although general procedures may be followed, they cannot be followed mindlessly. Students need to engage with the conceptual ideas that underlie the procedures in order to successfully complete the task and develop understanding
Procedures without connections tasks • Are algorithmic. Use of the procedure is either specifically called for or its use is evident based on prior instruction, experience, or placement of the task • Require limited cognitive demand for successful completion. There is little ambiguity about what needs to be done and how to do it • Have no connection to the concepts or meaning that underlie the procedure being used • Are focused on producing correct answers rather than developing mathematical understanding • Require no explanations or explanations that focus solely on describing the procedure that was used	Doing mathematics • Require complex and none algorithmic thinking (i.e., there is not a predictable, well-rehearsed approach or pathway explicitly suggested by the task, task instruction, or a worked-out example) • Require students to explore and understanding the nature of mathematical concepts, processes, or relationships • Demand self-monitoring or self-regulation of one's own cognitive processes • Require students to access relevant knowledge and experiences and make appropriate use of them in working through the task • Require the students to analyze the task and actively examine task constraints that may limit possible solution strategies and solutions • Require considerable cognitive effort and may involve some level of anxiety for the student due to the unpredictable nature of the solution process required

Table 11.5 Cognitive demands of mathematical tasks

Classification of tasks	All Tasks		Practice Tasks		Learning Tasks	
	Quantity	Percentage (%)	Quantity	Percentage (%)	Quantity	Percentage (%)
Doing Mathematics	22	17.1	1	1.6	20	42.6
Procedures with connections	32	24.8	13	21.3	19	40.4
Procedures without connections	62	48.1	45	73.8	8	17
Memorization	12	9.3	2	3.3	–	–
No mathematical activities	1	0.8	–	–	–	–

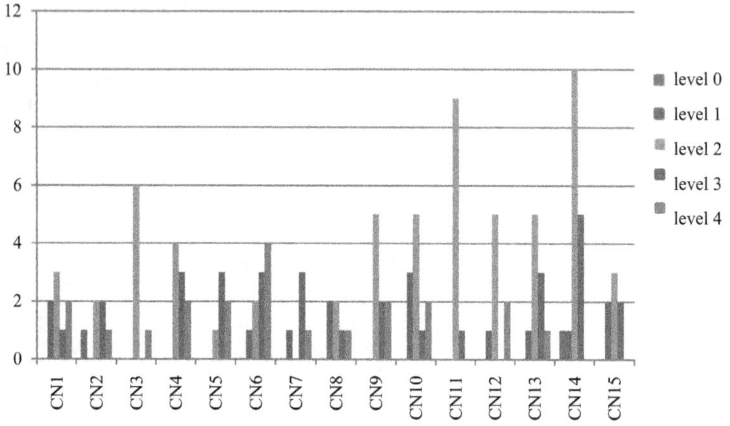

Fig. 11.3 Comparison of cognitive demand levels of tasks

11.2.3.5 Comparison of Cognitive Demand Levels Across Tasks

Table 11.5 shows the numbers and the percentages of each level of mathematical tasks in all the videotaped lessons.

From Table 11.5, it can be seen that the average cognitive demand level of all mathematical tasks in all lessons was 2.5, the average cognitive demand level of "learning tasks" was 3.3, and the average cognitive demand level of "practice tasks" was 2.2. It is apparent from the table that the cognitive demand level of the learning tasks was higher than that of the practice tasks.

From Table 11.5, it is also evident that the students in this Chinese sample engaged in more mathematical tasks involving "procedure without connection" (48.1% of the total tasks) as compared with "procedure with connection" (24.8%).

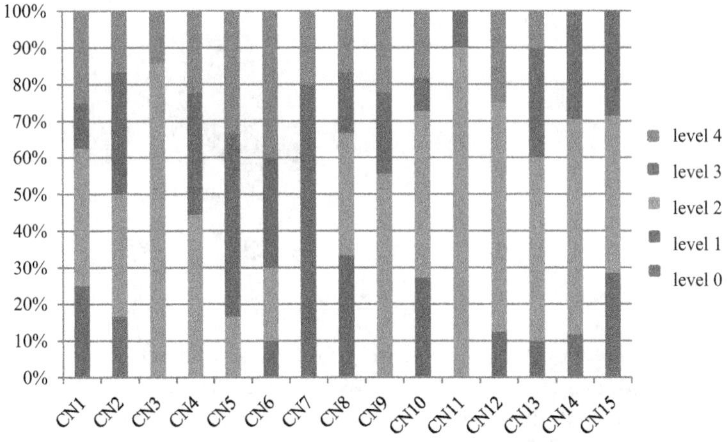

Fig. 11.4 Percentages of mathematical tasks by cognitive demand level

Fig. 11.5 Distribution of cognitive demand levels of "review tasks" in China

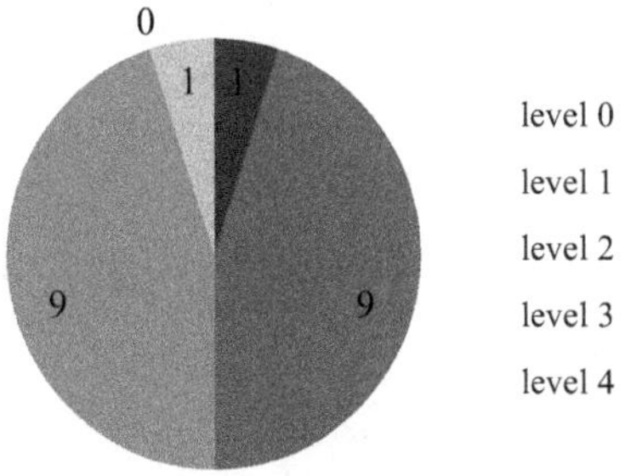

Fig. 11.6 Distribution of cognitive demand levels of "practice tasks" in China

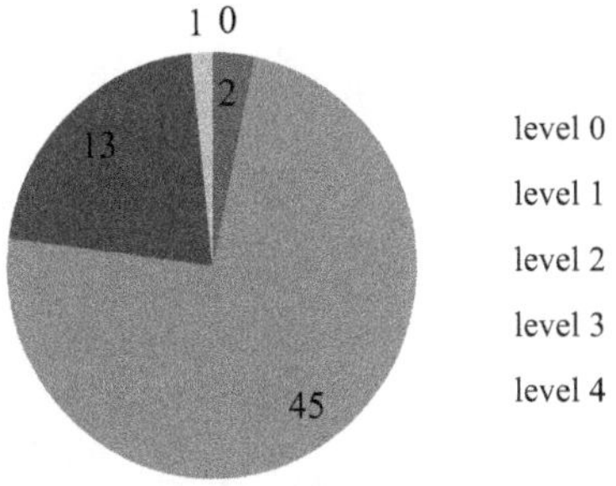

The fourth cognitive demand level, "doing mathematics," only accounted for 15.6% of the total.

We also collected information about the practice and learning tasks in order of the research questions presented in this chapter. There were 61 practice tasks in the videotaped lessons. From Fig. 11.6, it is evident that most of the practice tasks were at cognitive demand level 2, "procedure without connection" (73.8% of all practice

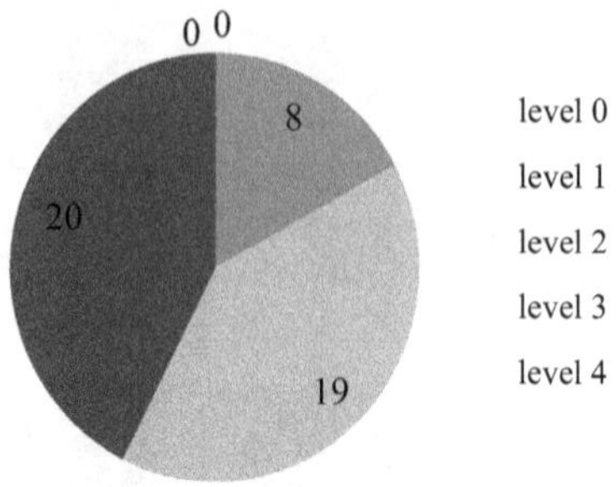

Fig. 11.7 Distribution of cognitive demand levels of "learning tasks"

tasks). This was followed by cognitive level 3, "procedures with connections (21.3%). The average cognitive demand level for practice tasks was 2.2.

There were a total of 47 learning tasks in all the lessons, of which 40.4% were at cognitive demand level 3, and 42.5% at cognitive demand level 4. The average cognitive demand level of all the learning tasks was 3.2. It is apparent that the majority of learning tasks in the classroom were high cognitive demand tasks.

11.3 Discussion and Conclusions

This chapter has described the features of mathematics tasks in a selection of routine but relatively high-quality mathematics lessons. This evidence of the nature of Chinese mathematical tasks can provide a very important reference for the reform of mathematics teaching in China and can also provide some useful information about Chinese mathematics instruction.

11.3.1 Reducing the Use of Low Cognitive Level Tasks

Nearly half of the tasks observed in the Chinese videos were at the second cognitive demand level, "procedure without connection," especially the practice tasks. Almost half (61 of 129) of the practice tasks were at level 2, and only one of them was concerned with "doing mathematics" at cognitive demand level 4. However, when the practice tasks were implemented, most of them were actually at a lower cognitive demand level. The main reason for this is that some types of tasks were repeated. Once students had engaged in the first task, then they could use it as an example to solve others of the same type; hence when they worked on these later, similar tasks were able to operate at a lower cognitive demand level. From this study, plenty of practice tasks were found to be at level 2 and to be repeated often. The results of the 2012 PISA showed that Chinese mathematics teaching focused on a high density of exercises and a lack of attention to applied mathematics. So, combining these two findings, it seems that it would be better to reduce the use of the low cognitive level tasks appropriately.

Is there too much investment in the "two basics" (basic knowledge, basic skills) in Chinese mathematics education? Can we reduce the relative number of exercises for basic knowledge and basic skills and increase the quantity of open problems in the mathematics classroom?

The overall object for mathematics learning stated in the *Compulsory Mathematics Curriculum Standards (2011 Version)* is that the mathematics curriculum should address basic knowledge, basic skills, basic mathematical activity, and basic mathematical thinking to enable students to adapt to modern everyday living and future development. We know that to master the basic knowledge and basic skills of mathematics practice is necessary, but to what extent? What is the optimal proportion of practice tasks in the overall task design? While these questions do not have a definitive answer, the results of this study do suggest that there are too many low cognitive level mathematics tasks in Chinese mathematics instruction, and hence it would be better for us to reduce the quantity.

11.3.2 Reducing the Quantity and Increasing the Cognitive Demand Level

In this study there were 129 mathematical tasks in the 15 videotaped lessons, with an average of 8.7 tasks per lesson.

Of these tasks, 48.4% were at level 2, "procedure without connection." There was one lesson that had 17 tasks, and 11 of these were at level 2, with another one being a memorization task at level 1. In this lesson, most of the tasks involved repeated practice of knowledge. The students in the class completed plenty of tasks, but there was a clear lack of mathematical thinking in the lesson. It would have been better for the students to ponder on the complex mathematical tasks, which would provide opportunities to develop their thinking. Reducing the overall quantity and increasing the cognitive demand levels of mathematical tasks, with more emphasis on students' thinking, would lead to better task design.

11.3.3 Solving Real-Life Problems

There is an entire series of descriptions and research about "situation," such as "situational teaching," "teaching situation," "situational importance," and "situational problems." This study also investigated the integration of real-life situations into mathematics tasks. There were very few tasks relating to real life. According to Table 11.1, there were only eight tasks of the 129 which involved solving real-life problems. Even so, a closer examination revealed that not all of these eight tasks really involved solving a real-life problem; five of them were used to introduce the lesson topic or some new knowledge. The new *Compulsory Mathematics Curriculum Standards (2011 Version)* proposed that it is better to choose content that is close to students' real lives and relevant to society, that can give them opportunities to reinvent mathematics by doing it. In the actual teaching process, the purpose of creating mathematical situations is to stimulate students' motivation to study and to self-learn. However, it seems that the terms "problem situation" and "mathematics" are mismatched in the Chinese mathematics classroom (Lv & Wang, 2001, 2006). It would be better to strengthen the application of the "problem situation." The tasks should be situated in students' real lives and draw on their real-life knowledge and experiences. But the important question is how to give students the "guided" opportunity to "reinvent" the mathematics by embedding it in a real-life context, not to isolate situation from the teaching content. This is well worth exploring in future.

References

Bao, J. S., & Huang, R. J. (2003). 变式教学研究 [The study of variant teaching]. *Mathematics Teaching, 1,* 11–12.

Cao, Y. M. (2006). 数学教学中的"生活化"与"数学化" ["Humanized" and "mathematicalized" in mathematics teaching]. *Journal of the Chinese Society of Education, 2*(46–48), 58.

Cao, Y. M. (2007). 中国数学课堂教学模式及其发展研究 [A study on the teaching model and development of Chinese mathematics classroom]. Beijing: Beijing Normal University Publishing Group.

Doyle, W. (1983). Academic work. *Review of educational research, 53*(2), 159–199.

Doyle, W. (1988). Work in mathematics classes: The context of students' thinking during instruction. *Educational Psychologist, 23*(2), 167–180.

Gu, L. Y. (2001). 教学任务与案例分析 [Teaching task and case analysis]. *Shanghai Research on Education, 3*(2–6), 36.

Guo, Y. F., & Shi, N. Z. (2012). 数学基本活动经验:提出、理解与实践 [Basic activity experience in mathematics: Presentation, understanding and practice]. *Journal of the Chinese Society of Education, 4,* 42–45.

Henningsen, M., & Stein, M. K. (1997). Mathematical tasks and student cognition: Classroom-based factors that support and inhibit high-level mathematical thinking and reasoning. *Journal for Research in Mathematics Education, 28*(5), 524–549.

Hiebert, J. (2003). *Teaching mathematics in seven countries: Results from the TIMSS 1999 video study*. Washington, DC: DIANE Publishing.

Hiebert, J., & Wearne, D. (1993). Instructional tasks, classroom discourse, and students' learning in second-grade arithmetic. *American Educational Research Journal, 30*(2), 393–425.

Lv, C. H., & Wang, B. Y. (2001). 论中小学 "数学情境与提出问题" 的数学学习 [On mathematics learning by using mathematical situations and posing problems in high school and in primary school]. *Journal of Mathematics Education, 10*(4), 9–14.

Lv, C. H., & Wang, B. Y. (2006). 论中小学 "数学情境与提出问题" 的教学 [On the teaching of mathematics situations and posing problems in high school and primary school]. *Journal of Mathematics Education, 15*(2), 74–79.

Ministry of Education, P. R. C. (2011). 义务教育数学课程标准*(2011年版)* [Full-time compulsory education mathematics curriculum Standard (2011 version)]. Beijing: Beijing Normal University Publishing Group.

Qi, C. X. (2007). 数学问题解决在中国的研究历史及其影响 [History and influence of research on problem solving in China]. *Curriculum, Teaching Material and Method, 12,* 32–35.

Shao, Z. H. (2014). 中美高质量初中数学任务的比较研究 [Comparative study of high quality mathematics task of junior high school in U.S. and Chinese classrooms.] (Unpublished master dissertation). Beijing Normal University, Beijing.

Shao, Z. H. (2015). 中美初中课堂中数学任务特征的比较研究 [Comparative study of mathematical task features in chinese and the U. S. classrooms]. *Comparative Education Review, 301* (2), 102–107.

Stein, M. K., Grover, B. W., & Henningsen, M. (1996). Building student capacity for mathematical thinking and reasoning: An analysis of mathematical tasks used in reform classrooms. *American Educational Research Journal, 33*(2), 455–488.

Stein, M. K., Smith, M. S., Henningsen, M. A., & Silver, E. A. (2000). *Implementing standards-based mathematics instruction: A casebook for professional development*. New York: Teachers College Press.

Tu, R. B., & Ning, L. H. (2002). 论数学活动的过程知识 [Process knowledge of mathematical activity and teaching reform]. *Journal of Mathematics Education, 11*(2), 9–12.

Yang, Y. D. (2012). 在学习目标导向下改进教学任务设计——以人教版小学数学五年级下册 人统计"计第一课时为例 [Improving teaching task design by formulating learning target—A case study on the first lesson of "Statistic" in Grade 5 of primary schools of PEP edition]. *Curriculum, Teaching Material and Method, 7,* 98–102.

Yuan, Z. L., & Lu, S. H. (2006). 高认知水平数学教学任务的特征分析 [Characteristics of high-level cognitive task of mathematics teaching]. *Journal of Mathematics Education, 15*(4), 24–28.

Chapter 12
The Design and Implementation of the Mathematics Teaching Goal

Dianshun Hu, Ke Ye and Jing Wang

Abstract Mathematical teaching goal is the intended learning outcome and standard in mathematical teaching and learning. It can be represented as the concrete description of students' learning results and behaviors, or an explanation of students' growth in knowledge and skills and other aspects at the end of teaching activities. During the thirty odd years, China's mathematical teaching goal has experienced a process of constant exploration and development, with a shift of focus from the evaluation of students' grasp of fundamental knowledge and skills, to the evaluation of the process of students' experience in mathematical activities and their solving of mathematical problems and their development mathematical literacy. The present teaching goal system can be seen as a three-dimensional one, which consists three aspects: knowledge and skill, process and method, emotions, attitudes and value. Under the new round of curriculum reform, the mathematical textbook editor and writer, academic experts, in-service teachers should work together in the designing, phrasing and final completing. The methodology for the design of mathematical teaching goal is an important technical problem, for which ABCD method and the method of combining internal and external are prominent. As for the realization of mathematical teaching goal, it shows both features of predetermination and estimated deviation.

12.1 The Outline of the Mathematics Teaching Goal

12.1.1 The Teaching Goal System

Teaching is a special kind of social, practical activity characterized by the participation of the key players, the teacher and the students, and the fact that it has clear goals. The teaching goal is a hierarchically structured system, which can be defined by five levels: educational purpose, training goals, teaching goals for a discipline,

D. Hu (✉) · K. Ye · J. Wang
Central China Normal University, Wuhan, China
e-mail: hdshsd@sina.com

Y. Cao and F.K.S. Leung (eds.), *The 21st Century Mathematics Education in China*, New Frontiers of Educational Research,
https://doi.org/10.1007/978-3-662-55781-5_12

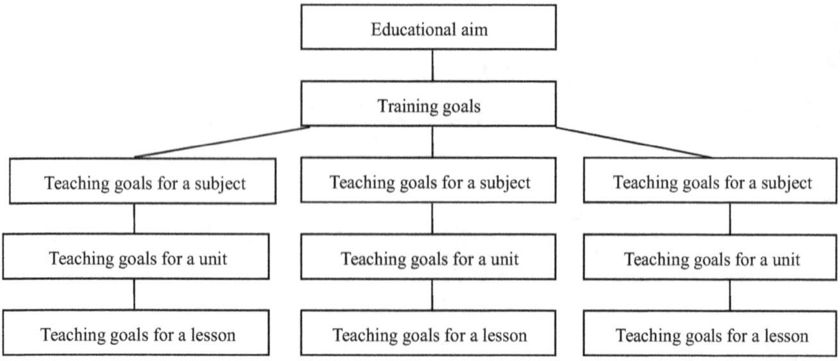

Fig. 12.1 System of the teaching goal

teaching goals for a unit, and teaching goals for a class period. These levels represent the transition from macro- to micro-views and from the abstract to the concrete (Fig. 12.1).

Educational purpose is the goal located at the highest level (tier 1). This refers to society's general conception that education is for the purpose of developing social individuals. Training goals (tier 2), based on the educational purpose, set out specific training requirements. Training goals in turn are realized in individual subjects; therefore, different subjects have particular teaching (or curriculum) goals (tier 3). In each discipline, the teachers then develop the subject teaching goals into teaching goals for specific units of work (tier 4). These goals in turn are developed further into goals for particular lessons/class periods (tier 5), these being the means by which the curriculum goal is achieved. The extent to which the general curriculum targets formulated by the new curriculum standards can be met depends on the clarity, appropriateness, specificity, and completeness of the design of the teaching goals for the class period. Throughout this chapter, the term "teaching goals" refers to the teaching goals for a class period.

12.1.2 The Mathematics Teaching Goal

The purpose of the teaching goal in mathematics is to direct the teaching activities and provide the basis for the teaching evaluation. Effective mathematics teaching goals emphasize the visibility, controllability, and measurability of mathematics teaching activities and their consequences, which are usually expressed as students' explicit, specific, and clear behaviors. However, our understanding of the mathematics teaching goal is developing and changing continually. It is generally believed that the teaching goal reflects the changes in students' behaviors brought about by cooperation between students and teachers leading to changes in the students' behavior because of the teaching. Therefore, the mathematics teaching goal determines the physical and psychological standards and requirements of the mathematics teaching activities (Xie & Zheng, 2003). It has also been suggested that the teaching goal is a subconcept of the task of mathematics teaching (Guan, 1992).

The teaching goal can also be described as the results and progress attained from doing mathematics activities. It can take the form of specific descriptions of students' learning achievements and eventual behaviors, or of the changes in aspects such as knowledge and skills.

12.1.3 Changes in the Concept of the Mathematics Teaching Goal

Over the past thirty years, with the mathematics educational reform,changes have occurred in our concept of the mathematics teaching goal. Between 1978 and 1982, attention was paid to teaching general knowledge, training in basic skills, nurturing students' intelligence, and developing their abilities. The *Mathematics Teaching Outline for the Ten Years of Full-Time Secondary School (Trial Draft)*, published in 1978, proposed that students should develop the necessary basic mathematics knowledge, the correct and fast operational skills, and certain logical and spatial imagination abilities, thus gradually developing the capacity to analyze and solve problems quickly. From 1983 to 1987, it was proposed that intelligence and capability should develop together; also independent thinking and the courage to bring forth new ideas were put forward as goals for the first time. In 1986, the *Mathematics Teaching Outline for the Full-Time Secondary School* pointed out that students should learn the necessary basic mathematics knowledge and skills and have command of operational, logical thinking, and spatial imagination abilities in order to develop the capacity to analyze and solve problems by applying mathematics knowledge. This document also described the importance of arousing students' interest in mathematics. From 1988 to 1992, in addition to the content mentioned above, the idea of promoting the healthy development of students' personalities was formulated for the first time; in fact, personality was confirmed as a new teaching goal. In 1992, *The Mathematics Teaching Outline for Nine Years of Compulsory Education Full-Time Junior High School* ruled that students must have a good command of the basic knowledge and skills of algebra and geometry. This document also referred to the development of operational and logical thinking abilities and spatial conception, solving simple practical problems by applying prior knowledge, and nurturing excellent personalities. During 1993–2000, the focus of primary and secondary school education changed from being examination-oriented toward the overall improvement of national quality. At this time, initiative and creativity were added as teaching goals in addition to the conventional goals (Tu, Yang, & Wang, 2011).

Since 2001, under the new round of curriculum reform, the mathematics teaching goals have been developed further. They have been developed from two categories to four and focus on three main areas, knowledge and ability, method and process, and emotions, attitudes, and values. To some extent, this means an expansion from education purely for knowledge to cultural education, emphasizing personal development, paying attention to social and practical skills, and highlighting the development of exploration and innovation (Tu et al., 2011). Thus, we can see that mathematics education has undergone an ongoing process of investigation and development. The changes, with additions and removals of foci, were in response to the changing social needs at certain times, thus presenting a trend toward more richness and diversification.

12.1.4 The Brief Introduction of the Mathematics Teaching Goal Under the Curriculum Reform

China began to be involved in research on teaching goals in the late 1980s (Wang & Lv, 2004). The teaching goals of different time periods have not really been the same. The two-category target of basic knowledge and basic skills has exerted a profound impact on the traditional mathematics teaching in China. The *Curriculum Reform Summary for Basic Education,* published by the National Ministry of Education in 2001, pointed out: "The national curriculum standards that are the basis of writing textbooks, teaching, assessment, examination proposition and the foundation of the national management and evaluation of courses should reflect such basic requirements as knowledge and skills, processes and methods, and emotions, attitudes and values that our country expects of students at different stages of studying." With the impetus of the new curriculum reform, the mathematics teaching goal is being transformed from a linear two-category target to a more encompassing three-dimensional, four-category one. According to the idea of the new curriculum standards, the mathematics teaching goals reflect the lowest levels of behavioral change or the learning levels occurring after a period of mathematics study, which is also regarded as the students' learning objective.

12.1.4.1 Dimensions and Levels of Teaching Goals Defined in the *Mathematics Curriculum Standards for Compulsory Education (2011 Edition)*

In the *Mathematics Curriculum Standards for Compulsory Education (2011 Edition),* the mathematics curriculum objectives were classified as general and specific to each learning stage. The general objectives can be illustrated as four goal fields (as shown in Table 12.1).

Table 12.1 Four goal fields in the "mathematics curriculum standards for the compulsory education (the 2011 edition)"

Goal field	Specific illustration
Knowledge and skills	• Experiencing the process of abstraction, operation, modeling, etc. In arithmetic and algebra, mastering the fundamental knowledge and basic skills of arithmetic and algebra • Experiencing the process of the abstraction, classification, discussion about the nature, movement, determination of the position, etc., of graphs; mastering the fundamental knowledge and basic skills of graphing and geometry • Experiencing the process of collecting and processing data, analyzing problems by applying data; obtaining information in the practical problems; mastering the fundamental knowledge and basic skills of statistics and probability • Participating in comprehensive practical activities; accumulating mathematics activity experience to solve simple problems by comprehensively applying mathematics knowledge, skills, methods, etc
Mathematical thinking	• Establishing number sense, symbolic consciousness, and spatial concepts; initially shaping geometrical visualization and operational capabilities; developing visualized thinking and abstract thinking • Feeling the meaning of statistical methods; developing the concept of analyzing data; understanding the random phenomena • Developing the ability of sensible reasoning and deductive reasoning in taking part in the mathematics activities of observing, experimenting, guessing, testifying, comprehensively practicing, etc.; clearly expressing thoughts • Learning to think independently; feeling the basic ideas of mathematics and mathematical way of thinking
Problem solving	• Initially learning to discover and put forward problems from the perspective of mathematics; comprehensively applying mathematics knowledge to solve simple practical problems; strengthening the sense of application; improving practical abilities • Acquiring some basic methods of analyzing and solving problems; experiencing the diversity of methods of solving problems; developing a sense of innovation • Learning to cooperate and communicate with other people • Initially shaping a sense of evaluating and introspecting
Emotions and attitudes	• Participating actively in mathematics activities; having curiosity and thirst for mathematics knowledge • Experiencing fun of success; exercising the will to overcome difficulties; establishing self-confidence in the process of mathematics learning • Feeling the characteristics of mathematics; understanding the value of mathematics • Developing learning habits of earnest and diligence, independent thinking, cooperation and communication, rethinking and casting; forming a kind of scientific attitude of seeking truth from facts

Table 12.2 Classification and implication of deed verbs in the *Mathematics Curriculum Standards for Compulsory Education (2011 Edition)*

Classification	Learning objective verbs	Descriptions
Result targets	Understand	Know about or exemplify the characteristics of objects from concrete examples; recognize objects from particular situations
	Comprehend	Describe the characteristics and origins of objects; illustrate the differences and relationships between objects
	Master	Apply objects to new situations on the basis of comprehension
	Apply	Draw on existing knowledge of objects to and create appropriate methods to solve problems
Process targets	Experience	Demonstrate perceptual knowledge in the specific mathematics activities
	Feel	Participate in specific mathematics activities; initiatively recognize or check the objects' characteristics, thus obtaining some experience
	Explore	Participate in specific mathematics activities independently or by cooperating with others; understand or raise problems, think mathematically to solve problems, discovering objects' characteristics and the differences and relationships between objects Demonstrate some rational knowledge

The four aspects of the general objective, which constitute an organic whole, should be taken into consideration concurrently when designing and organizing teaching activities. Such goals as mathematical thinking, problem solving, and emotional attitudes cannot be developed without learning the knowledge and skills needed to facilitate the realization of the other three goals. The overall realization of a goal is the indicator that students have received a good mathematics education and has an important significance to students' all-round, sustainable, and harmonious development.

There are two groups of learning objective verbs in the *Mathematics Curriculum Standards for Compulsory Education (2011 Edition)*; one group refers to learning objectives related to results, while the other describes process-related objectives. These two groups are described in Table 12.2.

Some illustrations of these verbs, using specific examples from the *Mathematics Curriculum Standards for Compulsory Education (2011 Edition),* are shown in Table 12.3.

Table 12.3 Specific examples from *Mathematics Curriculum Standards for Compulsory Education (2011 Edition)* to illustrate learning objective verbs

Learning objective	Equivalent verbs	Examples
Understand	Know, preliminarily recognize	Know the in-center and ex-center of a triangle. Preliminarily recognize decimals and fractions with concrete examples
Comprehend	Recognize, make	Recognize a triangle; make puzzles with rectangles, squares, triangles, parallelograms, or circles
Master	Recognize, signify	Recognize, read, and write the numbers to 10 thousand Signify the number of objects and their sequence and position
Apply	Prove	Prove the theorem: Two triangles that have two equal angles and the corresponding opposite sides also equal are congruent
Experience	Sense, try	Sense the meaning of large numbers in real-life situations Try to discover and put forward problems in real-life situations
Feel	Realize	Realize the meaning of arithmetic with integers by combining specific situations

12.1.4.2 The Dimensions and Levels of the Teaching Goal in *Mathematics Curriculum Standards for Ordinary High Schools (Experiment)*

The general objective of the document *Mathematics Curriculum Standards for Ordinary High Schools* is to enhance the mathematics qualities necessary for future citizenship so as to meet the needs of individual and social development. The concrete objectives are shown in Table 12.4.

The requirements of the general objective are divided into three correlated parts. When determining the content and scope of the mathematics teaching goal, all three fields should be taken into consideration, while each specific lesson should have its own focus.

The three categories of learning objective verbs and in the subcategories described in the *Mathematics Curriculum Standards for Ordinary High Schools* are shown in Table 12.5.

Table 12.4 Three goal fields in *Mathematics Curriculum Standards for Ordinary High Schools*

Goal field	Specific requirement
Knowledge and skills	1. Acquire the necessary fundamental knowledge and basic skills of mathematics, comprehend the basic concepts and the nature of mathematics conclusions, understand the formation and application of concepts, conclusions, etc., experience mathematical thoughts and methods underpinning concepts and conclusions, feel the process of making mathematical discoveries and creations through various forms of self-study and exploratory activities
Process and method	2. Expand the basic abilities of spatial imagination, abstract generalization, reasoning and demonstration, operational solving, data processing, etc. 3. Enhance abilities to propose, analyze, solve problems using mathematics (including simple practical problems) and the capacity for mathematics expression and communication, develop the ability to acquire mathematics knowledge independently 4. Develop mathematical thinking skills of applying and creating, endeavor to think mathematically, and make judgments about some mathematics applications in the real world
Emotion, attitude, and value	5. Increase interest to learn mathematics, develop confidence, perseverance, and a scientific attitude 6. Develop a mathematical sense, gradually come to understand the scientific, practical, and cultural value, form critical thinking skills, and sense the aesthetic aspect of mathematics, thus establishing a global outlook

Table 12.5 Learning objective verbs and subcategories described in the *Mathematics Curriculum Standards for Ordinary High Schools*

The goal field	Level	Verbs
Knowledge and skills	Know/understand/imitate	Understand, realize, know, distinguish, perceive, recognize, preliminarily learn, preliminarily comprehend, seek
	Comprehend/independently operate	Describe, state, express, formulate, show, depict, explain, guess, imagine, understand, generalize, summarize, abstract, extract, compare, contrast, judge, infer, seek out, can, apply, preliminarily apply, preliminarily discuss
	Master/apply/transfer	Master, educe, analyze, deduce, prove, study, discuss, choose, decide, solve problems
Process and method	Experience/imitate	Experience, observe, perceive, feel, operate, consult, aid, imitate, collect, retrospect, review, participate in, try
	Discover/explore	Design, sort out, collate, analyze, discover, communicate, study, explore, probe, search, solve, pursue
Emotions, attitudes, and values	Response/agree	Feel, recognize, understand, preliminarily realize, realize
	Grasp/internalize	Acquire, enhance, strength, form, cultivate, establish, exert, develop

12.2 The Design of the Mathematics Teaching Goal

12.2.1 Characteristics

To design mathematics teaching that can facilitate effective learning, careful planning is needed. This systematic planning is referred to in China as the teaching design; its primary component is to define a teaching goal that not only reflects the essence of mathematics teaching, but also conveys the intention of the teaching design and follow-up processes clearly. A well-designed teaching goal enables mathematics teachers to have a clear concept about "how students learn" and "how teachers teach." The design of the mathematics teaching goal in China has two important aspects:

12.2.1.1 The Theoretical Underpinning

Certain theories are involved in designing the teaching goal. Behaviorism, based on "stimulus-response," is concerned with foreseeing and controlling behavior. Cognitivism holds that teaching is a process of formation and reorganization that can improve learners' internal psychological structures and develop students' cognitive performance. Humanism considers not only students' behavior changes, but also the development of their emotions, attitudes, and values so that we should not raise the specific objectives in advance, but facilitate students to derive these objectives gradually from their experience. Constructivism is the theory that students do not acquire knowledge through the process of the teacher's teaching, but with the help of others (including teachers and learning partners), by using necessary learning materials and constructing meaning from certain situations. Constructivist teaching is directed by such processes as interactive teaching, "scaffolded" teaching, situational teaching, and random access to teaching (Yuan, 2004).

The various forms of teaching goal design all have advantages and disadvantages. The traditional teaching goal often just describes the teachers' teaching behaviors and teaching process, ignoring the fact that the essence and foothold of the teaching process are the learning behaviors and learning outcomes of the students. In view of this, teachers need to develop multifaceted teaching goals that consider not only their students' acquisition of knowledge and skills, but also their transformation in terms of emotions, attitudes, and values. These goals should also encompass extended teaching activities and embody the essence of the teaching process. The focus should not be only on the results, but should pay more attention to students, to studying rather than teaching, and to reflect the behavioral changes brought about by the teaching activities.

12.2.1.2 Changes in the Design Process

Traditionally, from the founding of the new China to the late twentieth century, the role of designing the teaching goal fell to experts. In the early days of the new China, the mathematics teaching plans, syllabi, and textbooks were uniform across the nation. Teachers were the passive "consumers" of the curricula and resources that experts had developed, and their work was merely to make some insignificant changes to the design provided by the experts, and then put it into effect (Zhou, 2009). Although the teachers implemented the teaching to address the goals, they were not the real "drivers"; these were the textbook authors, who often started with the logic of mathematics pedagogy but were not able to respond to the day-to-day needs of students in the actual classroom. With the new curriculum reform, the original mathematics syllabi evolved into the mathematics curriculum standards, giving teachers basic criteria to guide their teaching but allowing them the space and initiative to design specific teaching goals according to their students and their actual conditions. Teachers' voices in "designing the teaching for themselves" have become stronger and stronger, and the development of school-based curricula and the implementation of comprehensive, practical curricula have also provided the platform for them to do the design autonomously (Ren, 2009). As a matter of fact, the first-tier mathematics teachers are now the drivers of the teaching goal design, because they can be autonomous in designing the goals and guiding the classroom teaching practice, and their own development is an important focal point. The new curriculum reform also offers a platform for mathematics teachers in primary and secondary schools to communicate with experts on mathematics curriculum and teaching through various channels. With the help of the Internet, the first-tier teachers can engage in direct dialogue with mathematics specialists and learn about advanced theories, as well as exchanging experiences with other teachers.

12.2.2 The Basis of the Mathematics Teaching Goal

The mathematics curriculum standard is the fundamental basis of the mathematics teaching goal. During the time from the founding of new China to the late twentieth century, mathematics teachers often adjusted the teaching goals according to their experience or by virtue of copying ideas from some reference books or teaching plans; hence, the goals were often too general and complex and even detrimental to the teachers' effectiveness and development. The eighth curriculum reform clearly proposed that the teaching goal should be based on the mathematics curriculum standard. The requirements for specific goals were stated according to the learning stages of compulsory education and high-school education. Cui (2009) held that basing teaching on the curriculum would mean that teachers would carry out series of processes of designing and implementing teaching activities: determining the teaching goal, designing the evaluation, organizing the teaching content, carrying out the teaching, assessing the students' learning, and improving the teaching, in accordance with the learning outcomes expected by the curriculum standards.

The teaching materials are the important bases for the mathematics teaching goal. Both traditionally and in new courses, teaching is based on analyses of the textbook to identify opportunities for teaching mathematical knowledge and thinking methods, facilitating students to develop and improve their ability to learn mathematics, and developing a scientific spirit and affective qualities. In the mathematics discipline, the correspondence between the textbook content and the teaching goal is relatively obvious, hence what the focal points and difficult teaching points are, what the students should master, and at what level can be determined based on textbooks.

The learning conditions also play an important part in determining the mathematics teaching goal. Individual learning differences make it necessary for the teacher to analyze these conditions, including the students' previous level of knowledge, the level of psychological development, their maturity, and such personality factors as attitudes, interests, hobbies, and their inclination to learn. A teaching goal that is set too high or too low will negate the effectiveness of the teaching. For instance, when design the teaching goal for the monotonicity of functions, a goal such as "students can flexibly apply the monotonicity of functions to solve mathematical problems" may be too high, while one such as "students can utilize the graph of a function to determine the monotonicity of a function" may be too low (Pu, 2006). The ready conditions for learning and the common psychological characteristics prevailing in the whole class should be the main aspects taken into consideration when determining the mathematics teaching goal. Meanwhile, the students' individual differences should be fully taken into consideration when designing the target to frame the development goal at the corresponding level.

12.2.3 The Design Process

It is generally considered that the design of a mathematics teaching goal is a two-step process. The first step refers to analysis, mainly of the teaching content, understanding what knowledge the students should learn in the class, and the position and role of this knowledge in the whole textbook, how the students will acquire the knowledge, and which examples and exercises in the textbooks are suitable. The second step refers to transformation, that is, transforming the teaching requirements laid out in the curriculum standard into more specific and clear ones according to the analysis of the specific contents and the selection of learning objective verbs. At present, the process of designing a mathematics teaching goal represents a shift from a one-way process to a spiral one. The general process is shown in Fig. 12.2.

This process is not one-way. Reflection on the teaching goal also involves a review of the preparation, design, and implementation, leading to corresponding

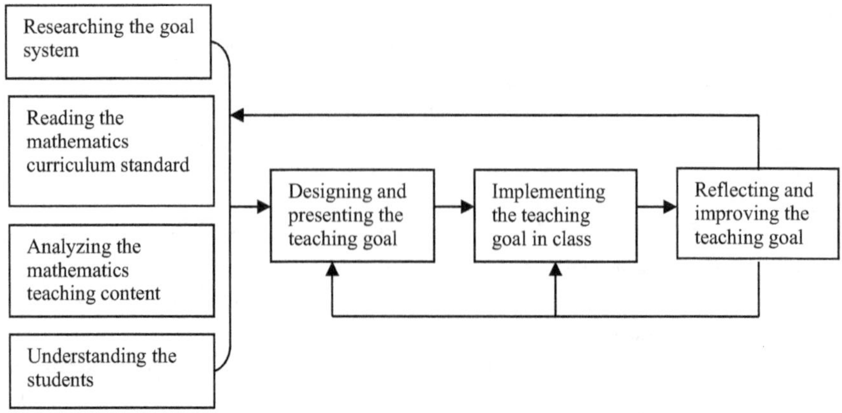

Fig. 12.2 Design process of the mathematics teaching goal

readjustments to the original teaching goal. The specification of each aspect of this process, shown in Fig. 12.2, is as follows:

1. Researching the specific tasks that should contribute to the goal: The mathematics teaching goal is the essential part of the whole goal system. Whether it is achieved or not will have impacts on further goals in the overall system. In turn, if we can look at the teaching goal on the basis of the overall goal system, we can locate the essential tasks more accurately. In this way, the design of the teaching goal is targeted, not divorced from the overall direction of the mathematics education activities.
2. Studying the mathematics curriculum standards: The mathematics curriculum standards embody our nation's basic requirements for students at different stages in such aspects as knowledge and skills, processes and methods, and emotions, attitudes, and values. It not only provides the nature, goal, and content frameworks for courses, but also gives guidelines for teaching and evaluation. It is important to study the curriculum standards and understand the society's expectations of students.
3. Analyzing the mathematics teaching content: The mathematics teaching content is not only the facts, concepts, generalizations, principles, and the theories relating to the discipline knowledge, but also includes how this information is organized. It refers to the information that students are required to master to realize the mathematics educational goal, as well as the way in which it is organized to arouse their interest. The concrete teaching content should be embodied in the design of the mathematics teaching goal, otherwise, the goal is not supported.
4. Understanding students' actual levels and developmental needs: The whole activity of mathematics education involves the provision of certain activities that will improve students' physical and psychological states to meet with social expectations. Therefore, only when we understand the students' current level of

development and knowledge structure, we can know what their zone of proximal development is and establish some goals that they will be able to achieve with the right support.

5. After completing the basic work mentioned above, the design process involves a substantial stage of proposing and classifying the goals. They need to be frames at different levels from different perspectives and according to different standards, and presented in an appropriate manner.
6. This step involves implementing the mathematics teaching practice in order to meet the teaching goals.
7. At this stage, the designer reflects on each step of the design process and makes amendments to the teaching goal in accordance with the actual effect of the mathematics teaching (Yan, 2010).

12.2.4 The Mathematics Teaching Goal Statement

12.2.4.1 The Basic Requirements for the Mathematics Teaching Goal Statement

After the mathematics teaching goal is determined, the key matter is how to express it clearly, accurately, and concretely.

Wang (2005) proposed some general requirements for a teaching goal statement: (1) It needs to state what the students can do after the lesson, not what the teachers can do after teaching. (2) We should state students' behaviors in as specific and observable a way as possible. (3) The learning content contained in each teaching goal should not be too limited, but should be comprehensive and wide. (4) Generally, we should not express the teaching goal as the learning activity or learning process. (5) For the presentation of the declarative knowledge, we had better not included the learning content.

Gu (1990) considered that a well-stated goal should have two basic traits: (1) including the detailed specifications of the specific content that is to be achieved, and (2) being able to depict the detailed specifications of the teaching to be obtained.

Zhu (1987) held that a good teaching goal should: (1) determine the behaviors that are evidence of the desired result, (2) confirm the occurrence of these behaviors, and (3) ascertain the desired standards of behavior.

The traditional mathematics teaching goals statements were relatively abstract and general and usually considered the teachers as the subjects, for example, how effectively they developed students' operational skills or logical thinking. However, these goals tended to be not clear enough, hard to measure and evaluate. General targets should be observable and measurable, illustrate what the students can acquire and what level they can reach after teaching, and state the behavioral changes expected of students. In summary, a clear, normative teaching goal should illustrate what (the required behaviors), and at what level (the levels and standards of the behaviors required) somebody (the learners) can do under what conditions (the specific and limited influential conditions).

12.2.4.2 The Frequently Used Ways of Presenting the Mathematics Teaching Goal

At present, there are three main arguments from a psychological perspective about the presentation of the teaching goal: the behaviorist view, the cognitive view, and the comprehensive view. The behaviorist view focuses on describing the observable and measurable learning behaviors; the cognitive view pays attention to changes in the learners' mentality; and the comprehensive view argues that overt behaviors and the process of the internal mentality should be taken into consideration when stating the teaching goal. Here are two common ways of expressing the teaching goal:

1. The presenting method of A, B, C, D

The subject of the behavior (A, for Audience) is the starting point. A behavioral goal describes the students' behaviors, not the teachers'. We cannot write such teaching goals as "teaching students …" or "the teachers will …". The goal should be expressed as "the students can …" or "the students will …" etc.

The behavior itself (B, for Behavior) is fundamental; it is necessary to state what standard the students should meet after the teaching process. Behavioral statements should be observable by using verb-object phrases. First, we should divide the specific curriculum content into different types according to content of lessons, then according to the types, choose verbs for the verb-object structure in presentation. Finally, we can consider the concrete learning content and action required in courses as the objects in the verb-object structure, for instance, compare the similarities and differences between a rectangle and a diamond in nature, and prove the Pythagorean Theorem using at least one method. The curriculum standard provides some verbs for each kind of action required for the instructional designers to refer to and select from.

The condition (C, for Condition) refers to the scenarios in which students exhibit the desired behaviors. In general, conditions include such factors as circumstances, equipment, time, information, and classmates or teachers, for example, understanding the quantitative relationship that the logarithmic function depicted through concrete examples and comparing the growth difference of the exponential function, the logarithmic function, and the power function by using computational tools.

The standard (D, for Degree) is the minimum measurable behavioral standard. Behavioral standards make the teaching goal measurable. The standards can be represented by using quantitative or qualitative methods or a combination of the two (Chen, 2001). The behavioral standards can usually be classified into three categories: (1) the time limit to complete the behaviors, for example, solving the problem within three minutes; (2) the accuracy, namely, the percentage or numbers of correct manipulations and operations, for instance, the percentage of correct answers is 90%; and (3) the characteristics of success, such as answering up to three decimal places. The curriculum standards provide a reference for the explanation of levels of behaviors, such as understanding and comprehending.

2. The method of combining internal and external

The A, B, C, D method is used by large numbers of mathematics teachers and teaching designers to develop the teaching goal. This kind of method, featuring clarity, observability, and measurability, is conducive to guiding and evaluating the teaching. However, it is difficult to present some psychological processes because of the complexity and diversity of the practical mathematics teaching. For example, only a few affective objectives can be observed and measured, and some objectives cannot be described with verbs. In order to overcome this disadvantage, Gronlund came up with a new method for observing and measuring psychological changes. Thus in defining the teaching goal, terms can be used that explain internal processes and then utilize the observable behaviors as examples. In other words, this is a combination of internal and external behaviors (Chen, 2001).

For example, the word "comprehends" in the phrase "comprehends the concept of exponential function" refers to an internal psychological process. Everyone's standard is unique and difficult to ob serve and measure. However, we can illustrate it by using behavioral examples that can reflect the "comprehension" level, such as: (a) paraphrasing the definition of the exponential function in ones' own words; (b) judging whether it is an exponential function or not according to the analytic expression of a given function; or (c) distinguishing the index function and the exponential function. Supplemented by these three examples, the teaching goal is no longer intangible as it can be described by the explicit external behaviors, "paraphrasing" "judging," and "distinguishing."

12.2.4.3 The Presentation of the Mathematics Teaching Goal Under the Curriculum Reform

Only when the mathematics teaching goal is presented in such a way as to combine the internal and external, we can describe all of the changes that occur during the teaching activities, thus providing comprehensive and effective information for evaluating students. At present, teachers can use various methods when presenting their mathematics teaching goals, specifically:

1. Divide the mathematics teaching goal into knowledge and skills, processes and methods, and emotions, attitudes, and values, based on the philosophy of the three-dimensional objective curriculum.
2. Divide the mathematics teaching goal into knowledge and skills, mathematical thinking, problem solving, emotions, and attitudes according to specific illustrations of the four aspects of the general curriculum goal according to the four categories in curriculum purpose.

3. Divide the teaching goal into knowledge, capability, and emotional goals by continuing to use the traditional design approach.
4. Directly list a few (e.g., 3–4) teaching goals, rather than establishing subobjectives (Luo, Xu, & Engstrom, 2014).

To illustrate the mathematics teaching goal, teachers should choose appropriate ways from the above four based on the specific teaching content. The method can be viewed as the embodiment of the curriculum philosophy. The second way is the specification of the general curriculum goal. The third way is to continue the traditional statement of the mathematics teaching goal. The fourth way does not provide the concrete name of the teaching goal. The latter two of these are less desirable because they are less clear and rational and do not keep pace with the changing times, which means that the teaching goal will not reflect the new curriculum. The second can contribute to a three-dimensional goal; therefore, increasing numbers of mathematics teachers are accepting using this approach.

The relationship among the three goal dimensions is an integral one, with all three realized concurrently in the same process. Knowledge is the purpose of cognition, which is in turn the means of learning knowledge. Attitude is the impetus of cognition. The emotional objectives are concerned with capturing the learners' interest and attention and creating cognitive conflicts. Thus, presentation based on the three-dimensional goal reflects the organic integration of knowledge and skills, processes and methods, and emotions, attitudes, and values from the three dimensions of knowledge content, cognitive manifestations of learning and attitudes, compliance with the laws of mathematics learning and teaching, and reflection on the teachers' scientific analyses and revisions of the teaching content. In addition, basing the teaching goal design on the presentation of the three-dimensional goal provides teachers with a clear direction to diagnose students' learning and evaluate their teaching in short term. The following is an example of a complete teaching goal statement:

The three-dimensional teaching targets are the topics of "intersecting angles between straight lines" (Grade 5, Chapter 5.1), "calculation of intersecting angles" (Chapter 2), and "spatial vector and stereoscopic geometry" (Mathematics Elective 2-1) of the Beijing Normal University textbook:

Knowledge and skills:

a. Exemplify the concepts of the intersection angle between two straight lines and the intersection angle between straight lines in different planes.
b. Calculate the degree of the intersection angle between straight lines with the spatial vectors.

Processes and methods:

a. Independently develop a method for calculating the size of the intersection angle between the spatial straight lines with the help of a pictorial diagram, spatial imagination, and/or vector operation.

b. Compare and analyze the concepts of the intersection angle between straight lines in the plane and the intersection angle between the spatial straight lines, drawing on the formula for the intersection angle between plane vectors and the formula for the intersection angle between spatial straight lines.

Emotions, attitudes, and values:

a. State the role that the spatial vectors play in calculating the degree of the intersection angle between straight lines.
b. Gradually establish a consciousness of drawing on and inter-relating geometric concepts and vector operations.

12.3 The Realization of the Mathematics Teaching Goal

Starting from a broad national educational purpose, the process of developing a teaching goal comes down gradually to different levels according to the course content, class time arrangements, the specific conditions for implementing, etc; at the same time, the goals at each of these levels are divided into different fields such as cognition, skills, and emotions. Specific objectives are then described. The teaching goal appears on the teacher's preparation plan for each lesson and is realized through the teaching and learning activities (Wang & Liang, 2007). Of course, even the most scientific and reasonable mathematics teaching goal is not necessarily realized; to ensure that it is, it is necessary to consider the links between the teaching design and each aspect of the classroom teaching.

12.3.1 The Realization of the Mathematics Teaching Goal in the Teaching Design

The design of the mathematical teaching goal is the first stage in mathematical teaching design, and its function is to give guidance for teaching, studying, and evaluation. Although the order of mathematics teaching design is open to debate, it is commonly agreed that the teaching goal design should be the departure point. The teaching goal gives the direction and the final purpose for the overall teaching design. It influences the selection of teaching methods and strategies, the selection and combination of the content, the application of the teaching media, and the evaluation of the teaching effects. Thus, the design and implementation of all aspects of the teaching must center on how to realize the teaching goal better (Pu, 2006).

The teaching strategy includes the teaching plan and its implementation in order to achieve the teaching goal (Yan, 2001). The teaching method encompasses the interactive teaching and learning activities that the teachers and students adopt to achieve the teaching goal and complete the teaching tasks (Li & Li, 1991). The teaching strategy and the teaching method are essential components of the teaching design, since both serve to achieve the goal but, at the same time, are constrained by it. As a result, it is necessary to choose strategies and methods according to the different levels and characteristics of the teaching goal. Experiments have shown that if the mathematics teaching goal focuses on the acquisition of knowledge and the learning outcomes, we should choose a teaching strategy based on meaning, and the most suitable corresponding teaching method is lecturing; if the goal centers on developing and improving skills, we should select a procedural teaching strategy and a teaching method that gives priority to training and practice; if the mathematics teaching goal lays stress on acquiring and exploring the knowledge experience and developing students' emotions, we should choose a strategy based on exploring the problems and hence the discovery or the task-driving method.

12.3.2 Realization of the Mathematics Teaching Goal

An important characteristic of mathematics teaching goal design is its "predetermination"; it is determined before the target groups engage in the activities and is therefore the mathematics teacher's subjective estimation of the teaching activities. The mathematics teaching is a purposeful, organized, and planned activity. It is necessary for the designer to be able to predict and set expectations well. However, it is not possible to preempt the many incidental and complex situations that can arise in mathematics teaching, thus there is a need to include provisions for "estimated deviation" in the design of the mathematics teaching goal (Long, 2008).

In mathematics teaching, teachers keep focusing on mathematical events. If unexpected events occur that were not foreseen in the design of the teaching goal, teachers need to show their professionalism and capture the resources available to them in order to refine the goal, thus producing teaching effects beyond the original teaching goal. In other words, the teaching goal is both predetermined and generative, and the latter aspect is an essential attribute.

With the mathematics curriculum reform, people have come to realize that, in actual practice, the preset mathematics teaching goal is far from able to address the actual classroom needs. It has been acknowledged gradually that the dual aspects of presetting and generation are both necessary. In other words, the realization of mathematics teaching goal depends both preplanning and improvisation. In particular, the uncertainty of the mathematics teaching practice requires that the supplements and adjustments are needed during the teaching process.

References

Chen, S. J. (2001). 课程与教学 [Curriculum and teaching]. Guilin: Guangxi Normal University Press.

Cui, Y. H. (2009). 课程实施的新取向:基于课程标准的教学 [The new orientation of curriculum implementation: the teaching based on the curriculum standard]. *The Educational Research*, 74–79.

Gu, M. Y. (Ed.). (1990). 教育学大辞典 [The great educational dictionary]. Shanghai: Shanghai Educational Press.

Guan, S. X. (1992). 教学论教程 [The tutorial for the teaching theory]. Xi'an: Shanxi Normal University Press.

Li, B. D., & Li, D. R. (1991). 教学论 [The teaching theory]. Beijing: People's Education Press.

Long, A. B. (2008). 教学目标的预设与生成 [The presetting and generation of the teaching goal]. *Modern Educational Science, 6,* 105–106.

Luo, X. B., Xu, H. Y., & Engström, A. (2014). 数学教学目标制定的问题及改进–以初中三维数学教学目标为例 [The problems and improvements of framing the mathematics teaching goal-taking for instance the three-dimensional mathematics teaching goal in the third year of junior high school]. *Journal of Mathematics Education, 5,* 23–26.

Pu, A. S. (2006). 新高中课程标准下数学课堂教学目标的设计 [The design for the mathematics classroom teaching goal under the new curriculum standard for senior high school]. *Journal of Mathematics Education, 1,* 92–94.

Ren, C. Q. (2009). 教师作为教学设计者的角色研究[A character study of teachers as the teaching designers] (Unpublished master dissertation). Sichuan, China: Sichuan Normal University.

Tu, R. B., Yang, Q., & Wang, G. M. (2011). 中国教学教学研究*30*年 [Thirty years of Chinese mathematics teaching research]. Beijing: The Science Press.

Wang, A. L., & Liang, C. F. (2007). 教学目标的具体化:一个复杂的过程 [The embodiment of the teaching goal: a complex process]. *Educational theory and practice, 4,* 57–60.

Wang, X. M. (2005). 教学论–心理学取向 [The teaching theory –the psychological orientation]. Shanghai: Shanghai Educational Press.

Wang, Y. L., & Lv, X. J. (2004). 论教学目标设计理论与实践的应用研究 [Talk about the theoretical and practical application research on the design for the teaching goal]. *Journal of Northeast Normal University (The Philosophy and Social Sciences Edition), 1,* 136–141.

Xie, L. M., & Zheng, B. W. (2003). 现代教学基础理论 [The fundamental theory of the modern teaching]. Shanghai: Shanghai Educational Press.

Yan, C. L. (2001). 课堂教学的策略、方式与艺术 [The strategy, way and art of classroom teaching]. *The Educational Research, 4,* 43–46.

Yan, Y. (2010). 课堂教学目标研究 [Objective research on classroom teaching] (Unpublished doctoral dissertation). Shanghai, China: East China Normal University.

Yuan, Z. G. (Ed.). (2004). 当代教育学*(*修订版*)* [The contemporary pedagogy (the revised edition)]. Beijing: The Education and Science Press.

Zhou, S. D. (2009). 论课程实施基本取向与教学设计的转变 [A discussion about the transformation of the basic orientation of curriculum implementation and the teaching design]. *Journal of China West Normal University (The Philosophy & Social Sciences Edition), 1,* 88–92.

Zhu, Z. R. (Ed.). (1987). 教育辞典 [The educational dictionary]. Nanchang: Jiangxi Educational Press.

Chapter 13
Integration of Information Technology (IT) and the Mathematics Curriculum

Nan Zhang, Rui Yang, Yanyun Liu, Jiushi Zhou and Guangming Wang

Abstract Since the beginning of this century, a new round of fundamental education curriculum reforms has been promoted widely. One key foundation of the curriculum reform has been to speedup the process of IT education and to promote the integration of IT into the mathematics curriculum. According to the *Basic Education Curriculum Reform Outline* (Ministry of Education of People's Republic of China, 2001a), *National Medium Long-term Education Reform and Development Plan* (2010–2020) (The State Council, 2010), *Education Information Decade Development Plan* (2011–2020) (Ministry of Education of People's Republic of China, 2012a), and *The National Education Development of The 12th Five-year Plan* (Ministry of Education of People's Republic of China, 2012b), it is necessary to continue collectively to advance the deepening of understanding, implementation, and promotion of IT to be used widely in the teaching process. To achieve the changes, it has been necessary to improve the presentation of teaching content, the methods of students' learning, teaching strategies, and the teacher–student interaction modes in order to fully implement the advantages of IT and provide a variety of educational environments and powerful learning tools for students' learning and development (Ministry of Education of People's Republic of China, 2001b).

Since the beginning of this century, a new round of fundamental education curriculum reforms has been promoted widely. One key foundation of the curriculum reform has been to speedup the process of IT education and to promote the integration of IT into the mathematics curriculum. According to the *Basic Education Curriculum Reform Outline* (Ministry of Education of People's Republic of China, 2001a), *National Medium Long-term Education Reform and Development Plan*

N. Zhang (✉) · R. Yang · G. Wang
Tianjin Normal University, Tianjin, China
e-mail: xiaoyu6109@126.com

Y. Liu
Meten International Education, Chengdu, China

J. Zhou
East China Normal University, Shanghai, China

Y. Cao and F.K.S. Leung (eds.), *The 21st Century Mathematics Education in China*, New Frontiers of Educational Research,
https://doi.org/10.1007/978-3-662-55781-5_13

(2010–2020) (The State Council, 2010), *Education Information Decade Development Plan* (2011–2020) (Ministry of Education of People's Republic of China, 2012a), and *The National Education Development of The 12th Five-year Plan* (Ministry of Education of People's Republic of China, 2012b), it is necessary to continue collectively to advance the deepening of understanding, implementation, and promotion of IT to be used widely in the teaching process. To achieve the changes, it has been necessary to improve the presentation of teaching content, the methods of students' learning, teaching strategies, and the teacher–student interaction modes in order to fully implement the advantages of IT and provide a variety of educational environments and powerful learning tools for students' learning and development (Ministry of Education of People's Republic of China, 2001b).

IT and mathematics curriculum integration does not simply involve adding the elements of IT into the traditional mathematics curriculum, but also the incorporation of essential changes which consist of three aspects:

- the formation of a new type of teaching environment (He, 2005; Wang, 2004; Zhang, 2007).
- the construction of a new teaching structure (He, 2005; He & Fan, 2007; Wang, 2004; Wang, Zhao, & Fu, 2006; Zhang, 2007).
- the formation of new ways of teaching and learning (He, 2005; He & Fan, 2007; Wang, 2004; Zhang, 2007).

He (2005) argued that these three basic attributes should be incorporated to understand and develop the integration of IT and mathematics—in particular, the incorporation of changes which can have significant impacts on the traditional teaching structure.

After fifteen years of curriculum reform, China has deepened the understanding of IT and mathematics curriculum integration. Practical experience and research results have accumulated in this field, and effective integration continues to be explored and developed.

Fig. 13.1 Number of computers per 100 students in Chinese schools

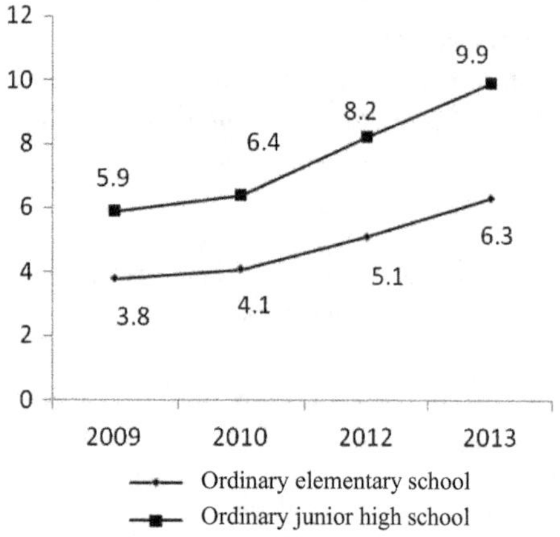

13.1 Practice Conditions for Integrating IT and Mathematics Curricula

13.1.1 IT Education Supported by China

With the advancement of national education information, compulsory IT education, and the availability of equipment, the quality of IT in classrooms has improved significantly in China. In 2009, the state elementary schools had 3.8 computers per 100 students, and the quota for the state junior high schools was 5.9 (Ministry of Education of People's Republic of China, 2011). By 2013, the primary schools had increased to 6.3 computers, while the junior schools rose to 9.9. By then, 68.1% of primary schools and 92.4% of junior schools were able to access the Internet (Ministry of Education of People's Republic of China, 2015), as shown in Fig. 13.1. The equipment was configured to support teachers to teach with IT.

In 2012, Vice Premier Liu Yandong delivered a speech to the national IT education television and telephone conference. She stated that in the government's twelfth five-year plan period, there would be broadband Internet access for schools to exchange information, for classes to share high-quality resources, and for learners to study and interact via the e-learning platform. In addition, the government planned to construct a public service platform for educational resources and educational management.

Furthermore, the Ministry of Education of the People's Republic of China has committed to implement a National Training Plan and Excellent Teacher Training Program in order to develop gifted teachers who produce high-quality lessons. This is to be coupled with a series of teacher education programs and activities, involving the Primary and Secondary Teachers' IT Competency Standards trial and the Teachers' IT Ability Training Curriculum Standards trial as well as the launching of the National Primary and Secondary School Teachers' IT Upgrade Project, to improve teachers' abilities to use IT and promote the depth of integration of IT and education.

At the national level, in order to support and promote education quality, the hardware and software for information technology and curriculum integration have been improved.

13.1.2 Conditions for Integrating Information Technology in Mathematics Education

Since 2001, the start of the current mathematics curriculum, IT and mathematics curriculum integration ideas had been formed step-by-step, and the proportion of IT usage in mathematics teaching has increased. Liu (2003) investigated mathematics teachers in Guizhou province and found that only 2.3% of them used computers often in their classes, 35.2% used them occasionally, and 62.5% never used computers. Hu, Wang, Wang, and Zhang (2011) investigated high school mathematics teachers in Shandong Province and found that more than 50% used information technology regularly and relatively frequently to show teaching content; teachers used information technology image drawing functions, explored the nature of relationships between functions, presented plane geometry and solid geometry, and provided image support. Liao (2006) found that 17% of Guangzhou City high school year 11 mathematics teachers used calculators often, and 66% used them for general activities. Wang (2009) found, in Shandong Province, that 93.7% of the teachers used IT in the open classes or competition courses, but less in normal teaching; also, older teachers used even less IT, down to only 6%. Guo, Cao, and Wang (2015) did research in China's north, northeast, southeast, and southwest regions. Their survey revealed that 68.5% of the teachers used IT at least once a week to assist their teaching, while 10% used it only monthly or never.

Despite the rapidly increasing availability of new ideas and technology to promote the mathematics curriculum, and teachers' access to IT to aid their teaching, classroom usage has not been uniform. There have been huge differences between regions, classes, and teachers of different age groups. In particular, in the rural western economic development areas, IT usage by teachers was found to be low; it was lower in normal classes than in public classes; and older teachers used it less than younger teachers. Therefore, balanced regional usage of IT development is an important issue that needs to be addressed.

13.1.3 Common Mathematics Education Software

Zhang and Peng (2009) divided the usage of IT in mathematics education into three categories:

- selective use of generally available IT, including e-mail, Internet, Chinese spell checker, and exchanges in the network forum community.
- IT commonly used in mathematics teaching, including dynamic geometry, dynamic curve mapping, dynamic measurement, symbolic computation, programming environment, random phenomenon simulation, statistical charting, fast formula editing, and courseware presentations.
- specific topics used for interactive teaching information technology, such as fractal design production and function fitting.

Wang, Zhang, and Kang (2015) described three main types of mathematics education software used commonly:

- mathematics software for uses including content-based professional software to facilitate students' abilities to explore and interact software to facilitate timely communication and feedback.
- general mathematics teaching software, secondary mathematics teaching multipurpose software which can be used in all disciplines, including courseware authoring software, knowledge structure, and after-school interaction.

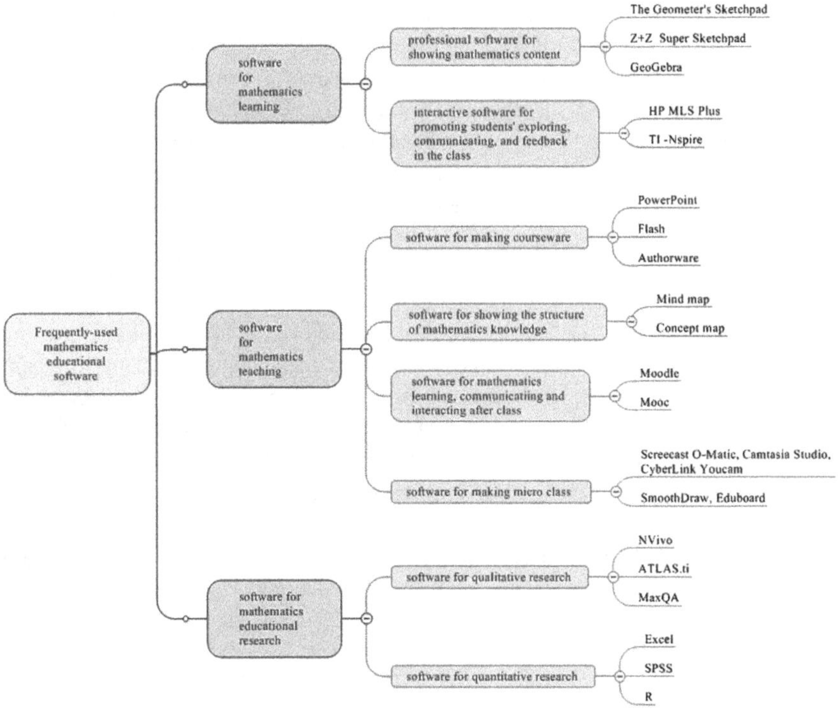

Fig. 13.2 Classification and examples of commonly used mathematics education software

- mathematics education research software, including quantitative and qualitative data analysis software.

Software categories and examples are shown in Fig. 13.2.

It is obligatory for mathematics teachers to master the functions and features of these educational software programs and to apply them flexibly to enhance the quality of classroom teaching and mathematics research.

13.2 Integration of Information Technology and Mathematics Curriculum: Research Status

13.2.1 Information Technology and Mathematics Curriculum Research Projects

Since the implementation of the new mathematics curriculum, a series of research studies has been conducted to promote the integration of information technology and the mathematics curriculum.

In 2001, the Institute of Mathematics Textbook Research and Development Center launched a project titled "High School Mathematics Curriculum Materials and Information Technology Integration Research." This was based on the *Ordinary High School Mathematics Textbook Experiment and the Ordinary High Schools Experiment Mathematics Textbook* (*Information Technology Integration Version*), and it was the first time that information technology and mathematics curriculum experiments were integrated. The research team carried out experiments in class teaching which involved more than 70 schools in Beijing, Guangdong, Yunnan, Zhejiang, and other provinces, with more than 16,000 students participating. The experiments involved changing the presentation of materials to improve teachers' and students' interactions with information technology; students' learning styles and teachers' teaching styles changed, and mathematics teaching, teachers' professional development, and the application of information technology in the mathematics teaching process were addressed (Middle School Mathematics Curriculum Materials and Integration with IT Research Group, 2004).

In 2008, the Ministry of Education Mathematics and Complex Systems Key Laboratory partnered with Beijing Normal University and the USA HP Company to do research about the viability of handheld technology in mathematics teaching in the middle school. The research established 10 municipal experimental districts, with more than 80 schools involved. It focused on six main topics:

- production of teaching plans, programs, and teaching cases to use handheld technology in the new middle school mathematics curriculum.
- the effects of handheld technology on different students' mathematics learning.
- impacts on handheld technology on students' mathematical understanding, ability development, and academic achievement.
- effectiveness of teaching with handheld technology.
- evaluation criteria and evaluation methods for using handheld technology.
- influence of new handheld technology and curriculum integration on teachers' professional development, etc.

In 2010, the HP e-Mathematics Laboratory and the TI-Nspire Wireless Classroom were introduced into Chinese mathematics classrooms. New technology tools such as graphing calculators were the core, and PCs, data collection, sensors, wireless interactive classroom management platforms for an experimental cloud service, and other educational technologies were used to provide opportunities for students to engage in mathematical experiments within an IT environment. With technology support, the research group gradually developed application-oriented classroom teaching materials, and experiments supported the experimental teaching curriculum design. As well, the group developed experimental teaching applications of a micro-curriculum, an experimental teaching service platform, and supported systems for schools and teachers to implement the concept of the mathematics laboratory in various ways.

In May 2011, the dynamic mathematics software GeoGebra, designed and developed by Markus Hohenwarter in the USA, was introduced into China. Beijing

Normal University established the GeoGebra Software Research Institute, and a similar institute was established at Tianjin Normal University. Both were committed to the usage of GeoGebra software and related applications of IT in classroom teaching research.

Based on research programs and projects, mathematics educators, teachers, and students deepened their understanding of integrating the mathematics curriculum with IT. This promoted an in-depth study of mathematics teaching, mathematics curriculum construction, and other aspects of mathematics learning through information technology environments.

13.2.2 Teaching Mathematics Within an IT Environment

13.2.2.1 The Functions of IT in Mathematics Teaching

Different kinds of IT have been used to carry out different mathematics teaching functions. Liao (2006) described four roles for IT in the classroom:

- as a presentation tool.
- to provide resources and environments.
- as a tool for inquiry and discovery learning.
- as a tool for information processing and knowledge construction.

The Textbook Information Technology Research Group for the Middle School Mathematics Curriculum and Integration (2004) stated another four roles for IT:

- as a powerful tool offering complex graph drawing, tedious calculations, and data processing—it can greatly improve the efficiency of the mapping, calculation, and data processing.
- to build "multiple contact representation" which means mathematical learning environment tools.
- to support mathematical experiments and practical mathematics activities.
- to promote the development of mathematical thinking.

Liu and Liu (2006) described three roles of IT in the integrated curriculum:

- providing a wealth of learning resources and helping to build immersive, multi-link mathematics learning environments.
- helping the discovery learning process.
- identifying and exchanging learning outcomes.

Overall, the research on the role of information technology in mathematics teaching could be categorized into three aspects:

- **IT support for visualizing mathematical concepts**
  Mathematics is abstract, so when IT is integrated into different domains like dynamic geometry, curve mapping, measurement, simulation of random

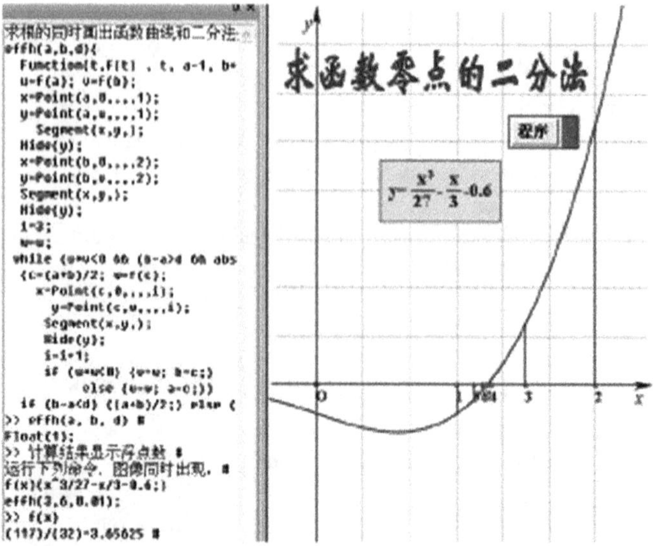

Fig. 13.3 Find zeros of a function

phenomena, or statistical charting, the visualization can be realized. Liao (2006) summarized the above process with the expression that "IT is a presentation tool" especially a tool for "applying software or graphing calculators to present a dynamic changing process of visualizing graphics." He also used a dichotomy equation method as an example to illustrate the necessity of using IT. Zhang and Peng (2009), by adopting the same example, said that IT is a must for teachers in programming and drawing images to visualize the mathematical nature so as to benefit teaching process as well as facilitating students' learning (Fig. 13.3).

- **"Multiple characterization" in mathematics was facilitated by IT**
  "Multiple characterization" refers to the use of a variety of methods to represent the same mathematical concept, with different representations focusing on different aspects of the concept being taught. Students need to be guided to combine several meaningful representations of information and to establish conceptual linkages that will deepen their understanding of the concept (Middle School Mathematics Courses Textbook and Information Technology Integration Research Subject Group, 2004).
  Mathematics software, such as Geometer's Sketch pad, Super Sketch pad, GeoGebra, and graphing calculators, is equipped to represent a variety of mathematical objects, including analytical style, images, graphics, and images. Liu and Liu (2006) believed that teachers can use IT to make mathematical object images, graphics, animation, and other forms of presentation. Zhang (2009) discussed a case study of sequences. The graphing calculator offered a general formula of sequence, sequence listing, sequence of images, and recursive sequences of four kinds of representation. Yang (2013) analyzed the

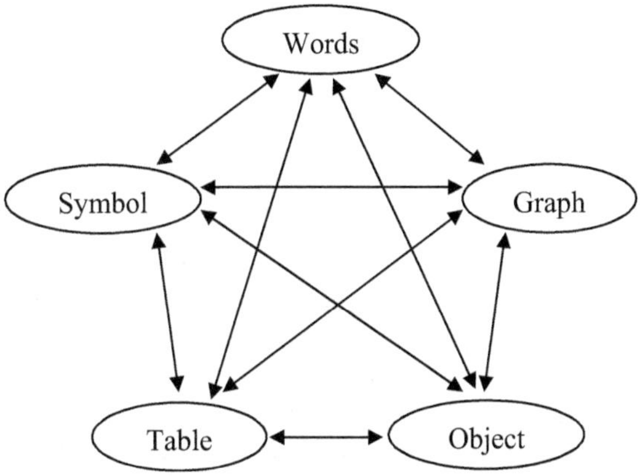

Fig. 13.4 Multiple characterizations of function

Five types representations of function $f(x) = -x^2 + 5x + 6$

Words: Function f(x)is minus x squared, then 6 is added5 times x. In the downward opening parabola, the symmetrical line is 2.5and the vertex is(2.5,12.25).

Symbol: $f(x) = -x^2 + 5x + 6$ Graph:

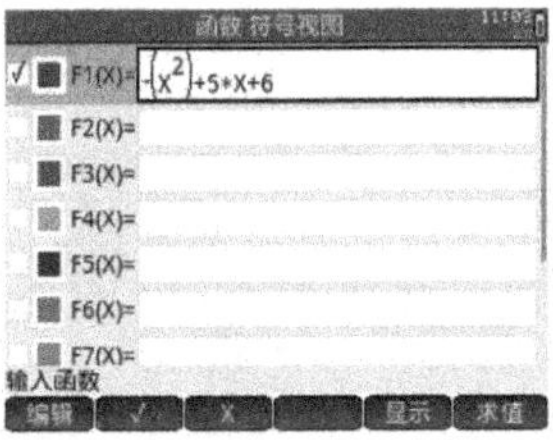

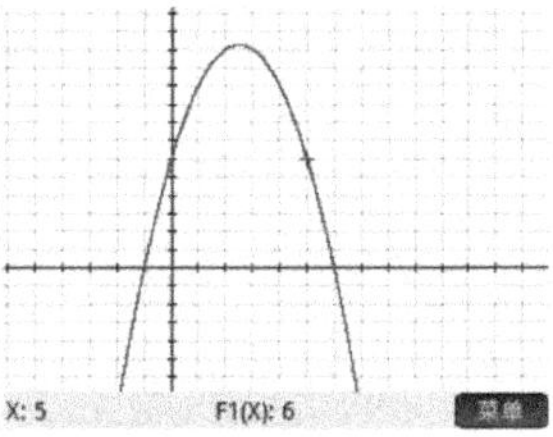

Table: Object:

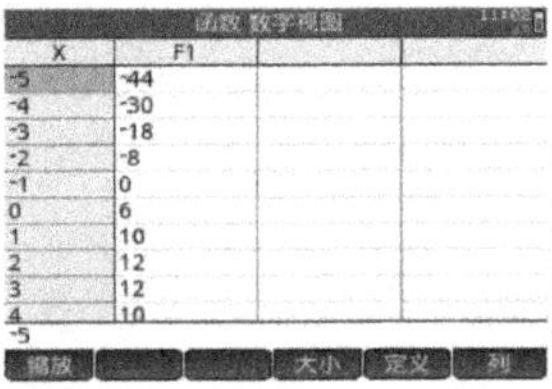

Fig. 13.5 Five types of representations of function

multiple representations of functions: the representation of a function expression, images, tables, text, and other physical objects (see Figs. 13.4 and 13.5). The same mathematical concept can be inducted from different angles, which might be able to help students to achieve a wide understanding of the learning object.

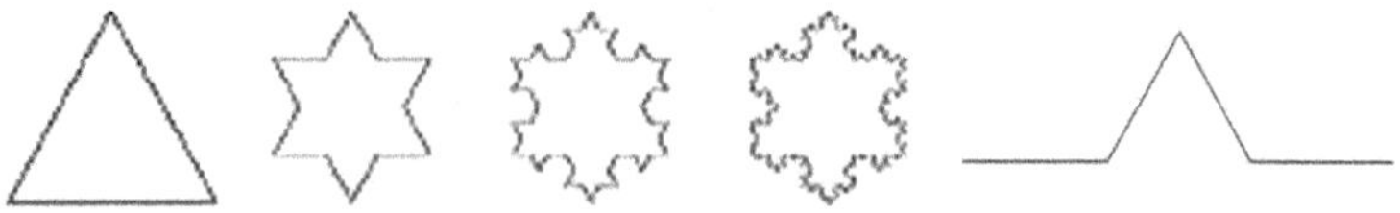

Fig. 13.6 Snowflake curve inquiry

- **Information technology helps students to explore mathematics**
  One of the highlights of new curriculum reform in mathematics was its student-centered philosophy, encouraging students to explore independently, have hands-on practice, and engage in cooperation and exchange in the process of developing an interest in learning mathematics, habits of independent and creative thinking, and active exploration. IT offers a platform for students to explore activities. It can point out the mathematical assumptions and reasoning, and then mathematical software can be used for verification, and the graphing calculator or software can be used to do experiments to find and summarize mathematical laws and mathematical phenomena (Liao, 2006).

An example is the use of GeoGebra, graphing calculators, and other integrated programming capabilities. Students can, for example, explore the snowflake curve after studying geometric sequences in high school. They can use the software to create snowflakes of various sizes (Fig. 13.6) and calculate the perimeters and areas. This allows students to experience through geometric sequence knowledge, to solve practical problems, to develop the algorithm, to gain a preliminary understanding of fractal geometry, and to feel the beauty of mathematics.

HP and TI-Nspire graphing calculators are equipped with data streaming technology; the probe has a strong function that allows data collection through measuring such things as light intensity, PH value, height, power, sound, and a true reflection of change in data availability, as well as allowing synchronous transfer to the graphing calculator, and the creation and display of graphs and tables. Based on these data, students can build mathematical models to solve practical problems. For example, in the study of the impact ionization equilibrium temperature, they can use the graphing calculator data stream to collect data, analyze them more efficiently, and manipulate variables easily, all of which stimulate students to explore scientific truth (Fig. 13.7).

13.2.2.2 Mathematics Teaching Design Based on IT

Yang (2013) and Yang (2013) designed the *Standards for Mathematics Learning and Teaching Based on IT*, as shown in Table 13.1. This was a guide for teachers not only to integrate IT, but also to give them a standard of reference to do self-evaluation and peer evaluation of their teaching.

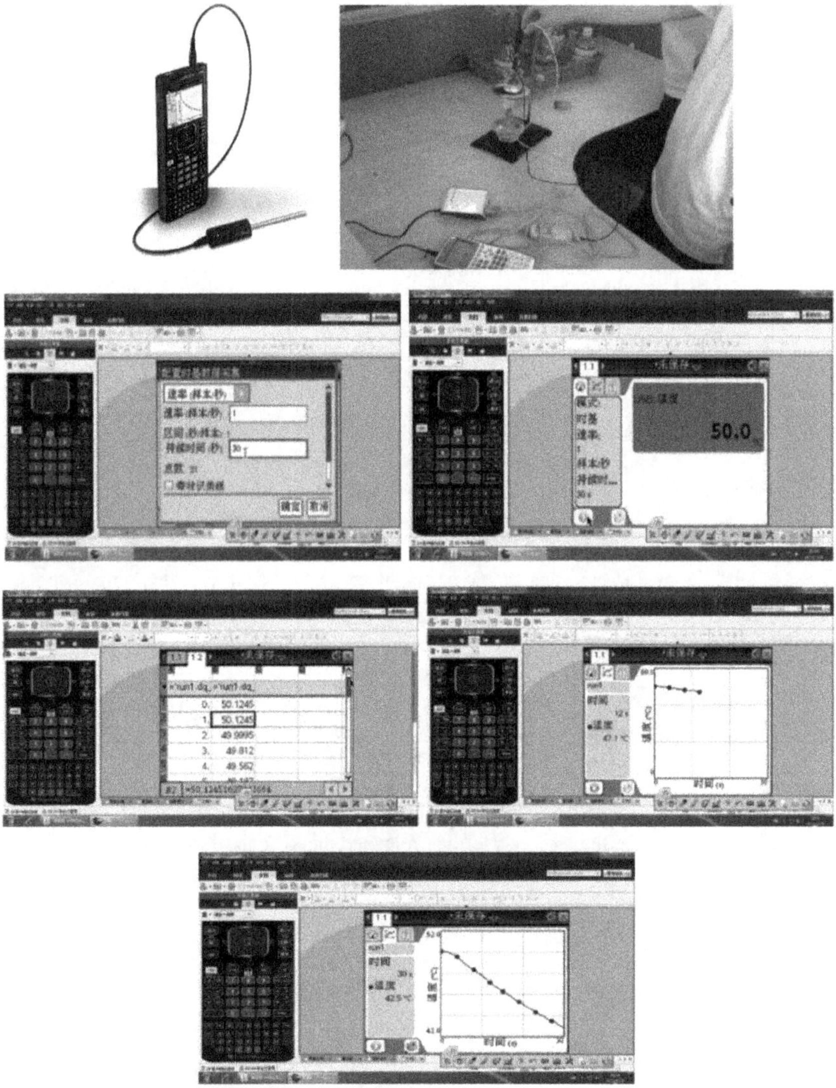

Fig. 13.7 Graphing calculator data streaming capabilities

13.2.2.3 Information Technology and the Mathematics Curriculum Teaching Mode

Chen, Lu, Xu, and Zhou (2005) focused on teaching for mathematical exploration. They argued that the growing diversity in the field of mathematical knowledge is dependent on choosing appropriate teaching strategies related to the teaching content and teaching objectives as well as students' ages, life experiences, and

Table 13.1 Standards of mathematics learning and teaching based on IT (Yang, 2013)

Elements of teaching design	Standards for student learning of mathematics based on IT
Materials analysis	1. Explain the meaning of current knowledge, explicit knowledge contained in the development process, and the cultural value of mathematics and knowledge
	2. Current knowledge about the status and role of mathematics courses
	3. IT integrated with the current knowledge and understanding IT in the teaching content
Analysis of learning	4. Students' mastery of existing knowledge and skills base, ability to draw on prior knowledge, and mastery of mathematical drawing, computing, IT, etc.
	5. Students' attitudes to IT, such as learning psychology and coping with emotions
Teaching goal	6. Target setting reflecting 4 dimensions (knowledge, thinking skills, problem-solving, emotion, and attitude);considering the overall integration of 4 dimensions; exact use of target action verbs
	7. Ability to link mathematics to real life, represent accurate pictures, create situations and dynamic presentations, carry out mathematical experiment
	8. Target setting should be combined with the analysis of teaching materials and teaching which should be clear, concise, and avoid abstractions
Teaching problem	9. Key points clear, specific, and properly explained
	10. Use of IT helps to address problems
	11. Teachers should adjust educational difficulties in accordance with students' conditions (levels, capacity to use IT technology, and mental states); clinging to tradition is a taboo for teachers
Teaching strategies	12. Identifying teacher's and students' roles in different aspects of mathematics teaching and learning, e.g., imparting new knowledge can be the teacher's leading role, and thinking and exploring should be student-centered
	13. Designing teaching methods and strategies in accordance with students' characteristics and cognition
	14. Teachers providing information resources to students for effective filtration, classification, to eliminate confusion, and for more effective use of information resources
The teaching process	15. Correct understanding of the auxiliary role of IT, improvements to the quality of mathematics teaching assistants—helping to promote students' understanding of mathematics and helping students to engage in mathematical thinking
	16 Encouraging student participation in the learning process to promote enthusiasm, initiative, and creativity
	17. Creation of situations suitable for high school students according to their life experience, interests, and hobbies
	18. Appropriate adjustment of content, examples, and exercises, and further help students better understand mathematics

(continued)

Table 13.1 (continued)

Elements of teaching design	Standards for student learning of mathematics based on IT
	19. Setting tasks of appropriate difficulty, encouraging student participation, creative thinking, confidence, enthusiasm, independent thinking, cooperation, and catering for a variety of learning styles
	20. Selection of information technology appropriate for teaching content, e.g., Geometer's Sketch pad and GeoGebra for geometry (2D geometry, 3D geometry, or trigonometry), graphing calculators and Excel for algebraic and statistical content
	21. According to the teaching goals and problems set, taking into account the principles of information technologies used in the process, such as necessity, balance, practicality, diversity, etc.
	22. Recognition that the process of knowledge generation, synthesizing, and summarizing students' own work; and plenty time to guide them
	23. Design of connecting and smoothly transitional teaching procedures with specified intention
	24. Delivery of well-structured handwriting on the blackboard and taking the integration of IT technology into consideration so as to focus on key points
Teaching evaluation	25. Supporting best possible use of IT to get good teaching results
	26. Merging IT technology into mathematics teaching to promote students' participation so as to enhance their enthusiasm and creativity
	27. Ensuring that mathematics objectives are addressed accurately

cognitive experiences. Accordingly, they proposed four modes of exploratory learning: "guide-the inquiry," "attempt–inquiry," "learning–inquiry," and "collaborative-research."

Zhang (2007) also proposed several modes for integrating IT and mathematics teaching: "demonstration–taught" mode, "discovery–inquiry" mode, "self-examination–mastery" mode, "dialogue–exchange" mode, and "cooperation–research" mode.

Wang (2007) focused on IT environments and identified several ways in which these were created: teachers teaching mainly through the conventional teaching mode; the student group project; teacher-based computer-assisted instruction (CAI); and the use of networked environments to explore cooperation through mathematical experiments. Wang found that the use of networked environments had the best teaching effects, with student performance slightly better than in the other teaching modes. The CAI mode was found to be slightly but not significantly better than conventional teaching. However, factors restricting the effectiveness of the CAI mode included a lack of proper mathematics teaching software and inadequacy of teachers' ability to use IT in their teaching. Wang also reported that mathematics teaching generally lacked evidence of understanding of interactivity, content courseware design, and presentation (text, color, and playback speed).

13.2.3 Study of IT Environments in the Mathematics Curriculum

There are some challenges to integrating IT and the mathematics curriculum, particularly relating to traditional course content and course design. Some mathematics education researchers have conducted studies to address these challenges.

Middle School Mathematics Curriculum and IT Integration Research Group (2004) edited an experimental textbook of ideas for ordinary senior and middle schools in which it explored four aspects:

- new ideas relating to the mathematics curriculum reform.
- changes to the textbook presentation.
- development of the "mathematical experiment" section.
- additions to teaching methods to include experiments, trials, simulations, conjecture, testing, control, operation, reasoning, proof, hands-on activities, and high-level mathematical thinking, as well as complex operations, difficult mapping, data processing of difficult issues, and real practical problem.

For example, Table 13.2 shows the "image and nature of the exponential function" as a textbook example of moving from integrating to inductive, presenting teaching materials for students to experience cognitive processes.

Guo and Cao (2012) assessed primary schools in China, Japan, Korea, Singapore, and 14 other countries, focusing on the standard textbooks for junior high school mathematics and analyzing the texts based on the national curriculum standards with emphasis on IT and its usage. The study found that several European countries (the Netherlands, UK, France, and Germany) placed more importance on IT usage, significantly more than the Asian countries. In France, Germany, and the UK, the IT usage varied; Asian countries mainly mentioned calculation tools, while Russia, South Africa, Finland, Japan, and the USA had much higher standards of IT technology. IT was used mainly in elementary or junior high schools, and the most common applications were in algebra, graphics, and geometric drawings.

Relatively, little research has focused on the mathematics curriculum standards. In the area of teaching material design, specifically the integration of IT and mathematics curricula, more needs to be done on course objectives, course content, and curriculum implementation.

Table 13.2 "Image and property of the exponential function": Comparison of teaching materials

Old version book	Integrating new book
1. x, y the table values () 2. Draw the image by using description method 3. List in $0 < a < 1$ and $a > 1$ both cases' images and their properties	1. Complete the form using information technology () 2. Draw image using information technology 3. Use of IT in the above image value of dynamic observation on function scope, monotonicity, etc. 4. Students choose several different exponential functions with different bases, draw and observe ranges, increasing and decreasing properties, etc. 5. Listed in $0 < a < 1$ and $a > 1$ both cases image and property

13.2.4 A Study of the Effects of IT on Mathematics Learning Environments

The ultimate goal of integrating IT and mathematics was to create better learning environments for students and hence to enhance their development in mathematics.

Zhang Jingzhong conducted many years of research and development in a project titled "Z + Z Intelligent Education Platform—Super Sketchpad." Zhang (2009) thought that the use of IT in teaching would help students to solve problems, develop their awareness of innovation, arouse their interest, and increase their in-depth understanding of mathematics by improving teaching efficiency and relating mathematics to real life and nature. The Middle School Mathematics Curriculum Materials and IT Integration Research Group (2002) suggested that more attention should be paid to students' understanding, with greater emphasis on laws of calculation. Zhang emphasized that the focus on operational skills could be reduced to save time for developing mathematical processes, understanding, and thinking. They proposed that IT could be presented as an illustrated, colorful, human–computer interaction, giving timely feedback and allowing students to simulate real situations to build their own internal and external problems and mathematical models and inquiry, which are more difficult to achieve in traditional mathematics teaching.

Wang, Yang, and Wang (2006) and Kui and Sun (2008) summarized the main mathematical inquiry learning styles in the IT environment, including the study of mathematical concepts, inquiry learning, and mathematical problem-solving. They designed teaching plans for the topics of "perpendicular" and "Napoleon's Theorem" as examples of mathematical inquiry. He, Deng, and He (2007) based an example on the "sides of a triangle relationship."

Guo and Cao (2015) selected 55 middle school mathematics teachers and tracked nearly 2000 students for a period of two years in three districts to explore how IT would impact on their academic achievements. The study took students' (2012) exam results as the dependent variable, while taking teachers' levels in using technologies like M-TPACK, IT, and students' academic achievements in 2011 plus tutoring times after class as independent variables to build a hierarchical linear model. The results showed that, in the IT environment, mathematics teachers' knowledge (TPACK) had a significant effect on students' academic performance; this had more influence in geometry than in algebra and geometry.

13.2.5 Factors Restricting Integration of IT with Mathematics Teaching

A variety of factors has been found to affect teachers' use of multimedia technology (Wang, 2009; Wen and Zhou, 2007; Xiong, 2014). These can be divided into internal and external factors (see Table 13.3).

Table 13.3 IT and Mathematics Curriculum Integration factors

Dimension	Description	Dimension	Description
Internal factors	Teachers' beliefs (students, self-efficacy, mathematics, teaching effectiveness, and external triggers) insufficient knowledge about information technology, inadequate teacher ability to integrate information technology	External factors	Inadequate funding for equipment, the lack of collaboration between colleagues and sharing of relevant teaching resources, lack of management, and insufficient rewards

Internal factors refer to IT and teacher beliefs, knowledge, and other related factors. Wang (2009) found that 46.7% of teachers cited difficulties in designing courses, and 78.2% said they did not have the time or energy to do so. Wen and Zhou (2007) believed that "teachers' knowledge, experience and motivation influenced their usage of IT". Xiong (2014) studied the influence of "teachers' beliefs on their integration of IT into the mathematics curriculum. They found that the teachers' mental structures, self-efficacy, teaching effectiveness, and external incentives had a significant influence on students. A two-year study conducted by the Middle School Mathematics Curriculum and IT Integration Research Group (2004) found that factors that affected the level of IT integration for teachers included their levels of IT usage, their capability to design and implement IT-integrated activities, and their ability to evaluate these activities.

External factors include hardware, software, funding, and school management. Wang (2009) argued that a lack of equipment, insufficient funds, insufficient attention to leadership, lack of cooperation, and exchange of environmental factors all influenced teachers' use of IT. Wen and Zhou (2007) suggested that relevant courses, teaching resources, school administrative support, and collaborative collegial environments, combined with the support of hardware and software, could provide an important platform for IT integration. Some studies (such as Wang, 2009; Wen & Zhou, 2007) showed that information technology exchanges and cooperation among teachers, school incentives, the accumulation of teaching resources, and other factors were more important than hardware configuration.

13.3 IT and Mathematics Curriculum Integration: Reflections and Future Directions

A review of the fifteen years of research on the new curriculum reform showed that experts have deepened their understanding of IT and mathematics curriculum integration. Even more gratifying is that the proportion of IT usage has risen. The rapid developments of IT have advanced its integration with the mathematics curriculum, accelerating the pace of the mathematics curriculum reform.

Despite an excellent start, there are still some problems that need to be addressed so that the use of IT in mathematics teaching will continue to develop.

- **Empirical research needs to be strengthened further**
  The research conducted over the 15 years of integrating IT and the mathematics curriculum under the new curriculum reform has indicated some promising results, such as those of Zhang Jingzhong, Wang Changpei, and Yiming Cao (Cao & Wang, 2009). However, these studies have been mostly quantitative comparative studies, and there have not yet been many case studies; the empirical research needs to be strengthened further.
  The questions still remain as to whether, in the use of IT, the quality of teaching has a positive or a negative effect on students' knowledge, understanding, and interest in learning mathematics. Therefore, further quantitative, theory-based experiments are necessary. Another need is to explore whether student's personality characteristics and cognitive levels require different approaches to use IT in mathematics.
- **Integration of IT and mathematics teaching should focus on essence of mathematics**
  The definition of the teaching mode is that it is a comprehensive system, including the teaching purpose in a certain period of time under the guidance of specific ideology and theory realized by adopting teaching links, methods, and process. Generally speaking, it is a practice of teaching theory as well as a summary of teaching experience. There are various technologies and methods featured in mathematics, but not all of them can meet the requirements of mathematics teaching since they have failed to consider the essence of mathematics, teachers' levels, students' conditions. etc. Hence, teachers should narrow down the range of model options. Pay more attention to the mathematics content and students' conditions when choose the teaching mode.
- **The effectiveness of integration or evaluation criteria needs further clarity**
  In practice, many teachers assume that integrating information technology and mathematics is merely the application of multimedia in their teaching. Hence, there has been a tendency for many teachers to use it for decoration but not for effective teaching purposes. Although some exploratory research has been done (Yang, 2013), in practice, there is also a need to conduct further in-depth investigation. Throughout the literature on the integration of IT and the mathematics curriculum, references to the effectiveness of the integration can be found everywhere. However, what is "effective"? In future studies, clear evaluation criteria need to be further defined.
- **Conduct a pilot reform of mathematical evaluation methods**
  Internationally, more and more countries are permitting the use of graphing calculators in important exams, while in China, only a few students are permitted to use them. If students only have access to experimental handheld technology in mathematics and elective courses, then the sense of affinity for technology will not be formed, not to mention their ability to learn and explore technology. At present, China has established school mathematics laboratories

in a number of provinces; some of our high school mathematics classes use graphing calculators as a teaching norm. In the next round of the high school mathematics curriculum reform process, it is recommended to reform the evaluation methods in a number of pilot areas first, by allowing students to use graphing calculators in exams.

- **Improve the hardware and software to improve teacher quality**
  While Guizhou, Guangzhou, Shandong, and other places surveyed have indicated that the proportions of teachers using IT are improving constantly; from a national perspective, the situation is still not satisfactory. First, the uneven development of education has led to a serious shortage of educational resources in many economically backward areas, hence the development of IT environments cannot be guaranteed. Second, even though relatively better educational resources are allotted to developed regions, if IT is only used in open or elite class competitions, the true meaning of integration will be lost. A survey of technology and mathematics teaching conducted by Shang and Huang (2005) found that, when asked about school leadership or related support with technology integration and teaching conditions, 90% of the teachers said that leadership support is important, but their perceptions of the actual support were not so optimistic.

It is important, as soon as possible, to narrow the education gap between regions, and education information construction is a priority. For ordinary teachers, updating their education concepts, improving their abilities to apply IT, and enriching and improving their information literacy are also very necessary.

At the time of issue of the *19th National Plan for Medium and Long-Term Education Reform and Development (2010–2020)* (Chap. 19), education was integrated into an overall national informational development strategy, which means the status of education had reached an unprecedented height. In the ever-changing IT revolution, China's educators must not be too conservative to embrace new technology, to emphasize the use of modern IT in the teaching of mathematics, to try to design and develop teaching materials, and to strengthen teacher training in IT. The new high school mathematics curriculum reform gives a high degree of attention to new technologies (Ministry of Education of People's Republic of China, 2001b; Ministry of Education of People's Republic of China, 2003) and also clearly states the requirements of new technologies in the curriculum standards. It is important to continue to develop the use of IT in mathematics, teaching, learning, and evaluation to improve the quality of mathematics education.

References

Cao, Y., & Wang, Z. (2009). MCL环境支持下的中学数学教学研究 [High school mathematics teaching in MCL environment]. *Journal of Mathematics Education, 18*(2), 56–58.

Chen, M., Lu, L., Xu, M., & Zhou, W. (2005). 信息技术与数学教学整合构建"探究"教学模式的研究 [Integration of information technology and mathematics teaching: Constructing an "inquiry" teaching mode]. *China Educational Technology, 216*(1), 52–55.

Guo, K., & Cao, Y. (2012). 数学课程中信息技术运用的国际比较研究——基于中国等十四国小学初中数学课程标准的研究 [International comparative study of information technology utilization in the mathematics curriculum: Based on 14 Chinese elementary and junior high school mathematics curriculum standards]. *China Educational Technology, 7*(108–113), 142.

Guo, K., & Cao, Y. (2015). 信息技术环境下数学教师教学知识调查研究与影响因素分析 [Analysis of teaching knowledge and factors influencing the mathematics information technology environment]. *Education Sciences Research, 3,* 41–48.

Guo, K., Cao, Y., & Wang, L. (2015). 教师信息技术使用对学生数学学业成绩的影响——基于三个学区初中教师的跟踪研究 [The effects of teachers' use of information technology on student academic achievement in mathematics—follow-up study based on three school district middle school teachers]. *Educational Research, 1,* 128–135.

He, K. (2005). 信息技术与课程深层次整合的理论与方法 [Deep theory and methods of information technology and curriculum integration]. *E-educational Research, 1,* 7–15.

He, Q., & Fan, D. (2007). 信息技术与高中数学课程整合的实践研究 [Pilot study of information technology and mathematics curriculum integration]. *China Educational Technology, 9,* 73–76.

He, P., Deng, P., & He, P. (2007). 信息技术环境下数学探究式学习的个案研究 [Cases study of mathematical inquiry learning in the IT environment]. *The Chinese Journal of ICT in Education, 10X,* 41–43.

Hu, F., Wang, W., Wang, S., & Zhang, S. (2011). 高中数学教师使用信息技术的现状研究 [Status of high school mathematics teachers using information technology]. *E-educational Research, 2,* 36–43.

Kui, K., & Sun, K. (2008). 如何利用信息技术开展数学探究性学习 [How to use information technology to carry out mathematics exploration]. *The Chinese Journal of ICT in Education, 10,* 35–37.

Liao, Y. (2006). 高中新数学课程使用信息技术的调查与思考[Investigation and Thoughts on the integrating Information Technology into the new mathematics curriculum of senior high school]. *Journal of, Mathematics Education, 15*(03), 83–86.

Liu, L. (2003). 贵阳地区信息技术在中学数学教学中应用的调查与思考 [Investigation and thought of the application of information technology in middle school mathematics teaching in guiyang]. *Journal of Mathematics Education, 2003*(03), 46–50.

Liu, X., & Liu, Z. (2006). 论信息技术与中学数学课程整的意义和存在的问题 [The significance and Problem of the integration of information technology and mathematics in secondary schools]. *Curriculum, Teaching Material and Method, 26*(2), 64–68.

Middle School Mathematics Curriculum Materials and Information Technology Integration Research Group. (2002). 中学数学课程教材与信息技术整合的思考 [Thoughts on high school mathematics curriculum materials and information technology integration]. *Curriculum, Teaching Material and Method, 10,* 51–55.

Middle School Mathematics Curriculum Materials and Information Technology Integration Research Group. (2004). 高中数学课程教材与信息技术整合研究与实验 [Research and experiment on high school mathematics curriculum materials and information technology integration]. *Curriculum, Teaching Material and Method, 24*(3), 47–52.

Ministry of Education of People's Republic of China (2001a). Basic Education Curriculum Reform Outline (Trial). Retrieved from http://www.gov.cn/gongbao/content/2002/content_61386.htm.

Ministry of Education of People's Republic of China. (2012a). 教育信息化十年发展规划*(2011–2020*年)[*Education informatization development plan in next decade (2011–2020)*]. Retrieved from http://www.edu.cn/zong_he_870/20120330/t20120330_760603.shtml.

Ministry of Education of People's Republic of China. (2015). *National educational development situation in 2013*. Retrieved from: http://www.gov.cn/guoqing/2015-02/09/content_2816730.htm.

Ministry of Education of People's Republic of China. (2001b). 全日制义务教育数学课程标准*(试验稿) [Mathematics curriculum standards for full-time compulsory education (trial)]*. Beijing: Beijing Normal University Press.

Ministry of Education of People's Republic of China. (2003). *Mathematics Curriculum Standards for High School (trial)*. Beijing: People's Education Press.

Ministry of Education of People's Republic of China. (2011). *China education profiles: National educational development situation in 2010*. Retrieved from: http://www.gov.cn/test/2011-10/31/content_1982280.htm.

Ministry of Education of People's Republic of China. (2012b). 国家教育事业发展第十二个五年规划 *[National education development twelfth five-year plan]*. Retrieved from http://www.moe.gov.cn/publicfiles/business/htmlfiles/moe/moe_630/201207/xxgk_139702.html.

Shang, X., & Huang, T. (2005). 关于技术与数学教学整合现状的调查与思考[Investigation and thoughts about the current condition of integrating information technology into mathematics teaching and learning]. *Bulletin des Sciences Mathematics*, 05, 14–16.

The State Council. (2010). 国家中长期教育改革和发展规划纲要*(2010–2020*年*)* *[National medium long-term education reform and development plan (2010–2020)]*. Retrieve from: http://www.moe.edu.cn/publicfiles/business/htmlfiles/moe/moe_838/201008/93704.html.

Wang, X. (2004). 信息技术与数学学科教学整合的尝试与思考 [Conformity of information technology and mathematics teaching]. *Journal of Mathematics Education, 13*(2), 97–98.

Wang, D. (2007). 信息技术环境下数学教学模式的实验研究 [An experimental study of mathematics teaching in an information technology environment]. *The Chinese Journal of ICT in Education, 8,* 69–70.

Wang, A. (2009). 现代信息技术在数学教育中的应用与现状调查研究 [Research about modern information technology in mathematics education]. *Journal of Mathematics Education, 18*(2), 69–71.

Wang, G., Zhao, X., & Fu, D. (2006a). 信息技术环境下的数学探究学习 [Mathematics inquiry learning in an IT environment]. *China Educational Technology, 231*(4), 48–51.

Wang, H., Yang, W., & Wang, J. (2006b). 信息技术与数学课程整合的探讨 [Approach to the integration of information technology and mathematics curriculum]. *Journal of Shanxi Normal University (Natural Science Edition), 20*(1), 36–39.

Wang, G., Zhang, N., & Kang, Y. (2015). 数学教师培训视域下的教育软件培训内容研究 [Educational software training content: perspective of mathematics teachers]. *Theory and Practice of Education, 35,* 35–37.

Wen, Y., & Zhou, Y. (2007). 影响教师将信息技术整合于数学教学的因素分析 [Analysis of factors affecting teachers' integration of ICT in mathematics instruction]. *Journal of Mathematics Education, 16*(3), 44–48.

Xiong, B. (2014). 教师信念对信息技术与数学课程整合的影响研究 [Influence of teachers' beliefs on information technology and mathematics curriculum integration]. *E-Educational Research, 5,* 87–90.

Yang, R. (2013). 基于信息技术和学生学习的数学课程教学设计案例研究 [Research based on information technology and student learning of mathematics: A curriculum design case of Tianjin] (Dissertation). Tianjin Normal University.

Zhang, R. (2007). 信息技术与数学课程整合的教学模式探讨 [Teaching mode in the context of information technology and mathematics curriculum integration]. *E-Educational Research, 12,* 69–72.

Zhang, N. (2009). HP图形计算器在高中数列教学中的应用示例 [Application of HP graphing calculators in high school sequence teaching]. *Journal of Mathematics Education, 18*(2), 63–65.

Zhang, J., & Peng, X. (2009). 深入数学学科的信息技术 [Discipline-based Information Technology for Mathematics]. *Journal of Mathematics Education, 18*(5), 1–7.

Chapter 14
Chinese Mathematics Teaching Methods Reform in the 21st Century

L. Wu, J. Zhou and G. Wang

Abstract Curriculum reform is the key to basic education reform, just as teaching reform is fundamental to curriculum reform. Since the beginning of the twenty-first century, based on the "Two Basics", the emerging mathematics teaching approach and the pedagogy of variation, a great many outstanding achievements in the field of mathematics teaching reform have shown in China.

14.1 Background

Curriculum reform is the key to basic education reform, just as teaching reform is fundamental to curriculum reform. Since the beginning of the twenty-first century, based on the "Two Basics", the mathematics teaching approach and the pedagogy of variation, a great many outstanding achievements in the field of mathematics teaching reform have shown in China. These include the GX Experiment conducted by Chongmu Chen and Naiqing Song from Southwest University, Mathematics Teaching Efficiency carried out by Guangming Wang from Tianjin Normal University, Mathematical Situations and Problem-Posing proposed by Chanhan Lv and Bingyi Wang from Guizhou province, and Implementing Mathematics Methodology and Enhancing Student Quality (also called the MM experiment) conducted by Liquan Xu from Jiangsu and Shangdong Dulang Kou. In addition to these experimental studies, Chinese students' excellent performances in the mathematics Olympics and TIMSS, Shanghai students' first ranking in PISA have all attracted worldwide interest in mathematics teaching in China (Kan, 2015). As mathematics educators from other countries become interested to know more about Chinese mathematics education, it is important to document its development.

L. Wu (✉) · G. Wang
College of Teacher Education, Tianjin Normal University, Tianjin, China
e-mail: 273664122@qq.com

J. Zhou
Mathematics Department, East China Normal University, Shanghai, China

Y. Cao and F.K.S. Leung (eds.), *The 21st Century Mathematics Education in China*, New Frontiers of Educational Research,
https://doi.org/10.1007/978-3-662-55781-5_14

14.2 Inheritance and Innovation: The Mathematics Teaching Reform Experiment

14.2.1 Mathematics Teaching Reform Based on Traditions

Traditionally, the arrangement of textbooks follows the sequence of definition-theorem-proof-examples; in line with this, the teaching model follows the sequence of organization-review-teaching new lessons-conclusion-assignment, which is typical Kairov's "Five-Step Teaching". This teaching model has been adopted by many Chinese primary and middle school teachers and is recommended highly by educators in China. The advantage of "Five-Step Teaching" lies in that it is easy to carry out in real classrooms, can be combined with mathematical deduction and is welcomed by teachers. However, some problems will arise if focus too much on the teachers' teaching and ignore students' learning.

Seeing the potential problems, many schools and teachers are trying to make some changes. Some typical teaching methods introduced since 1990 are the Self-Study Guide Pedagogy proposed by Zhongheng Lu, Qingpu, the Teaching Experiment by Lingyuan Gu from Shanghai, the GX Experiment conducted by Chongmu Chen and Naiqing Song from Southwest University, and "Self-Study-Discussion-Guidance" put forward by Yunan Li.

14.2.1.1 Self-study Guide Pedagogy

The Self-Study Guide Pedagogy was developed by Zhonghe Lu, a professor from the Chinese Academy of Sciences Psychology Department, in 1958. The experiment was carried out in more than 30 provinces around China. The teaching model consists of five parts: enlighten, read, practise, understand and conclude. Teachers help all students with the "enlighten" and "conclude" phases at the beginning and end of the class, each part lasts for 10–15 min. The students spend the remaining 30–35 min in the middle of the class reading, practising and understanding the mathematics content by themselves. The underpinning philosophy of this model is to develop students' self-study abilities, enhance both teachers' and students' enthusiasm for study, build students' learning confidence and develop good self-study habits.

14.2.1.2 Qingpu Teaching Experiment

In 1977, after a three-year survey, one year of selection, three years of scientific experiment and seven years of application, Qingpu Teaching Experiment was introduced as a new teaching pedagogy. Two books were published, *Learning to Teach Based on the Qingpu Experiment* (Mathematics Reform Group from Qingpu Shanghai, 1991) and *Teaching Experiment—the Method and Themes of the Qingpu Teaching Experiment* (Gu, 1994).

There are four principles underpinning this model: motivation, sequence, activities and feedback. To be precise, the motivation principle is the reason for students to study, while the sequence principle is for a structure for the teaching content and process. The activities principle refers to the teaching pedagogy and the feedback principle is concerned with performance. This model has six main steps: arouse students' interest by creating problem situations; try to explore knowledge; form a conclusion; practise with variation; organize thoughts and regulate feedback.

14.2.1.3 Self-study-Discussion-Guidance Pedagogy

There are three tenets to this pedagogy. Self-study is the base, while discussion is the bridge and guidance is essential. Combined, these three contribute to building students' learning ability.

14.2.1.4 GX Experiment

The GX experiment was concerned with enhancing the efficiency of middle school mathematics classes. It was the work of Professor Chongmu Chen and Professor Naiqing Song from Southwest University in 1992. The main purpose of this experiment is to relieve the pressures of study while increasing the learning quality at the same time. The big idea can be underpinned by the following words "moving forward positively, circulation rise; stressing the nature of mathematics rather than the form, coming straight to the point and concentrating on proper issues, studying prior to teaching, and co-performance of the teacher and students" (Chen, Zeng, & Song, 1994). The experiment was carried out in hundreds of schools across 14 provinces and was welcomed by students and teachers (Pang, 2007).

The sequence of the GX model is to focus on problems, then followed by practice in class, give feedback in relation to the practice, and reflect and carry out an assignment. It is based primarily on lecture-type teaching, but also combined with guided-discovery and activity teaching. It is suggested that teachers relate mathematics content to students' real lives. Xiaoda Zhang commented that "If we sum up GX into one sentence, it is that it encourages using the least time to learn more useful mathematics" (Zhang, 1995). The essential part of the model is teaching meaningfully while learning productively.

14.2.1.5 MM Experiment

MM is short for the Mathematical Methodology Education Pattern, which was proposed by Liquan Xu from the Jiangsu Education Research Institute. This experiment started in 1989, requiring teachers to carry out technical and cultural education in their classes and teaching students how to prove, guess and do research. The teacher's role was to help the students to improve their scientific and social literacy. The model was based on eight themes, returning to nature, aesthetics, discovery method, moral quality, history, deduction, plausible reasoning and general problem-solving (Xu & Yang, 2002).

14.2.2 Mathematics Teaching Reform Based on Learning Cases

The Learning Case approach emerged in China in 1997. Now, it has come to be a leading teaching reform and is well accepted by almost all primary and middle schools in China. There have been a great number of research studies and publications relating to this model. The Jinhua No. 1 Middle School in Zhengjiang was the first school to adopt the learning case model, and Donglu Middle School in Zhengjiang reported remarkable results from it. The essential theme of this model is for students to self-study first, then be taught by their teachers. In 1999, the Hujian Middle School proposed a five-step process: preparation-self-study (students)–discussion (students)–inspiration (teacher)–extension (teacher) (Wan, 2015).

In 1997, the Jinhua No. 1 Middle School followed three stages in the process of developing the model: first the teachers in the school worked together to develop after-class activities; second, they developed activities to be completed before the subsequent classes and third, a uniform teaching plan was developed. In 1998, the Dulang Kou School modified this model by proposing three parts, preview, demonstration and feedback, and six steps, preview and communication, clarification of goals, group cooperation, demonstration, reinforcement and assessment (He & Xu, 2009). This model enables students to enjoy a more open study environment. The concept of "students teaching students" also motivates them. However, there are some disadvantages of this philosophy of "teach less but learn more". Some of the "lecture and study" activities are not particularly meaningful, and the idea of students teaching students can result in that they learn less knowledge and may not use their study time efficiently.

Influenced by the Dulang Kou model, many areas started to develop their own variations in efforts to achieve high results. Due to the range of experiments that ensued, there are now many variations of the learning case model. For example, Jinfeng No. 3 Primary School in Ningxia used a highly efficient model which advocated "studying first, teaching later", and "How to teach depends on students' learning results". This model resulted in the development of seven teaching models and more than 20 teaching themes. The Shiji Middle School in Zibo set up a new teaching model, with five steps. The first step is to guide students in self-study according to the textbook and other related materials. The following four steps are demonstration, practice, assessment and feedback (Hu, 2008).

The DJP teaching model was proposed in the Long Quanyi District in Chengdu, Sichuan province. This consists of three parts: study guiding, explanation and assessment (Wang & Wang, 2013). Study guiding is the fundamental base,

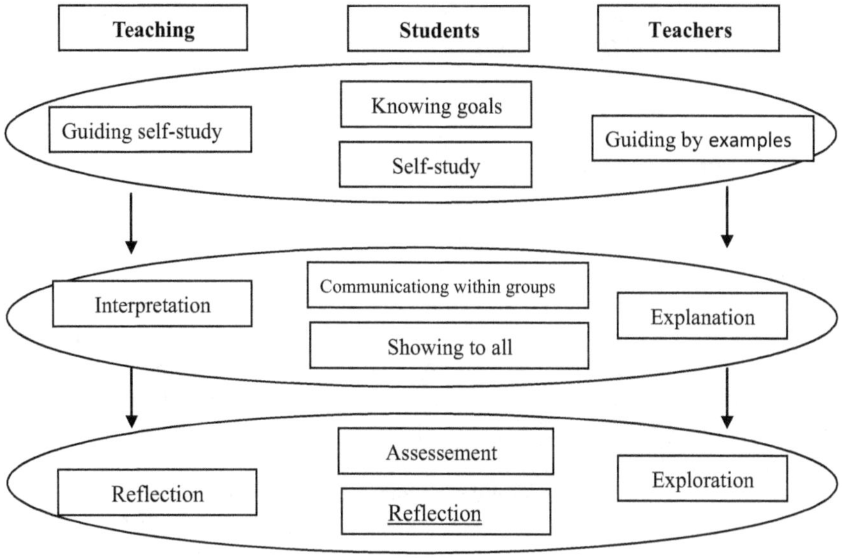

Fig. 14.1 Teaching process of DJP

focusing on "what to learn", while explanation is concerned with "how to learn", which involves both teachers and students, and assessment relates to the achievement of the teaching goals, "how is the learning". It is only real DJP teaching if the three parts come together. The teaching process is summarized in Fig. 14.1. This model was developed over a period of six years, with 3274 teachers in 23 schools.

14.2.3 The Situational Teaching Method

Linking mathematics to real life is one of the most important ideas of the curriculum reforms of the twenty-first century. In 2000, Chuanhan Lv and Bingyi Wang, from Guizhou Normal University, proposed the situational teaching method that aims to develop innovative thinking and practical abilities. The experiment was introduced into schools in 2001 and continued with some outstanding achievements for the following five years (Lv & Wang, 2006).

This model has four steps, set up mathematics situations, introduce mathematics problems, solve the problems and apply the knowledge.

In situational teaching, teachers help their students to develop strategies dealing with real-life problems and posing new questions. The strategies of setting mathematics situations include: (1) Game situation, (2) Practical situation, (3) Real-life situation, (4) Process situation (5) Suspense situation, (6) Competition situation, (7) Analogy and guessing situation, (8) Argument situation, (9) Constructive situation, (10) Dynamic situation (Wang, 2014). Six strategies of posing mathematics questions include: cause and effect, comparison, generalization, limit, change and reverse strategy (Zhu & Zeng, 2004).

14.2.4 Mathematics Teaching Efficiency Experiment

Even though many new mathematics teaching theories and models have been developed this century, they need to be efficient; if not, they can lead to discouragement and fatigue on the parts of both teachers and students (Liu, 2006).

A well-known contribution to mathematics teaching efficiency was the study conducted by Guangming Wang, from Tianjin Normal University. His book *Mathematics Teaching Efficiency* compared high-achieving and average high school students in terms of cognition structure and understanding the factors the students themselves believed could affect their learning efficiency (Wang, 2006). The research suggested that teaching efficiency depends on two factors. One is the time that students spend in full, active study. The other is the results of learning, such as scores, self-perceptions of efficiency, cognitive structure and mathematics learning competence (Wang, 2005). The book also pointed out the characteristics of high teaching efficiency: focusing on teaching thoughts, understanding problems and building cognition structures (Tu, Wang, & Yang, 2011). Their psychological model for enhancing students' learning efficiency had four dimensions: reasons, senses, cognition and learning competence. Xinmin Wang modified the model by categorizing content into quantitative and qualitative efficiency and classifying knowledge into explicit and tacit efficiency (Wang, 2006).

A national project titled the Teaching Efficiency Study of Fundamental Education, conducted by Guangming Wang, suggested that highly efficient teaching should focus on scientific, intellectual and aesthetic aspects. To be more concise, being scientific means teachers can establish reasonable goals for a lesson and know its level of difficulty. They should then try to help students with their understanding and cognitive structures. Teachers should be interested in building and knowing various learning methods. The intellectual aspect refers to choosing suitable content and teaching methods and also organizing the classroom effectively. Aesthetics refers to the aesthetics of teaching gestures and language, and even classroom organization (Wang, 2011). In short, the most efficient teaching methods are those that trigger highly efficient study (Wang & Wang, 2011).

14.2.5 Class Selection System

As modern technology develops, it is becoming more common for students to have more flexibility in selecting their classes. For example, in 2004, the Shenzhen Middle School in Guangdong Province offered a combination of compulsory and elective courses, in which students had individual class schedules. In September

2009, Qingdao No. 2 Middle School in Shangdong introduced this system, providing their students with 8 study areas, 14 subjects and 87 modules. Similar systems have also emerged in the Affiliated High School of Peking University, the Affiliated High School of Renmin University and the No. 11 Middle School in Beijing. Beijing No. 11 Middle School won a national teaching achievement in 2014 for its related research, and now has some 4000 course schedules for its 4000 students, which means that almost each student has his or her own course schedule.

The aim of class selection system is to stimulate students' enthusiasm to study and make the classroom a favourite place. Like Dulang Kou and Longquan Middle School, many other schools in China are using group work in classes to enable students to communicate and cooperate.

14.2.6 Teaching Reform Based on Technology

As technology has developed so quickly in the recent decades, many teaching methods have changed to keep up. Teachers are required to combine technology suitably with their teaching (Liu, 2015). The most popular methods of doing so are digital narratives, e-schoolbags, electronic whiteboards, flipped classrooms and micro classes.

14.3 Final Thoughts

14.3.1 Focus on Teaching Integrity

In all of the teaching reforms described here, it is important to keep in mind the importance of a complete teaching process. Even though students' learning is the core, teaching cannot be ignored, since no high-level studying happens without high-level teaching. The argument about whether learning or teaching comes first is meaningless because the two should be happening simultaneously (Cheng, 2015). Attention should be paid to the relationship between teaching and learning in teaching reforms. It is easy for mathematics learning or the role of teacher to weaken in these forms. How to achieve a balance between them is still an issue for teaching reform (Cong, 2008).

14.3.2 Student-centred Teaching and Learning

Professor Sato Manabu from Japan expressed the belief that the problems of mathematics teaching in Japan lie in students depending too much on their teachers teaching and a lack of initiative on the part of the students (Manabu, 2013). Similar problems also exist in China. However, since Yelan (1997) suggested, "Let the class be more energetic", some mathematics teachers have started to move away from lecture methods and towards more student-centred ones. Methods such as the Yangsi experience, Dulang Kou experiment and DJP model focus on students' learning and developing. The purpose of teaching is to motivate students to learn (Guo, 2008). Students' cooperating in their study and teachers' classroom organization to facilitate this are essential themes of the reform (Wang & Wang, 2015). Teaching reform should not be hampered by a fear of students performing poorly in the beginning as long as educators have clear reform maps in their minds (Tian, 2015).

14.3.3 Appropriate Attitudes to Teaching Models

Generally, each teaching method has its own advantages, so learning from different models is necessary; however, it is important to avoid applying them mechanically. The teaching model directly affects the study efficiency and achievement (Cao, 2007). Whether teaching is successful or not depends on the students' knowledge and motivation, and also the teacher's professional background, attitudes and personality. If students are to develop their own study styles it is inappropriate to choose just one teaching model and use it all the time. Even the same teaching model may be used differently by different teachers or schools. Highly efficient teaching should focus on scientific, intellectual and aesthetic aspects (Wang, 2011). Teachers' creativeness should be encouraged; as Zunshan said, "Teaching is an art. A good artist is not only familiar with all fundamental knowledge, but is also creative and shows his style constantly. Similarly, an excellent educator will not constrain himself into one teaching method, but constantly creates new personal ones" (Cao, 2002, p. 2).

14.3.4 Handling the Relationship Between Teachers and Students, Courses and Classroom Culture

There are six kinds of relationship between teachers and students, courses and classroom culture. Two of these are between teachers and courses, one between students and classroom culture, another between teachers and classroom culture, the next between students and students, and the last is between teachers and students (Hao, 2005). To some extent, teaching is a battle between teachers and students to grab discourse power. It is suggested to give more discourse power to students in order to make independent thinking spaces for them. The key core of education and study is to gain the ability to learn by oneself, to develop independent, creative thinkers (Zhi, 2007). In the ideal classes, students should try to discover problems, pose their own problems and analyse them, solve them, and apply the knowledge independently. Teachers should leave more room for students to find solutions, patterns, conclusions and themes and build their own thinking; the teacher's role is only to offer the necessary help (Shi & Lai, 2008).

14.3.5 Reforms Affected by Examinations

Since 1980, in order to satisfy the entrance examination, a model of "using two years to teach three years' content and leaving one year to review" has been very popular around China. Even now, examinations still mainly test students' knowledge and skills to solve problems, which limit students' development. When introducing a new reform into schools, teachers on the one hand apply it to meet the policy needs, while on the other hand, they still try to help students get higher scores (Shao & Zhou, 2006). As in China, almost any reforms are affected by the examination-oriented system, so it is important that any reform will not affect students' examination performances.

14.3.6 Avoiding Removing Mathematical Features

Some teaching reforms do not place enough emphasis on the nature of mathematics, and this affects teaching efficiency to some extent. In one way, it weakens the teaching of content, even though it does strengthen discussion and exploration (Guo, Peng, & Yang, 2007). On the other hand, it reinforces the theme that the teacher is in charge of the class (Yu, 2005). Mathematics teaching reformation needs to be carried forward, developed and innovated unceasingly. As Dianzhou Zhang said, "Mathematics is the result of thinking, so for mathematics learning, communicating and discussing is quite important. However, it always requires much more time for one to think. I doubt that the primary and middle classroom could offer enough" (Zhang, 2009, p. 158).

14.4 Conclusion

Philosopher Caracon once said, "It is unimaginable for a community totally become separated from its previous culture. Change without considering cultural issues only incurs tragedy" (Yu, 2003, p. 48). Learning from previous experience is of high necessity.

Throughout the twenty-first century, Chinese teaching reforms have required teachers to update their thoughts and change their teaching models from a single value orientation to multiple ones. It is highly recommended that teaching methods are able to keep students' desire for knowledge alive, make schools places to "learn" rather than "teach" and turn classrooms into places where teachers can really communicate with their students.

References

Cao, Y. M. (2002). 数学教学模式导论 [Mathematics Teaching Methods Guide]. Beijing: China Federation of Literary and Art Circles Publishing House.

Cao, Y. M. (2007). 中国中课堂教学模式及其发展研究 [Chinese Middle School Teaching model and its Development]. Beijing: Beijing Normal University Press.

Chen, C. M., Zeng, Z. J., & Song, N. Q. (1994). 减轻负担、提高质量——GX (提高课堂教学效益)实验简介 [Relieve burden and enhance quality ——GX, brief introduction]. *Journal of Mathematics Education*, 3(2): 1–4, 61.

Cheng, S. R. (2015). 回到教学的基本问题上去 [Going back to the basic problems of teaching]. *Curriculum, Teaching Material and Method, 35*(1), 21–28.

Cong, L. X. (2008). 讲授法的合理与合法 [The rationality and validity of the lecture method]. *Educational Research, 7*, 64–72.

Gu, L. Y. (1994). 教学实验论——青浦实验的方法学与教学原理 [Teaching Experiment-the Method and Themes of Qingpu Teaching Experiment]. Beijing: Education Science Press.

Guo, J. P., Peng, M. H., & Yang, L. Y. (2007). 正反例在概念教学中的研究和应用 [Using positive and negative examples in concept teaching]. *Journal of Educational Studies, 6*, 21–28.

Guo, S. L. (2008). 杯子边上的智慧 [Wise on the side of cup]. *People's Education, 3*(4), 19–21.

Hao, Z. J. (2005). 探究性教学的实质:一种复杂性思维视角 [Essence of inquiry teaching: A complicated thinking perspective]. *Educational Research, 11*, 66–70.

He, S. L., & Xu, W. B. (2009). 东庐中学"讲学稿"教学改革探索与启示 [Donglu middle school teaching reform]. *Journal of the Chinese Society of Education, 10*, 46–48.

Hu, J. H. (2008). 为学生的终生幸福奠基——山东省淄博市桓台县世纪中学办学侧记 [Shandong Shiji middle school teaching reform]. *People's Education, 2*, 54–59.

Kan, W. (2015). 专题:基础教育课程改革的走向 [Primary and middle school education reforms]. *Journal of Educational Studies, 11*(2), 62.

Liu, S. F. (2006). 厚积薄发 引领数学教学新潮——评《数学教学效率论》 [Mathematics teaching——comments on mathematics teaching efficiency]. *Maths Teaching in Middle School, 23*, 57–58.

Liu, B. Z. (2015, May 21). "互联网 + 教育"将让"个性化"教育成为现实 [*Technology make individual education possible*]. China Education Newspaper.

Lv, C. H., & Wang, B. Y. (2006). 中小学数学情境与提出问题教学研究 [Research on Situational Problems in Primary and Middle School]. Guiyang: Guizhou people's Publishing House.

Mathematics Reform Group from Qingpu Shanghai. (1991). 学会教学 [Leaning to Teach]. Beijing: People's Literature Publishing House.

Pang, K. (2007). GX实验的再研究——GX教学模式的建构 [Restudy of GX Experiment—the construction of GX instruction model] (Unpublished doctoral dissertation). Southwest University, Chong Qing, China.

Sato Manabu. (2013). 学校的挑战 [The Challenge for School: Build Learning Community]. Shanghai: East China Normal University Press.

Shao, C. Y., & Zhou, M. (2006). 试论内容标准、表现标准的特点与关系——基于评价与标准一致性的角度 [Discussion of the relationship between content standard and performance standard]. *Contemporary Educational Science, 10*, 37–40.

Shi, J. M., & Lai, P. G. (2008). 让课堂活起来的力量 [Make classes more energised]. *People's Education, 1,* 44–46.

Tian, H. S. (2015). 落实立德树人根本任务，全面深化课程教学改革 [Implementing the fundamental task of morality education; deepening the curriculum teaching reform]. *Curriculum, Teaching Material and Method, 35*(1), 3–8.

Tu, R. B., Yang Q., & Wang, G. M. (2011). 中国数学教学研究30年 [Research on Chinese Education in 30 Years]. Beijing: Science Press.

Wan, W. (2015). 三十年来教学模式研究的现状、问题与发展趋势 [Reviews on teaching methods in 30 years]. *Journal of the Chinese Society of Education, 1,* 60–67.

Wang, F. Y., & Wang, X. M. (2013). 让知识在对话交流中生成——DJP教学中知识生成的过程与理解分析 [Make teaching happen in discussion-case of DJP]. *Journal of Chinese Mathematics Education, 21,* 3–6.

Wang, G. M. (2005). 重视数学教学效率，提高数学教学质量——"数学教学效率论"课题简介 [Value mathematics teaching efficiency, advance mathematics teaching quality]. *Journal of Mathematics Education, 14*(3), 43–46.

Wang, G. M. (2006). 数学教学效率论 *(理论篇)* [Mathematics Teaching Effiency]. Tianjin: Xinlei Publishing House.

Wang, G. M. (2011). 高效数学教学行为的特征 [Features of effective mathematics teaching behavior]. *Journal of Mathematics Education, 20*(1), 35–38.

Wang, G. M., & Wang, Y. (2011). 高效与低效数学课堂导入的案例比较 [Case study between high efficiency class and lower one]. *Journal of Teaching and Management, 1,* 49–51.

Wang, J., & Wang, M. D. (2015). 高效课堂的建构及其策略 [On the construction of efficient classroom and its strategy]. *Educational Research, 3,* 112–118.

Wang, X. H. (2014). "递进式-四环节"小学数学课堂教学模式研究 [Primary school mathematics teaching model study]. *Teacher's Journal, 14,* 5–6.

Wang, X. M. (2006b). 关于数学教学效率及其效率意识的分析 [Analysis on mathematics teaching effect and its effective consciousness]. *Journal of Mathematics Education, 15*(3), 16–18.

Xu, X. Q., & Yang, S. M. (2002). 数学知识的两种形态与数学教学 [Two forms of mathematical knowledge and mathematics teaching]. *Journal of Mathematics Education, 11*(2), 71–73.

Ye, L. (1997). 让课堂焕发出生命活力——论中小学教学改革的深化 [Let the class be more energetic-on deeping the teaching reform of basic education]. *Educational Research, 9,* 3–8.

Yu, W. S. (2005). 新课程教学改革的成绩与反思 [Achievements and problem reflection of the new curriculum teaching reform]. *Curriculum, Teaching Material and Method, 5,* 3–9.

Yu, Y. S. (2003). 中国思想传统的现代诠释 [The New Interpretation of Chinese Traditional Education]. Nanjing: Jiangsu People's Publishing House.

Zhang, D. Z. (2009). 我亲历的数学教育 [My Own Mathematics Education Experience]. Nanjing: Jiangsu Education Publishing House.

Zhang, X. D. (1995). GX是可能的——写在《GX》第一轮实验完成之际 [GX is possible]. *Correspondence of the Teaching of Mathematics, 4*(2), 42.

Zhi, T. J. (2007). 为思维而教 [Teaching for thinking]. *Educational Research, 10,* 44–48.

Zhu, Y. L., & Zeng, X. P. (2004). 中小学数学创设情境与提出问题的策略 [Strategies of math creating situation and posing problem in the elementary schools junior and senior high schools]. *Journal of Mathematics Education, 13*(4), 93–94.

Part IV
Mathematical Learning

Chapter 15
A Case Study of Transformation of Students with Mathematics Learning Difficulties

Yufeng Guo and Guowen Yu

Abstract The term "learning difficulties" was first proposed by American scholar Kirk in the 1960s to describe students who have normal intelligence but lag in academic achievement. China began to study the issue of students with learning difficulties at the end of the 1980s. In 1996, the Research Group of the National Education Administration described students with learning difficulties in a document titled *Study of the Education of Students with Learning Difficulties in Junior Middle School.* This was the earliest reference to this group of students in the current literature.

The term "learning difficulties" was first proposed by American scholar Kirk in the 1960s to describe students who have normal intelligence but lag in academic achievement.

China began to study the issue of students with learning difficulties at the end of the 1980s. In 1996, the Research Group of the National Education Administration described students with learning difficulties in a document titled *Study of the Education of Students with Learning Difficulties in Junior Middle School.* This was the earliest reference to this group of students in the current literature.

According to a Japanese survey conducted by Matsuhara Dachiya in 2009, about 20% of students had learning difficulties. Chen Yingsan also identified about 20% of students with learning difficulties in Taiwan. China's Huang Jiafen and Xu Min administered questionnaires to schools in Shanghai and found 20.04% in primary schools, 20.50% in junior high schools, and 28.14% in high schools (Li, 2009).

In 1995, professor Du Yuxiang from Tai'an normal college, Shandong Province, together with other researchers investigated about 6000 junior high school students from nearly 20 middle schools and found students with learning difficulties in mathematics in the country they studied accounted for about 22% and in the township accounted for about 29% (Dai, 2001).

These statistics suggest that there is indeed a problem with learning difficulties in mathematics.

Y. Guo (✉) · G. Yu
Beijing Normal University, Beijing, China
e-mail: guoyufeng@bnu.edu.cn

Y. Cao and F.K.S. Leung (eds.), *The 21st Century Mathematics Education in China*, New Frontiers of Educational Research,
https://doi.org/10.1007/978-3-662-55781-5_15

15.1 Existing Research on Mathematics Learning Difficulties

15.1.1 Causes of Mathematics Learning Difficulties

Foreign studies have pointed out a number of reasons why students have mathematical learning difficulties. One US study found that most fourth to eighth grade students' mathematics achievement was at or below the basic level (Kiuhara & Witzel, 2014). About 26% of 15-year-old students in the USA showed low levels of mathematical reasoning in PISA2013.

Some overseas researchers have suggested that the root cause of students' learning difficulties is school education allowing them to have the experience of failure (Wang, 2007), in addition to problems with society, family, and students' personal reasons, findings which coincide with those of Chinese studies.

Soviet scholars have expressed the belief that students' mathematical learning difficulties are due mainly to their lack of education. Ki conducted a study in which educational factors accounted for 70% of learning difficulties, and Linsky suggested that learning difficulties are caused by the social environment, family, and school education, rather than by problems with the individual students.

Chinese scholars have also conducted in-depth explorations of the causes of difficulties in learning mathematics. Their findings can be summarized in two categories: One is the factors relating to students themselves, and the other is external factors.

The main factors relating to students are:

- A lack of self-confidence and initiative; even if they accept that their intelligence is equivalent to their classmates', they do not believe they have the ability to make progress, to catch up with other students;
- Resistance to some practices by teachers or parents that the students do not like;
- A lack of interest in learning mathematics, inability to stay focused in the classroom, and lack of motivation;
- Poor learning habits and self-learning ability which can lead to low motivation;
- A lack of basic mathematics learning methods and skills, or inability to understand the crux of a problem;
- A lack of basic moral judgment which can cause some students to be misled into inappropriate behaviors that affect their learning;
- Teacher–student relationship tension, or tension with peers, which means they will not seek peer assistance or consult the teacher when having a problem.

In Tan Feng's *Investigation and Research on Intervention Measures of High School Students' Mathematics Learning Difficulties* (2012), it was claimed that nearly 60% of the students with mathematical learning difficulties thought their mathematics learning was not good because the teacher or learning environment was not good, or even because of bad luck. This concept of "incorrect attribution" gives another perspective for the causes of learning difficulties in mathematics.

Another viewpoint was put forward by Chen (2006) that learning difficulties in mathematics can be affected by reading comprehension, understanding, cooperation and communication, independent thinking, reflective thinking, and the ability to explore and apply new knowledge to solve problems (Chen, 2006).

The following sections describe the main external reasons that have been put forward.

15.1.1.1 Lack of Family Education

The lack of proper love and responsibility can be harmful to children. Spoiling a child can lead to a lack of autonomy, responsibility, and motivation. Other issues can include poor family atmospheres, poor relationships between family members, unsound family structures such as single-parent families or children left behind while parents move away for work, or lack of examples set by parents to have courage to solve problems.

In addition, some studies have pointed out that parents having inconsistent concepts about education is one of the leading causes of students having difficulties in learning mathematics, since the children do not know which advice they should take.

15.1.1.2 School-Related Factors

Many teachers lack confidence to deal with students who have learning difficulties, which in the course of time makes the students lack confidence and willingness to change. Some teachers do not treat students with mathematical learning difficulties fairly, which can cause them to lose confidence. Teachers' teaching methods can also mean that some students are unable to understand and master the knowledge. Some teachers cannot lead by example, so they cannot inspire their students to have passion for mathematics learning, enthusiasm for problem solving, or reasonable planning of their own development. If teachers give heavy workloads, other students may be able to cope, but students with learning difficulties will waste a lot of time and energy. Another reason can be teachers' evaluation strategies and the messages they give students about standards of excellence.

The term "teacher-originated students" has been used to describe students whose learning difficulties have originated from teacher-related reasons. Schools and society have a basic moral obligation to address the problems of teacher-originated students.

According to Gardiner's theory of multiple intelligences, students who are not achieving highly in mathematics may have strengths in other areas.

15.1.1.3 Adverse Impacts of Society

Unwholesome social tendencies, such as too much publicity about successes in speculation, can cause students to think that success does not necessarily require painstaking effort. The rapid spread of unhealthy content of film and television drama, books, and magazines has adverse effects on children, and now the Internet has exacerbated this problem.

15.1.2 Students' Characteristics

In recent years, some scholars have described the characteristics of students with mathematical learning difficulties: cognitive and behavioral disorders; reading comprehension difficulties; background knowledge, low vocabulary learning efficiency, limited background knowledge, and poor memory; inability to make learning plans; and poor self-control of their learning (Kiuhara & Witzel, 2014). These authors have suggested that students' mathematical learning difficulties are characterized by cognitive obstacles, difficulty in understanding, poor learning quality, and poor attitudes to learning.

The related research from the Soviet Union described the characteristics of students with learning difficulties as having defects in the development of thinking, the basic skills of learning, practical knowledge, and factors not related to intelligence (Wang, 2007).

The following summarizes related research on the psychological and academic performance characteristics of students with mathematics learning difficulties.

15.1.2.1 Psychological Characteristics

Students with learning difficulties in mathematics often exhibit not only serious inferiority, but also aspirations and ambitions. Their levels of curiosity and ambition are significantly lower than those of gifted students, and they typically demonstrate a lack of independence. They generally have higher levels of anxiety than other students and can become restless and irritable more easily. As well, they tend to lack concentration and self-control and cannot focus on mathematics learning for very long.

Students with learning difficulties in mathematics lack good learning habits. They are unable or unwilling to take the initiative to learn, and lack the desire to learn.

After experiencing failure, they no longer believe they can learn mathematics. This suggests a lack of self-confidence and low self-efficacy.

Even though, as mentioned above, students with learning difficulties in mathematics may have strengths in other disciplines, their failure in mathematics still makes them feel inferior.

Even though these students admit to having difficulties in mathematics learning, they are not willing to make efforts to solve the problem. It is difficult for them to find the motivation and action to support their efforts.

15.1.2.2 Academic Performance Characteristics

Reading and writing difficulties can often lead to difficulties in understanding the meanings of questions and problems. It is therefore difficult to present their understanding methodically. Another performance characteristic is that they can be more prone than other students to make calculation mistakes.

A lot of students with learning difficulties find it more difficult to solve more advanced mathematical problems. Their spatial organization can lead to difficulties in learning geometry that is highly dependent on spatial knowledge (Xie, Su, & Xu, 2009).

Other studies have pointed out that the students who have difficulty in learning mathematics have the characteristic of "poor memory" (Zhang & Liu, 2009). Indeed, a part of the problem can be difficulty in memorizing mathematics concepts and methods, but I think the reason does not lie in the poor memory but rather in their fear of mathematics.

On the contrary, a lot of students with mathematics learning difficulties can perform well in Chinese, English, and other subjects, which need even more memorization. They can easily grasp the knowledge, which suggests that poor memory is not the problem.

15.1.3 Suggestions for Supporting Students with Mathematics Learning Difficulties

To support students with learning difficulties, it is important to solve the root cause, to find a solution starting from the reasons. It is important that the students no longer suffer repeated setbacks, but get love and self-value.

American scholars have given some suggestions for supporting students with mathematics learning difficulties. For example, 26% of the US 15–year-olds in PISA2013 showed poor mathematical reasoning ability. There are six suggestions that have been put forward to help develop mathematical reasoning ability in students with learning difficulties through improving teachers' training: give them clear guidelines; pay attention to their verbal reasoning; give them visual representations of concepts; use heuristics; give appropriate examples; and encourage peer-assisted learning (Kiuhara & Witzel, 2014).

The heuristic teaching method is one that cannot be ignored in China and has also received the attention of scholars abroad. For example, Quine and Douglas (2009) suggested some steps to support the teaching of the number: start with real illustrations of a number; encourage students to use oral language to describe the number; and then introduce formal symbols to give them a need to use this number (Quine & Douglas, 2009).

Traditionally, there was an idea of teaching students in accordance with their aptitudes. Al-Makahleh (2011) reported an American study, which found that direct mathematics instruction given to fourth and fifth grade primary school students with mathematics learning difficulties had positive effects on their basic skill achievement and attitudes toward mathematics.

As well, there have been studies exploring how to change students with learning difficulties. These studies have focused mainly on changing the students themselves, and changing home education, school education, and social education. The following lists suggest how these aspects may be addressed.

15.1.3.1 Students Themselves

- Students should develop interest in mathematics study and develop faith in their abilities;
- Develop strong motivation and study actively and consciously;
- Develop good study habits and positive attitudes;
- Acquire and use appropriate study methods.

15.1.3.2 Home Education

- Transfer the idea of home education, educate children using scientific methods, develop harmonious personalities, and communicate with them equally.
- Create a democratic home atmosphere. The fundamental home condition to change students with learning difficulties is to let children study and grow up in a healthy home atmosphere.
- Change the methods of home education. Do not beat or punish children, do not blame and scold, do not spoil, and do not let them develop without guidance.
- Appreciate education. Parents should learn to praise their children, to find their strong points and give them consistent and equal care.
- Parents need to cooperate with the school and provide role models of appropriate study behaviors.

15.1.3.3 School Education

- Emotions. Teachers should communicate frequently with students about their difficulties. It is important that teachers respect students' personalities and protect their self-respect.

- Appreciation. Learn to find these students' strong points and praise their strengths in order to inspire their interest, prevent them from feeling ashamed of themselves, encourage them to face difficulties, and care about the development of students as human beings.
- Promote the professionalization of teachers. Teachers should change their teaching methods to enable the students to keep pace with the class; teachers should guide students, especially those with learning difficulties, to develop good study habits; teach them in accordance with their aptitudes; give them opportunities to participate; and inspire a sense of achievement and protect their self-respect by changing the methods of evaluating.
- Create a healthy and active atmosphere for these students.
- Communicate with parents in order to understand the students' home conditions.

Provide different learning activities to cater for students with different ability levels, so that all students are dealing with knowledge that they are able to digest. Variations of these ideas have been discussed by some Chinese scholars. He and Sun (2008) proposed using the theory of Multiple Intelligences to understand students with learning difficulties and work on the strengths as well as the weaknesses of students with learning difficulties. This is consistent with the scholar Chen Youke, who said that the student is his own master and we should promote the lifelong development of students.

15.1.3.4 Social Education

- Emotion. Eliminate unhealthy aspects of the school, including negative unsafe factors. This should be extended to entertainment places and Internet cafes around the schools.
- Focus on developing leisure activities for students that offer informal learning along with entertainment.

Individualized education has gained attention from many schools and teachers in China as a way of helping students with mathematics learning difficulties, as it is easy to operate.

Beijing 166 School is a typical case of the implementation of individualized education for the transformation of students with learning difficulties. The remainder of this chapter will give details about how Beijing 166 School implements this type of teaching.

15.2 Study Design

The *Mathematics Curriculum Standards* propose that mathematics is for all students and that different people should be able to develop differently. This leads to the need for individualized education, including emphasis on students' differences and respect for these differences.

The observations reported here focused on students with mathematics learning difficulties from Grade 8 of Beijing 166 School. The purpose was to investigate any transformation of students with learning difficulties as a consequence of teaching at different levels (hereafter referred to as "level teaching").

A pre-test was given to 334 students and they were allocated to levels based on this. Different teaching methods were allocated to students of different levels; these included different learning objectives, learning requirements and assigned work. The observation was conducted over one semester, and at the end of a cycle the students took a post-test to examine any transformation of students with learning difficulties.

15.2.1 Study Process

15.2.1.1 Pre-test

Based on the pre-test and their previous overall mathematics achievement, the 334 Grade 8 students were grouped into A, B and C levels. They were then allocated to two level A classes, four level B classes and two level C classes which respectively had 96, 175 and 61 students.

15.2.1.2 Teaching

The students were only grouped by levels for mathematics classes. For other subjects they remained in their original classes. Each teacher was assigned to teach two classes of two different levels. All levels use the same syllabus and textbooks and follow almost the same teaching schedule, except that there are different specified targets and different degrees of difficulty. The aim is to ensure all students can learn mathematics within their own abilities.

This kind of teaching requires more from the teachers. They not only need to prepare cross-level lessons, but also to grasp the differences between the different levels so as to implement individualized education. What is particularly important is that the teachers of Level C need to observe and regulate the mental states of the students and their parents, and to gain the understanding and support from them, which is one of the prerequisites to making the level teaching work.

15.2.1.3 Post-test

The students took a test in the end of the semester to measure the effectiveness of the level teaching, particularly for those with mathematics learning difficulties. The test paper is the Final Examination paper of Dongcheng District.

Overall, five tests were taken by the students, from pre-test to post-test.

15.2.1.4 Conclusions of the Study

Differences across levels

The differences between students from different levels were obvious. We used five tests during the period of the implementation of level teaching to describe the trends in changes in the students' mathematics achievement. The two variables **intercept** and **slope** can be used to describe the change of students' achievement (Tables 15.1 and 15.2).

According to the tables above, there were significant differences between levels before the teaching intervention (**intercept**); a similar conclusion can be drawn from the students' average scores. However, no significant differences were found through comparing the **slopes** of the different levels. This indicates that there was a difference between levels before the level teaching, but the trends were approximate.

After the teaching intervention, level A and level B maintained a significant upward trend, while the students from level C declined; level B gained the largest increase. Does this suggest that the level teaching did not work in the transformation of students with mathematics learning difficulties? We know that there must be an increase or a decline of achievements within a single school, and the increases of levels A and B will surely indicate a decrease of level C. But our tests were conducted within a district, so it was necessary to compare with students from other schools.

Table 15.1 Levels distribution before level teaching

	Level A	Level B	Level B	Level C	Difference
Intercept	0.9891	0.3362	−0.3494	−1.5007	4 < 3 < 2 < 1
Slope	−0.2905	−0.0521	0.2407	0.1898	No difference

Table 15.2 Levels distribution after level teaching

	Level A	Level B	Level B	Level C	Difference
Intercept	0.5560	0.2498	−0.6278	−0.3571	4, 3 < 2 < 1
Slope	0.1717	0.1881	0.2784	−0.7262	4 < 3, 2, 1

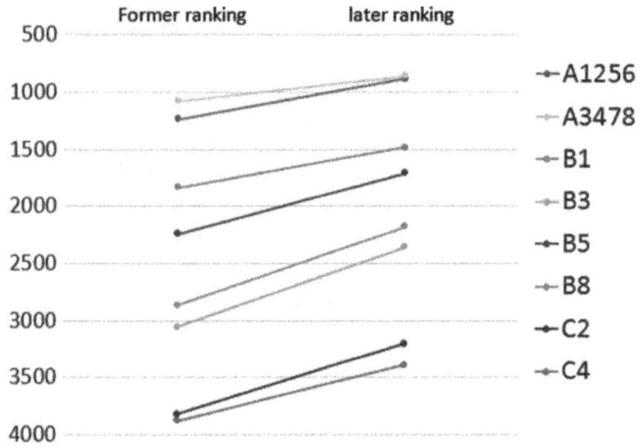

Fig. 15.1 Ranking among the schools after level teaching

Comparisons within the district

At district level, before Beijing 166 School implemented the level teaching, the students' test scores were arranged according to their mathematics rankings. It should be noted here that this was not based on the pre-test that was used for Beijing 166 School. What we were observing here was the impact on the overall mathematics academic achievement and the academic achievement of students with mathematics learning difficulties after the implementation of the level teaching at Beijing 166 School, so we just needed to provide two scores before and after the implementation of level teaching.

The results show that the rankings of students from levels A, B and C of Beijing 166 School all increased after the level teaching (Fig. 15.1).

The result fully corroborates that the level teaching played a significant role in the improvement of teaching effectiveness and in the transformation of students with mathematics learning difficulties.

As we know, level teaching is based on individualized learning, hence the transformation of students with mathematics learning difficulties was affected positively by individualized learning. If we really pay attention to each student, and provide each student with appropriate experiences to learn at their own levels and arouse their interest, then this should lead to their transformation.

15.2.2 Some Precautions Regarding the Transformation of Students with Learning Difficulties

Based on the literature, there are some precautions suggested when attempting to transform students with mathematics learning difficulties.

First, the students' emotions must come first; care and love are important. According to the Pygmalion Effect, we need to give students more positive expectations.

Second, the teacher is the key factor during the transformation of students with learning difficulties, and the peer group also plays an important role.

Third, only if schools, families, and society form a unified force can we transform students with learning difficulties.

Finally, the transformation of students with learning difficulties cannot be rushed, because the process is not easy.

15.2.2.1 Teachers' Thoughts About Transforming Students with Learning Difficulties

The results of the study reported here on students with learning difficulties reflects the positive response to the idea that mathematics is for all students and that people should have the opportunities to develop differently, as proposed in the *Mathematics Curriculum Standards*. We also need to pay special attention to the causes, characteristics and transfer strategies of the group of students with mathematics learning difficulties in order to convey the view that mathematics is for all students.

Foreign studies, especially the American studies, have focused more on transforming students with mathematics learning difficulties through the use of specific teaching strategies. The Chinese studies have been broader, paying attention to the potential impacts of the students themselves, their parents, schools and teachers as well as social factors. As those who are the closest to the students with mathematics learning difficulties, it is important to gain deeper insights into student-teacher interactions in individualized learning. The following section reports a case study of one student and his individualized interactions with his teacher.

15.3 Case Study

Design

As mentioned above, the teacher has an important role in helping students with mathematics learning difficulties to gain confidence and put more effort into their mathematics study. In addition, the success of level teaching lies in the implementation of individualized education. Each student has his best-fit learning methods under which he can achieve academic progress. Under this premise, this study attempted to explore if we could support students with mathematics learning difficulties successfully by giving special attention to every student.

We selected student H from Class 2, Grade 7 of the High School Affiliated to Beijing Normal University to be the subject of a two-month classroom observation and follow-up study. Student H was not selected randomly, he was one of the groups of students with mathematics learning difficulties and his mathematics academic performance usually placed him 32nd–40th of the 40 students in his class.

He had most of the characteristics of students with mathematics learning difficulties described earlier in this chapter, especially poor self-control and inability to focus on learning for any length of time.

The researcher observed H's behavior and participation in the mathematics classroom, and communicated with him after class, answering his questions, and pointing out his poor performance in the classroom so that it could be corrected, as well as praising him for good performance. This was to help him to build confidence, enhance his strengths and overcome his weaknesses, and help him to build an interest in mathematics learning gradually. He was interviewed after the classroom observation to explore the deep-seated reasons for his difficulties in mathematics learning, through thorough questioning about himself, his family, his school experiences and his teachers' specific performance characteristics.

15.3.1 Interview

The aim of the interview was to understand more about H's mathematics learning difficulties and society. He was also asked about his expectations for future studies, to explore whether his motivation changed after enhancing academic achievement.

15.3.1.1 Earlier Period

Homework

When contact was first made with H, according to his homework (Fig. 15.2) and the test volume (Fig. 15.3), his mathematics performance was not good and there were problems with his attitude.

In this homework, there was no "solution" written for every answer, which shows that he had not mastered basic mathematics specifications and that his work was arbitrary. In the 13th question, H crossed out the wrong part randomly and did not make any serious attempt to modify it, indicating that his mathematical learning attitude was poor. In the 14th question, he wrote 15 = CN + AM, while the right answer was 5 = CN + AM, but he still got the right answer in the end. Whether it was a careless typo or copied from another student was not clear. The 15th question was an extension of the 14th but H wrote nothing, which means a lack of mathematical thinking, and an ability to understand, analyze and combine the numbers and figures. In the 16th question, his solution process showed now evidence of following any procedure and no causal link. It looks as if he might have copied from another student.

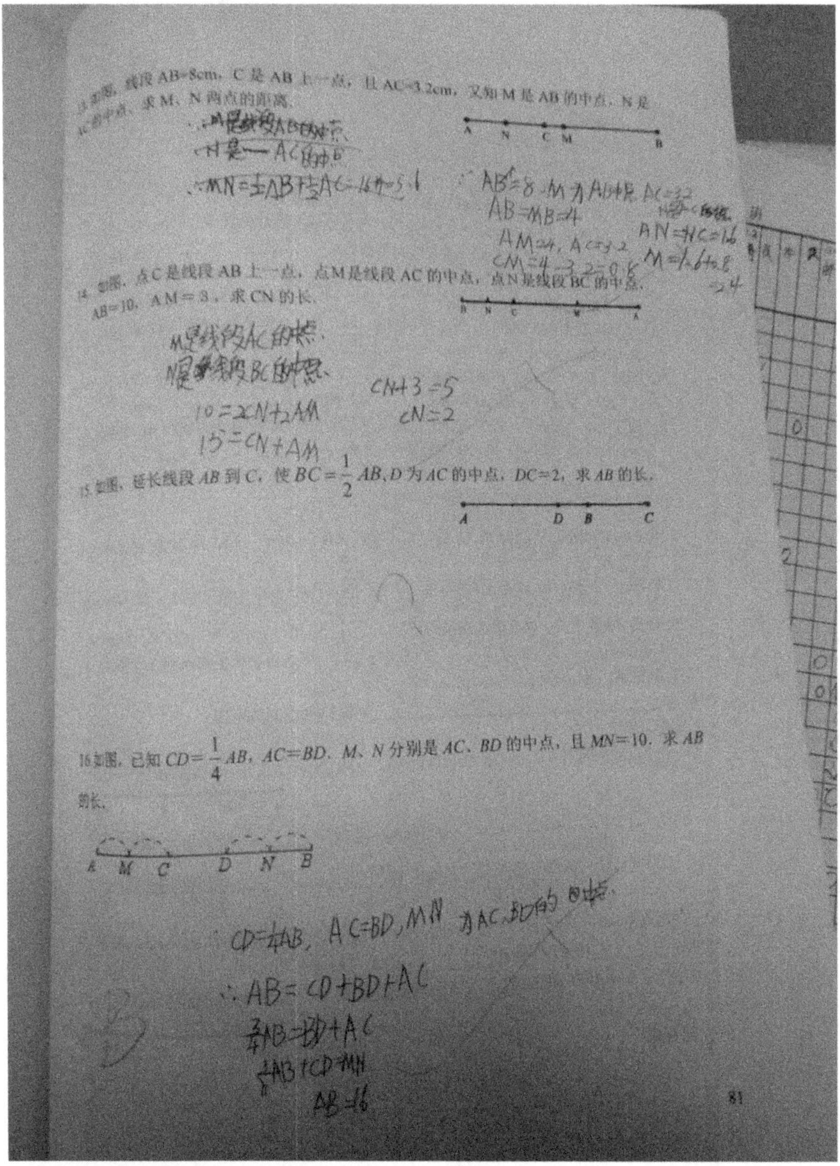

Fig. 15.2 Homework of H

In summary, although this was only one piece of homework, the problem above was also reflected in H's other mathematics homework. As a student with mathematical learning difficulties, H exhibited many characteristics of a student with mathematical learning difficulties as previously mentioned: reading difficulties, difficulties in understanding the problem, and lack of mathematical abstraction.

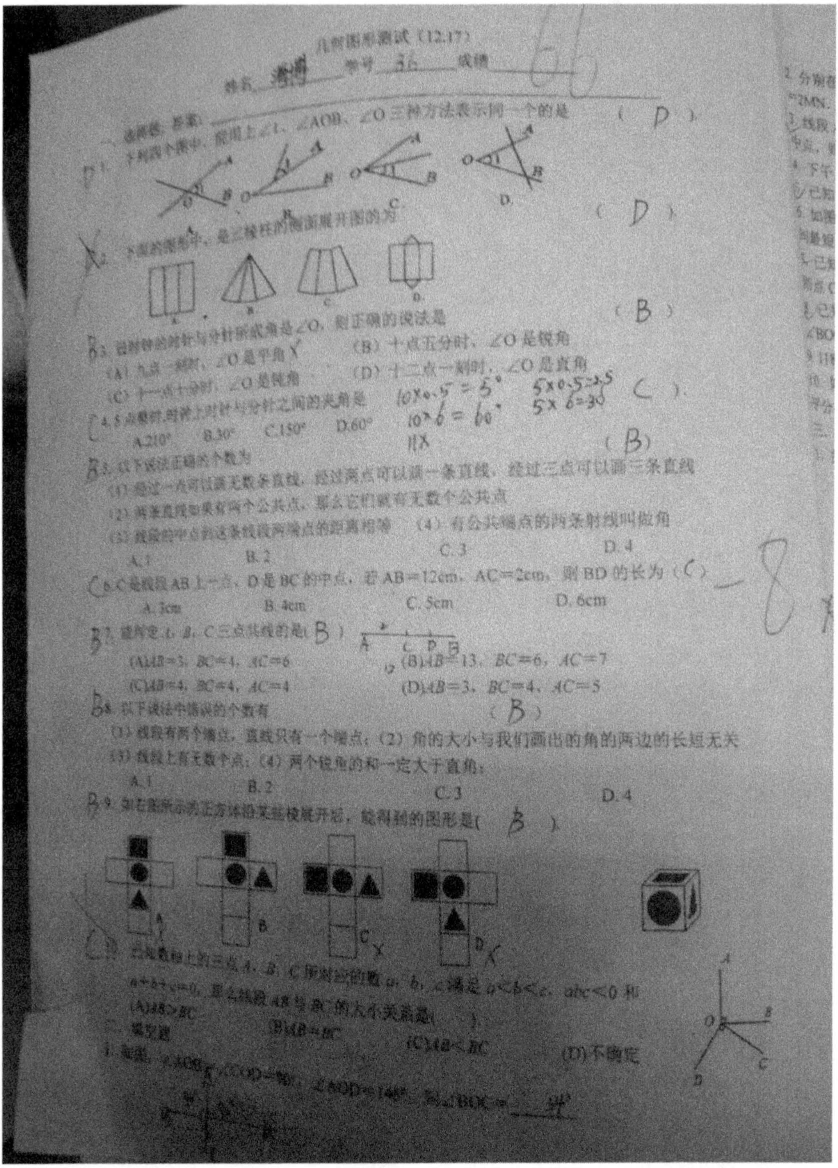

Fig. 15.3 Test paper of H

Test

This was a geometry test, mainly of the common knowledge and questions to which junior grade 1 students were exposed, such as the segments and cube nets. According to the existing research, the student with mathematical learning difficulties is characterized by weak spatial organizational ability. H, as a student with significant mathematical learning difficulties, demonstrated some difficulties in the completion of geometry exercises.

From this geometry test, it can be seen that H only scored 66 points in the relatively easy papers (ranked 36 in a class of 40 students), indicating a lack of spatial imagination. The second question involved a side expanded view of a triangular prism, but H selected the expanded view, indicating that his moderation was unclear and his attitude was not careful enough. The tenth question indicated his lack of ability to combine numbers and figures and to use symbolical mathematical language to show the subject content clearly. H did not write anything here, noting that he had no idea about this problem, which can be attributed to a lack of mathematical thinking, not doing well in finding the right method to solve the problem and an inability to link the request with the knowledge.

Classroom observation

As an example, the following record of H's classroom performance illustrates his ordinary classroom performance.

From the beginning of class, H concentrated on playing with his pen then his ruler and did not seem to be interested in the lectures. Then he was copying things from the blackboard to the paper while the teacher explained. He had been very bored for the first 18 min of the class. The teacher asked the students who had made errors to take notes but he did not respond, and his notebook was not open. At 22 min into the lesson, he turned back to look at his classmates' books and tables. At 26 min, he was scratching on his desk. When the teacher talked about the previous day's test score and praised students who had made progress, he stopped to listen, but soon was busy scratching again.

That he stopped to listen for two minutes shows his caring about mathematics progress, and suggests that he could be transformed successfully with proper methods.

Subsequently, 30 min into the class, he said some words. When the teacher asked if this question had been done, he looked at the subject. The teacher asked the class to write, and he began to do so. The teachers found some students to explain the solution on the blackboard; he did not look up at the blackboard at all. When the students finished speaking, he looked up to copy everything down.

In the break, I asked him to show me his last examination paper, and he avoided doing so again and again. Once he said he had forgotten, other time he said the teacher had taken it. I knew he was ashamed to let me see his score. Every break when he saw me walking toward him, he began to slump on the table and not move, waiting for me to ask him questions.

From the description of H's classroom observations (which were typical of his performances in mathematics classes), it can be seen that his involvement and enthusiasm were not high and he did not participate enough. He was not in unison with the other students and often switched off.

The comprehensive early observation suggested that H displayed some performance characteristics of a student with mathematical learning difficulties. And H's other characteristics as a student with mathematical learning difficulties will be illustrated later in relation to his interviews.

15.3.1.2 Later Period

Toward the end of the two-month observation period, H realized he was concerned about giving better presentations in class. Of course, we hope that students will be able to take the initiative to modify their behavior without outside interference, but if the outside pressure can encourage them to do so, it is not a bad idea.

Earlier he had played with stationery or sat in a trance for about 30 min every lesson. Then a gradual downward trend happened. He often stopped playing suddenly and maybe realized that someone was watching him. His class participation gradually increased. When the teacher asked to see the blackboard or to do the subject, he was more able to synchronize with other students.

Once the teacher asked the students to complete a few questions in five minutes. When the time was up, the teacher asked the students who had finished to put up their hands. Some raised their hands, some stopped writing, but H was still writing. His attitude was remarkable compared with earlier; an obvious difference was that he was no longer self-defined as "a poor student". Being given appropriate encouragement and shown some concern had profoundly inspired his inner motivation and self-confidence.

Although he was still unwilling to show me the previous test paper, he was prepared to share his current one. He was also willing to share his notebook with me because it was more detailed and careful than the previous one (Fig. 15.4), even though it was not the best in his class.

In these notes, there are emphases and difficulties labeled with red pen and also calculus done in class. Although these marks make the notes appear not clean, they do show that H was engaging in his own thinking in the process of doing the exercises and had a high degree of involvement in the class.

Before every mathematics lesson, I went to H to give him some positive message, check his class readiness and remind him to listen carefully in class. After each lesson, I sat next to him. In the early days, every time he saw me, he slumped on the table and did not move. Later, he started to turn around to look for me after class and even came to me to ask questions or tell me about his classroom performance initiatives. Whether my questions were about mathematical knowledge, or about his activities in the classroom, he was able to answer calmly and confidently and if something was unclear he would ask humbly. It is apparent that his self-confidence had been greatly improved.

With the in-depth observation and communication, gradually, H became more willing to spend more time and energy on his weak mathematics learning. In the class, his performance was no different from the others and his homework was done more seriously (Fig. 15.5).

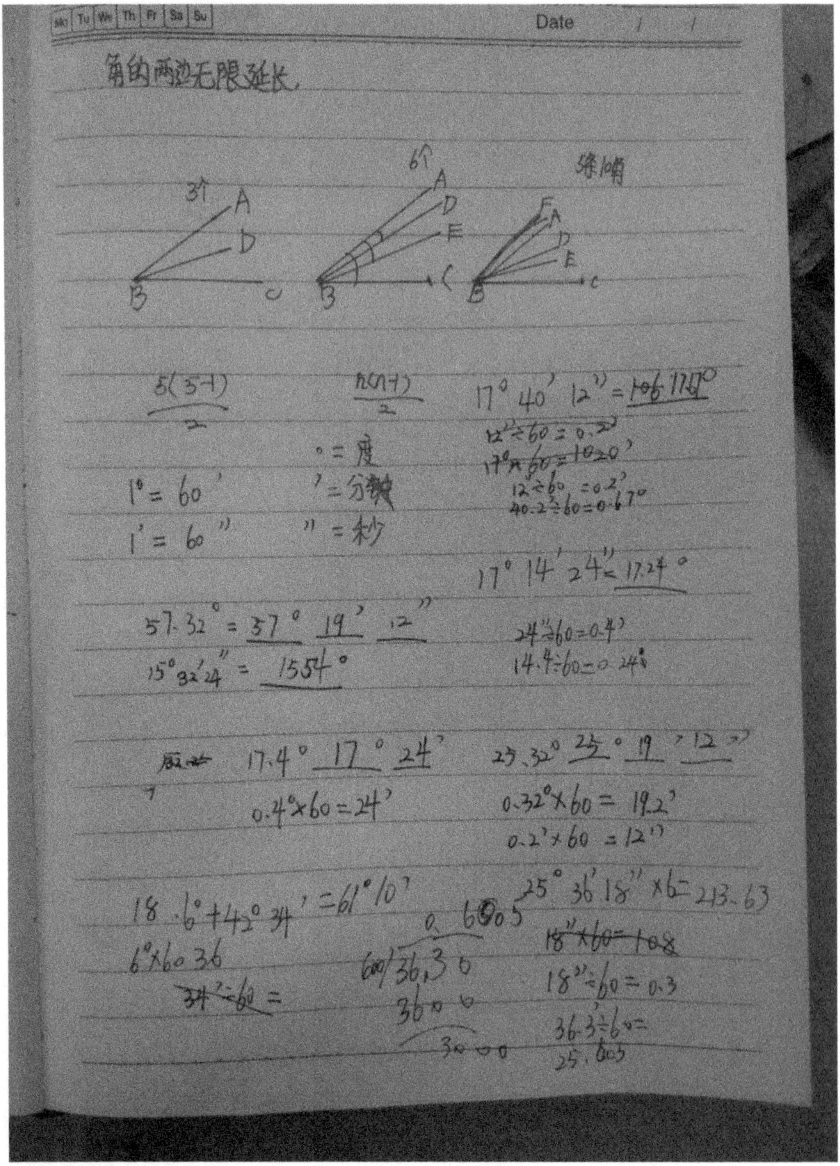

Fig. 15.4 Notebook of H

In this piece of homework, it is noted that in the first question, he wrote "solution", though this was lost in the third question. He had improved a lot in his problem-solving skill and was worthy of recognition.

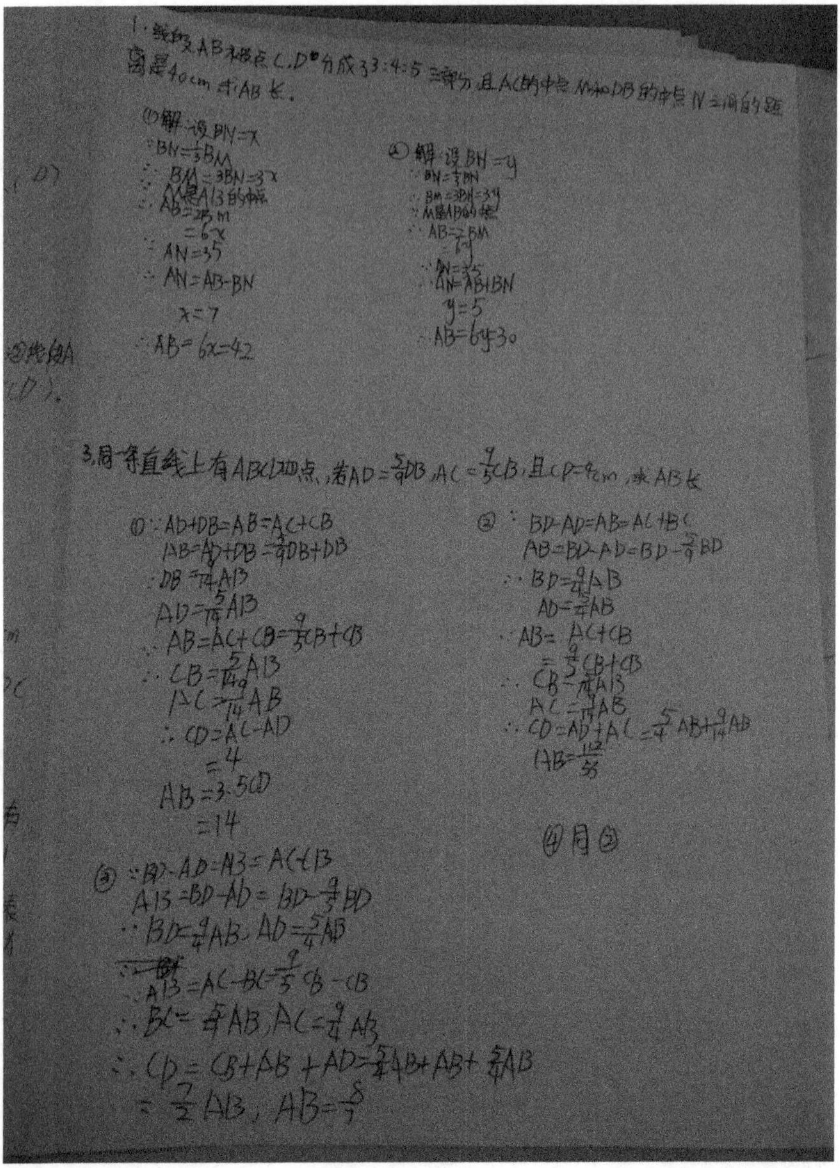

Fig. 15.5 Exercise of H

15.3.1.3 Interview

Later, we interviewed the student H in order to find out the underlying causes of his various previous difficulties in mathematics learning and trying to dig out his new attitude and orientation to mathematics or himself after his mathematics academic achievement had improved.

First, H was asked to recall the knowledge that he had learned at the beginning of the term. He represented the concept and the solution method for inequalities in one unknown.

A detailed record of the core interview is as follows.

Researcher (hereinafter referred to as "R"): How do you feel about mathematics learning now?

H: Good.
R: This state, good, is too vague. How do you feel compared with the previous semester?
H: Simpler than the previous semester.
R: Has there been there any test? Can you talk about it?
H: Once 83, sometimes more than 70.
R: Where did you rank in your class?
H: In the middle of it.
R: Can you be specific?
H: More than 20.
R: 28 or 22?
H: About 22 or 23.
R: Are you satisfied about this position?
H: Yes.
R: No more demands?
H: The students before me are the best students that are always good at mathematics.
R: I think your progress is quite obvious. You had always played with ruler and pen on class before. What about now?
H: I do not play.
R: Why did you play before?
H: I do not know.
R: Do you think you like mathematics now?
H: I quite like it.
R: Do you think you can keep up with the teacher's rhythm?
H: Yes. Completely.
R: How about the homework?
H: I can finish all of it.
R: How about attitude? How much do you score yourself?
H: The teacher will do it. Usually A, and sometimes B+.
R: Very good. Why do you think your mathematics was not particularly good before?

H: Sometimes I did not understand what the teacher said. Then I did not want to hear. Last semester, the application problems were more difficult to understand.
R: Is there a mathematics tutor to help you?
H: No, I learn by myself.
R: How is your relationship with the other students?
H: Good.
R: Do you communicate with your teachers often?
H: Often, because my hand is injured, many teachers are more concerned about me.
R: What is your general arrangement after school?
H: I hurry to go home and do homework.
R: Oh, hurry to finish homework, and then hurry to watch TV?
H: I do not watch TV.
R: Play I-pad or something? Play on your phone?
H: I do not play. Saturday and Sunday I play, my mother manages it.
R: Only Saturday and Sunday? How long can you play?
H: Half an hour, and then rest. But I can play about two hours, two days together.
R: That's not too much, very good. Now can you briefly introduce your family? What do your parents do?
H: My mother is a waitress, and my dad is a driver.
R: Can they help you solve learning problems?
H: Yes.
R: Can they help you with the specific mathematical problems?
H: Sometimes they can, but sometimes too hard.
R: Some topics are really hard. How do you think they are the greatest help to your study?
H: They can supervise me. If I have a lot of work, they will not sleep and stay awake with me.
R: Wow, good parents. So do you feel particularly warm in your heart?
H: I just want to finish it quickly so that they can go to bed earlier.
R: Really sensible. You've worked so hard and next time I hope to hear that you are into the top 20, top 15 and even top 10. As I told to you, H, you need not think this is impossible and hard work can create a miracle. As long as you are making progress, even if slow, it is very good. If we have nothing else to say, can you make a mathematics learning goal? I will look forward to when you can achieve it.
H: Same as you said, things are slowly going better.
R: In this semester, what is your goal?
H: I think I could be in the top 20.
R: Within 20, right? I like that! That is no problem, thank you! Goodbye!
H: Goodbye.

From observing the performance and characteristics of H, some interesting things emerged. His inability to understand what the teacher was talking about and what the problems meant is consistent with other studies that students with mathematics learning difficulties may have understanding disorders or dyslexia. H's parents were not educated well enough to be able to provide significant help to him;

however this was not the main reason for his learning difficulties. His concentration and self-control were poor, but there were no signs of inferiority or other psychological manifestations. In other words, he did not display all of the characteristics of a student with mathematics learning difficulties described in the literature.

An examination of the causes of H's learning difficulties suggests that it was mainly of his own cause. His family was stable and he had been receiving concern and attention from teachers because of his injured hand. His teachers did not give up on him because of his bad scores or absences. He was never absent without reason, and did not leave the school or his family to participate in social games or go to Internet cafes. So there did not appear to be any societal reasons for his substantial learning difficulties. So the only reason was himself, as the interviews and classroom observations indicated at the beginning of the project, his attitude, understanding and classroom involvement needed to improve.

15.3.2 Case Tracking Study Result

Academic Achievement

In the midterm examination before the observation, H was ranked 36th in his class of 40 for his aggregate score and 36th for math as well. Generally his test rankings had been between 32nd and 40th, and occasionally in the top 30. In the final examination, after the two-month observation and exchange, his aggregate score was ranked no. 29, and mathematics no. 31 of the 39 students present in the examination. In the quizzes held during the two months, he progressed from the original 32nd–40th to 20th–32nd, and only occasionally 35th or 36th. Undoubtedly, H became the student with the most significant advancement in his class.

Learning Attitude

It can be seen from the homework and notes in the late period that, with concern and intervention, H's attitude toward learning had improved significantly. Compared with previously, he rarely became distracted in class and could basically follow the teacher's rhythm. When the teacher issued instructions, such as to complete a problem, he could always be as fast as the other students and completed the task properly. His homework was completed more clearly and seriously and his notes became cleaner, with emphases marked.

Interest in Learning

Although H still lacked mathematical understanding and had personality traits such as lack of attention, he had begun to develop intrinsic motivation and to be involved more actively in mathematics and to be more confident. He was beginning to show signs of transformation. By slowly raising his interest and confidence, he was able to reach a good state in his self-feelings.

15.3.3 Discussion

The case study above can fully illustrate that H had made significant improvement that was due largely to the teacher's concerns and reminders. Teachers paying attention to every student, in fact, is the aim of individualized and hierarchical teaching programs. A lesson can be learned from this: In the process of boosting the transformation of a student with mathematics learning difficulties, the teacher's effect cannot be ignored and is in fact paramount. If teachers really do pay attention to and help every student, every student should be able to succeed and make progress in academic achievement.

The research about students with mathematical learning difficulties still has a long way to go. It is rooted deeply in the practice of basic education highlights that students should get different opportunities and support for their development in mathematics, as proposed by curriculum standards. There is still a need to explore more about how to transform students with mathematics learning difficulties. However, we know that exploring and implementing individualized and people-oriented programs can help students with learning difficulties in mathematics to transform.

References

Al-Makahleh, A. A. A. (2011). The effect of direct instruction strategies on the mathematics achievement of primary 4th and 5th grade students with learning difficulties. *International Education Studies, 4*(4), 199–205.

Chen, Y. K. (2006). 初探高一数学学困生的成因及转化策略 [Reasons and strategies for students with mathematics learning difficulties in Grade 10]. *Math Education Communication, 2,* 14–16.

Dai, F. M. (2001). 苏北农村初中数学学习困难学生的情况分析和对策研究 *[A study of analyses and strategies for students with mathematics learning difficulties in junior high schools in the north of Jiangsu Province] (Doctoral Dissertation).* Yangzhou: Yangzhou University.

He, X. N., & Sun, F. (2008). 多元智能视野下的数学学困生转化初探 [An exploration of the transformation of students with mathematics learning difficulties, using the theory of multiple intelligences]. *Modern Education Science, 2,* 79–80.

Kiuhara, S. A., & Witzel, B. S. (2014). Focus on Inclusive Education: Math Literacy Strategies for Students with Learning Difficulties. *Childhood Education, 90*(3), 234–238.

Li, Y. Y. (2009, April 22). 小学学困生的原因分析及教育策略 *[Reasons and educational strategy for students with learning difficulties in primary school]* [Electronic version]. Retrieved from http://www.fjzzjy.gov.cn/newsInfo.aspx?pkId=51717.

Quine, D. D. (2009). Why is Mathematics so Hard for Some Children?: The Nature and Origins of Mathematical Learning Difficulties and Disabilities. *Childhood Education, 85*(3), 200–202.

Tan, F. (2012). 高中数学学困生的成因调查与转化措施研究 [Cause Investigation and Research on Intervention Measures of High School Students' Mathematics Learning Difficulties]. *Education for Chinese After-school (Theory), 3,* 24–25.

The National Education Management Research Group. (1996). 初中学习困难学生教育的研究 [A study of education for students with learning difficulties in junior high school. *Educational Research, 8,* 40–46.

Wang, Z. Y. (2007). 关注每位学生的发展——"学习困难学生"转化的理论与策略 *[Pay attention to each student's development—theory and strategies for the transformation of students with mathematics learning difficulties]*. Shandong: Shandong Education Press.

Xie, M. C., Su, S. D., & Xu, Y. (2009). 数学学困生的转化 *[The transformation of students with mathematics learning difficulties]*. Shanghai: Huadong Normal University Press.

Zhang, H. M., & Liu, Y. (2009). 教师如何做好学困生的转化 *[What teachers can do to transform students with mathematics learning difficulties]*. Tianjin: Tianjin Education Press.

Part V
The Professional Development of Mathematical Teachers

Chapter 16
School-Based Professional Development of Mathematics Teachers in China

Yiming Cao and Xinlian Li

Abstract Begin with a part introducing the general system of qualifying and assessing teachers in China, this chapter then mainly investigates the school-based activities that affect the professional development of mathematics teachers, including: the teaching research group, the lesson plan group, class observation and assessment and the master-apprentice relationship. Detailed Reviewing of these different activities helps to cast light on the professional development of mathematics teachers in China.

16.1 Introduction

16.1.1 An Overview of Mathematics Teachers in China

16.1.1.1 Qualifications

In China, a teacher qualification examination system was established to ensure rigorous levels of professionalism would be achieved. To become a teacher at the basic education level it is necessary to pass the primary and secondary school teacher qualification examination to obtain a teacher qualification certificate. According to the *Teachers' Law of the People's Republic of China* (Standing Committee of the National People's Congress of China, 1993), school leavers need to have achieved minimum school diploma levels. The Secondary Normal School Diploma is the minimum prerequisite to become a primary school teacher. A junior college diploma from a normal college or other kind of junior college is necessary to be a secondary school teacher. A bachelor degree or higher from a normal university or other type of university is required to become a high school teacher.

Since the start of the twenty-first century, qualifications of primary and secondary school teachers have improved along with the development of the economy,

Y. Cao (✉) · X. Li
School of Mathematical Science, Beijing Normal University, Beijing, China
e-mail: caoym@bnu.edu.cn

Y. Cao and F.K.S. Leung (eds.), *The 21st Century Mathematics Education in China*, New Frontiers of Educational Research,
https://doi.org/10.1007/978-3-662-55781-5_16

society, and education in China. In 2013, the number of primary school teachers was 5,584,644, with those at postgraduate level accounting for 0.3%, those with undergraduate qualifications accounting for 36.9%, those with junior college education accounting for 50.1%, and those with high school education accounting for 12.7% (Ministry of Education of the People's Republic of China, 2015) The total number of secondary school teachers was 3,480,979 with postgraduates accounting for 1.3%, undergraduate qualifications for 73.6%, junior college education for 24.4%, and high school education or below for 0.7% (Ministry of Education of the People's Republic of China, 2015). The number of high school teachers is 1,629,008 (postgraduates 5.8%, undergraduates 91.1%, junior college education 3.1%, high school or below 0.08%) (Ministry of Education of the People's Republic of China, 2015).

Thus, most primary school teachers have junior college education level or above, which account for 87.3%. Most of the secondary and high school teachers have undergraduate degrees or above, which account for 74.9 and 96.9%, respectively. From primary school to high school, teachers' qualifications are improving. Mathematics teachers are important members of teaching teams, hence they tend to have above-average qualifications.

16.1.1.2 System of Evaluating Mathematics Teachers

A unified system of teaching positions has been established for compulsory education teachers, and the upgrading of these positions is considered as important for encouraging teachers. The teaching positions are divided into three levels: junior, medium, and senior. For secondary and high school teachers, the teaching positions are first-level, second-level, and third-level teacher of secondary and high schools. For primary school teachers, as the positions are senior teacher of secondary and high school, senior teacher of primary school, first-level teacher, second-level teacher, and third-level teacher. The most senior are the senior teacher positions, followed by first-level teachers. The second and third levels are junior positions. In order to encourage and commend teachers who contribute distinctively to education and teaching, the honor of distinguished teacher is awarded. Holding a senior professional position is the basic qualification for applying to become a distinguished teacher (Standing Committee of the National People's Congress of China, 2006).

In 2013, the distribution of primary school teachers was as follows: senior teachers of secondary and high school (2.1%), senior teachers of primary school (52.3%), first-level teachers (33.7%), second-level teachers (3.3%), third-level teachers (0.2%), and teachers without professional position (8.6%) (Ministry of Education of the People's Republic of China, 2015). The distribution of position of secondary school teachers was as follows: senior teacher of secondary and high school (16%), first-level teachers (43.2%), second-level teachers (32.4%), third-level teachers (1.6%), and teachers without professional positions (6.8%) (Ministry of Education of the People's Republic of China, 2015). The distribution of positions of high school teachers was as follows: senior teachers of secondary and high school (26.4%), first-level teachers (36.1%), second-level teachers (29.7%), third-level teachers (0.7%), and teachers without professional positions (7%) (Ministry of Education of the People's Republic of China, 2015). Comparatively, the proportion of senior positions was the lowest in primary schools and the highest in high schools. The proportion of medium positions was the highest for all three types of schools. Additionally, there were still some teachers without professional positions in China.

16.1.2 Main Forms of Professional Growth of Mathematic Teachers in China

Teachers are responsible for teaching and educating students, cultivating the building of society and improving the quality of the nation (Twelve Key Normal Universities, 2002). The Chinese new curriculum reform advocates that teachers should be organizers, guides, and collaborators, having important effects on students' emotional attitudes and acquisition of knowledge and skills, as well as the quality of the subjects they teach. Teachers' professional levels are related to the quality of education. Improving teachers' professional level is not only necessary, therefore, for their own self-growth, but also the improvement of education. Thus, it is important to examine the most effective approaches to developing teachers' professionalism.

China started a new round of curriculum reform at the beginning of the twenty-first century. In line with this, the Ministry of Education produced the *Outline of Basic Education Curriculum Reform (Trial)* (hereinafter referred to as the *Outline*), the *Mathematics Curriculum Standards for Compulsory Education (Experimental Manuscript), Mathematics Curriculum Standards for Compulsory Education (2011 Edition),* and other documents. They provide guidance, in the context of the curriculum reform, to help teachers improve their professional skills, renew their teaching ideas and change their teaching methods; through pre-service training and in-service training they gradually mature in these aspects of their development. In-service training can be both part-time training and school-based training (Twelve Key Normal Universities, 2002), and school-based teacher growth development is an important component of in-service training.

Teachers' professional development is a process of continuous socialization and individuation. Studies have shown that the good qualities of many excellent teachers in high school tend to be accumulated and developed gradually throughout the teaching practice, for example the ability to deal with teaching content, use teaching methods and approaches, and organize and manage teaching. About 65.31% of these abilities are obtained during the post-service period (Twelve Key Normal Universities, 2002). School-based professional development is related closely to teachers' daily work, which is an effective way for them to make progress.

The school-based development of mathematics teachers involves various group institutionalized and non-institutionalized activities; this is not the same as continuous professional development happening in a school context. The school is not only the teachers' work place, but also the place to improve their professional qualities. Chinese mathematics teachers spend twenty-five to thirty-five years working and living in schools, from recruitment to retirement. The institutionalized approaches of school-based growth mainly include activities of teaching research groups, teaching preparation, class observation and assessment, and mentoring designing for newly recruited teachers. The non-institutionalized mainly include self-initiated communication, inquiry, and discussion between teachers.

16.2 Activities of the Teaching Research Group

16.2.1 Formation of the Institution of Teaching Research

The systematic and complete system of teaching research is a characteristic and strength of Chinese education. The Chinese teaching research system was founded in 1952, at the time of the *Provisional Regulations of Primary School (Draft)* and the *Provisional Regulations of Middle School (Draft)*. The names of teaching research organization in different form area to area. In 1957, the *Work Regulations for the Research Group for Secondary School Teaching, the Ministry of Education (Draft)* was the first formal document to use the name "teaching research group." From 1985 to 1999, teaching research groups were established widely in primary and secondary schools, at the same time, institutions of group teaching preparation and institutions of group learning and discussion were also established. A four-level (province-city-district (country-school)) system of teaching research was established covering different sections and subjects, and this became a uniquely Chinese teaching administration model (Liang, Lu, & Huang, 2010). Since the Chinese curriculum reform for basic education began in 2001, the function of teaching research organizations at different levels developed into teaching research and teaching guidance based around curriculum and textbook reforms, and the school-based institution of teaching research was developed.

The Chinese teaching research institution has provided an excellent platform for communication and cooperation between teachers. It plays an important role in the development of the mathematics teachers. Some scholars have considered from the perspective of Western sociology, that is, teacher cooperation is a kind of social activity which starts from establishing identity and obtains social capital by social interaction (Li & Xiong, 2013). Others have pointed out that the essence of teacher cooperation is for social capital. Teachers obtain valuable resources through intentional interaction in the organizational structure. Interaction and collaboration between teachers are beneficial for building teams broadening their knowledge, and developing their teaching skills (Li & Xiong, 2013), thus accelerating their professional growth and lightening their burdens. Through peer cooperation, mathematics teachers carry out professional dialogue, share experiences, learn from each other, support each other, and develop together (Yu, 2003). Teachers can share and complement knowledge through collaboration, since there are so many individual styles of knowledge structure, thinking modes, and cognitive styles (Zhang & Zhang, 2011). Some scholars have pointed out the need for research on local and original teaching issues, such as school-based teaching research activities, which can not only improve the teaching of mathematics in China, but also contribute globally to mathematics education, integrating international perspectives and research methods (Huang, 2004).

16.2.2 Survey of School-Based Teaching Research Activities

A school-based teaching research group is the grassroots unit for educational research and teaching research. Groups usually conduct regular seminars, which are attended by all the mathematics teachers of a section. Some combine junior and senior high schools and some have separate groups for them. In order to understand the teaching research activities, we sampled 190 teachers from urban schools of four cities from Eastern, Western, Southern, Northern, and Central China.

16.2.2.1 Frequency of Meetings

Meetings are held regularly and hosted by the group leader. All mathematics teachers are expected to participate in these meetings, and the vice principle in charge of teaching or academic dean sometimes attend as well. The duties of the group leaders usually include observing, inspecting, monitoring, and advising. They also have to offer comments and suggestions on teaching research, put forward the latest school requirements, announce recent activities, and so on.

From our survey, it was found that the majority of teaching research groups meet fortnightly. The next most common is weekly meetings, although some meet monthly, once or twice a semester, or less frequently. Generally speaking, the schools with regular and frequent meetings hold larger scale and better quality meetings than others.

The durations of meetings also vary. The most common is to take up one or two class periods. However, in many cases the meeting will continue until the research questions have been solved, without a time limit.

16.2.2.2 Content

The contents of teaching research meeting are varied, since schools, meeting time, and teaching processes are different. However, there are several common practices.

The first of these is the lecture. Schools sometimes employ experts in the field of mathematics education, education, or mathematics, teaching researchers from their school districts, or excellent teachers from similar schools to make special reports. These lectures may focus on experiments, changes to the mathematics curriculum standard, knowledge of advanced mathematics, theory of education and teaching, learning psychology, typical cases, and outstanding experiences in education. Professional books, journals, and curriculum standards are important supplements to activities of this kind.

The second is work arrangements and summaries. This type of meeting is usually held in the beginning, medium, and end of a semester. The meeting held at the beginning of the semester deliberates on teaching process and teaching schedule. In the middle of the semester, the focus is on developing test papers, test preparation, test schedules, test result analysis, and adjustments to teaching schedule for the next semester. In the end-of-semester meetings, teachers deliberate on an overview of the semester's teaching, arrangements for the semester test, and distribution of assignments for the vocation period.

The third type of activity is open-class assessment. Generally, open classes can be divided into several categories. The first one is the report class, which is often implemented by young teachers. Some schools require young teachers to implement report classes periodically, and the role of the teaching research group is to give feedback and suggestions. The second category is the presentation class, which is implemented by excellent teachers and backbone teachers in the teaching research group. The observers learn about the teaching design, blackboard-writing design, class introduction, and teacher–student interaction as well as the overall teaching presentation. The third is the competition class, which refers to school-level teaching competitions or participation in teaching competitions at district, municipal, provincial or national level. The teaching research group organizes school-level teaching competitions occasionally; this stimulates teachers and leads to improvements in their teaching. Often, other teachers will work with the contestants to develop excellent teaching design and implementation. During the interview, some teachers told us that the professional growth they achieve from participating in a teaching competition is almost equal to that gained from several years of regular teaching.

The fourth type of teaching research meeting is the most common and the most closely aligned with daily teaching, focusing on the teaching itself. The problems of teaching and students are discussed and solved; examples include any mismatch between teaching progress and the teaching schedule established at the beginning of a semester, the selection of teaching methods for particular content, breakthroughs in teaching challenges, class administration, dealing with teaching emergencies, creating a culture of excellence, supporting students with learning difficulties, and research on graduation tests (some schools even organize teachers to develop items for graduation test periodically).

There are many types of teaching research group activities, which cover all aspects of mathematics teaching. The teaching levels and professional qualities of new teachers, and experienced teachers are improved through conducting various activities. The activities of teaching research groups provide macro-guidance to solve problems in the implementation of teaching, provide an important platform for exchange, and learning for mathematics teachers of different grades, and thus ensure smooth and effective school teaching.

16.2.2.3 Effects

The periodical teaching research meetings serve, in effect, as a "service station" for mathematics teachers, providing important support for their teaching. The strength lies in the collective wisdom of the group. A famous Chinese educator, Confucius, said "*Among any three people walking, I will find something to learn for sure*"; mathematics teachers build learning communities which make them communicate with each other and learn from each other, with the resultant effect that the whole is greater than a part. According to the *Professional Standards for Kindergarten Teachers (Trial)*, *Professional Standards for Primary School Teachers (Trial)*, and *Professional Standards for Middle School Teachers (Trial)*, one index of "professional capability" is "communication and corporation," which requires teachers to cooperate, communicate, share experiences and resources, and develop together with colleagues. The effects of participating in teaching research group activities can be summarized as follows:

1. Planning. The teaching research group activities conducted at the beginning of the semester aim at designing and planning the semester's work according to the needs of the students in their classes. Notices, new documents, and important schedules are disseminated at these meetings.
2. References. During the interview, almost all of the mathematics teachers admitted that their teachings were influenced by the activities of the research group to some degree. For example, one young teacher said, "*I am young, I have little work experience, and I lack teaching experience. I also have difficulty grasping some of the key points of teaching... the depth of knowledge....It was difficult for me to teach the relationship of sides of triangle, so I valued the experiences of other teachers about the examples and exercises they used.*" Inexperienced teachers can learn from experienced teachers, while experienced teachers can reference each other.
3. Communication. Most teaching-related problems can be resolved through communication and sharing of resources between teachers in the teaching research group.

> Each teacher's teaching style is different from others. When I listen to other teachers, I find their teaching methods, examples, and exercises to be useful. The teaching research group has inspired me to analyze the textbook, prepare my teaching, and research my teaching more deeply.

4. Promotion. Teaching research group activities promote professionalism and teaching quality. Mathematics teachers are able to enrich themselves, solve problems, learn from each other, and think deeply about their education and teaching. "*The leadership of experts, including teaching research group leaders and backbone teachers, as well as peer coaching, can help me broaden my*

horizons and improve my teaching.... The activities of the teaching research group make me pay more attention to teaching research and to reflect on my teaching."

16.3 Activities of Lesson Plan Group

The lesson plan group is usually a subsidiary of the teaching research group. Teachers meet by grade level. The purpose of this group is to plan lessons together. In our study, 190 teachers in municipal schools from East, West, South, North, and Central China were interviewed about their roles and activities in lesson plan groups. The following sections describe the analysis of the interview data.

16.3.1 Frequency, Time, and Materials Needed

The group leader plans all the teaching content for the following week and presents it to other group members for their review. One leader explained, "*I am a member of the lesson plan group for grade two of a secondary school, as well as a academic dean, so I attend every lesson plan group meeting, because I am responsible for teaching two mathematics classes. As well as attending the lesson plan meetings, I always make some suggestions about the lesson plans, inspect other group members' work, and transmit the ideas of the New Curriculum Reform, notices of the Department of Education and the school's arrangements for the next period of time.*"

Most of the schools hold their lesson planning meetings once a week. The durations of activities vary; some are one hour, some are two hours, and the most common is the duration of one lesson.

Lesson plan groups include various materials, including textbooks, teaching reference books, all kinds of supplementary materials, and resources issued within the district such as textbook analysis materials, teaching design for a whole textbook chapter, students' typical errors, items for final examinations, curriculum standards, journals, books, and previous lesson plan groups' lesson plans.

16.3.2 Content of the Activities of the Lesson Plan Group

A typical meeting starts with a weekly work report and summary. Teachers discuss whether their teaching went according to plan, and any problems are discussed. They also discuss ideas about how to adjust their teaching based on feedback.

Sometimes the work of gifted students is discussed; this makes the teachers feel happy and inspired. As well, teaching schedules are planned for the semester and changes are discussed according to needs that arise. One teacher commented, "*We can see in the lesson plan group activities the process, extent of teaching, and coordination and unification of the teaching progress.*"

Another activity is to design the teaching for the next teaching content. Topic introduction, selection of examples and exercises, and design of teaching activities need to be negotiated and debated. The group leader arranged work, with one member developing the lesson plan, and others giving feedback and suggestions.

A fourth task is to identify teaching difficulties and key points. Over the past sixty years, mathematics teachers in China have developed a framework of teaching design, classroom observation, and teaching reflection, with three points: teaching emphasis, difficulties, and key points. The first of these refers to the core mathematics content. The second refers to the cognitive difficulties encountered when students learn the points. Key points of teaching refer to how to help students to overcome difficulties and grasp the emphasized points (Li & Huang, 2013).

The group is also concerned with the selection of examples and exercises. This is particularly important for mathematics teachers and often challenges new teachers. The group sets regular examinations and quizzes. This includes determining the scope of the examination questions. After the examination, the results are analyzed items by items, focusing on the students' responses, their common errors, and reasons for these errors. At the end of each semester, a review plan is developed.

If some group members are going to attend lessons competitions, the lesson plan group will help them to develop their lessons, from the aspects of selecting topics, designing teaching sections, introduction, selecting examples, and exercises. The competitors combine other teachers' advice and their own thinking. Other teachers observe their teaching and give feedback and advice on which revisions are made. One teacher who has received awards in competitions told us, "*Rather than saying I won the prize, it is a prize of my lesson plan group; my achievements carry the collective wisdom of the team.*"

16.3.3 Effects of the Lesson Plan Group Activities

Teachers speak favorably about the role of lesson plan activities in professional development, from different aspects. They believe that it is an effective approach to improve teaching skills and professional qualities. One of the key reasons for this is the degree of supervision. One teacher takes the role of designing a sequence of lessons and the external pressure to do this well for the sake of the group prompts them to high performance levels. One teacher commented, "*When I attend the group lesson planning meeting and it is my turn, I will make my lesson plans more precise and more fulfill, so that I can present my best. My group leader is very strict and has high requirements for lesson plans.*"

Communication is also important; even though teachers vary in age, experience, and quality, they can still learn from each other. The new teachers can learn from the experienced teachers about classroom management, teaching instruction, teaching language, and selecting examples and exercises; at the same time the experienced teachers can learn from the teachers about initiative and passion to teach. Teachers can reflect and compare the strengths and weaknesses of their teaching designs. The process of thinking and brainstorming contributes to new progress. As one teacher said, "*In the lesson planning activities, some problems I have never found are presented, and I will pay more attention to these from then on.*"

Another said, "*Although we teach the same grade, the students of each class are quite different, I will put forward a discussion on the problems in our class.*" A third said, "*Before the class, teaching design is just a kind of presupposition, there is a gap between it and the actual class. Sometimes students in my class reflect that my teaching is not good, then I will discuss with the teachers in our lesson plan group so I can establish whether the problem is my fault or if it is a common difficulty for students in this grade.*"

The leader of lesson plan group usually divides the teaching content for a semester into several components, with each team member in charge of a part, in turn. Other group members comment and give feedback. In fact, teaching design is the collective wisdom of the whole lesson plan group. Regardless of effectiveness or effect, the collective lesson plan is superior to an independent one. One teacher said, "*The lesson plan group helps us to overcome our own defects and draw on the good experiences of others with team power.*" Another said, "*I remember how worried I was when I attended the school-level teaching competition for the first time, but my confidence grew after I negotiated with other group members.*"

16.4 Class Observation and Assessment

Class observation and assessment is an important part of the school-based research activities; this is different from the teaching research meeting. The organizational form of teaching in China is that students belong to a fixed classroom, teachers move between different classrooms, and the mathematics teachers who are of the same grade or subject share an office. Most schools have rigorous rules for class observation and assessment of mathematics teachers, and assessment becomes the routine for mathematics teachers.

Class observation and assessment is defined as a kind of practical activity whereby teachers observe other teachers' classes, assess the classes observed, and discuss with the demonstration teachers (Zhao, Pei, Feng, Cheng, & Jin, 2013). The following describes the analysis of the interviews with the 190 mathematics teachers.

16.4.1 Categories of Class Observation and Assessment

16.4.1.1 Open Class

The open class or observation class can be defined broadly as one that is observed or commented by colleagues, peers, leaders, or experts. (Yu, 2006). The purpose of this activity is to study the teaching content, format and methods or to implement an experiment relating to the teaching reform, to assess teachers' professional levels, and promote their professional development (Zhang & Yu, 2010). In the open culture background of our country, open classes are very common. Usually, these classes are taught by new teachers with the aim of developing their professional skills. Most teachers agree that this is one of the most effective ways to accelerate professional development (Yang & Ricks, 2013).

There are five categories of open classes.

1. The standard-meeting class is suitable for new teachers with less teaching experience. The new teacher needs to meet the basic level of school's standard in a certain period. The teachers are assessed by the members of teaching administration department at school.
2. The communication-focused class is a situation in which teachers observe and assess each other's classes, communicating and studying with each other, and working together to improve their lessons.
3. The study-focused class is one in which the teacher describes any existing problems in relation to teaching ideas or approaches. Other teachers who observe the class will engage in discussions about the most efficient teaching strategies, approaches, and methods to address the specific problems.
4. In an example class, an excellent teacher or one who is outstanding in one particular aspect of teaching demonstrates the essence and rules of teaching, which others are inspired to imitate.
5. A Competition-focused class is one that engages in teaching competitions organized within the school or a cluster of schools. The aim is to understand the school's teaching approaches and improve the teaching. In this activity, the classes are assessed by experts (Yu, 2008).

Mathematics teachers observe both the open mathematics classes and those of the other subjects.

16.4.1.2 Regular Classes

The regular class refers to the daily classroom teaching. This differs from the open class in that the teachers do not study it repeatedly to make sure it is perfect. Mathematics teachers observe and assess regular mathematics classes as well as other subjects.

16.4.2 The Objective of Observing and Assessing Classes

In the interview, almost all of the mathematics teachers mentioned that their schools had fixed numbers of observation classes, ranging from at least 8 per semester to at least 20. As a special case, one mathematics teacher who is also the education director of his school said that he had observed and assessed at least 60 classes in one semester. On average, a mathematics teacher observes 18 classes per semester. The majority of teachers explained that they are not forced by their schools to take part in observing and assessing activities, but they are eager to do so. Usually they observe more than the minimum required number of classes; they are generally only restricted by timetable clashes with their own classes. Most mathematics teachers said they prefer to observe mathematics classes of the same grade levels that they teach, as well as the open classes at school level for any subject.

There are three kinds of teachers who are observed: excellent teachers, including leading teachers, experienced teachers, teaching research group leaders, lesson plan group leaders, "master" teachers and others with rich experience in teaching mathematics; mathematics teachers in the same lesson plan group, who have similar teaching levels and experience; and other teachers including those who teach different subjects at the same grade level. The teachers interviewed reported that they have different preferences for the types of teachers they choose to observe, but the first and second of those described above are the most common.

In the course of the class, the mathematics teachers usually use a standardized class observation record book, to improve the efficiency of the observation and ensure that they are all observing the same aspects. As well, some teachers will bring corresponding lesson plans, learning materials, and so on. When mathematics teachers observe a colleague's class they will record background information about the subject, topic, teacher's name etc. They also write extensive notes about their observations, for the purpose of giving feedback or for their own reflections.

16.4.3 Aspects to Focus on When Observing Class

Mathematics teachers with different teaching styles and teaching experiences indicated in the interviews that they focus on different aspects when observing a class. However, the most common aspects were identified, as summarized in the following table.

Category	Item
Teacher-related	Example exercises
	Exercises
	Teaching structure
	Key points, points of difficulty, and critical points

(continued)

(continued)

Category	Item
	Teaching method
	Teaching language and gestures
	Introduction of lesson
	The design of blackboard
	The teaching of mathematics content and mathematics ideas
	Use of the textbook
	Teacher's questioning
	Classroom management
	Dealing with unexpected events
	Achievement of teaching aims
Student-related	Students' activities
	Students' questions
	Students' cooperation with the teacher
	Students' responses to teacher's questions
	Students' enthusiasm and participation in class
	Students' understanding of the content
Class environment	Interaction between teacher and students
	Classroom atmosphere

The three categories above are the main aspects of mathematics class observation. The items related to teachers are the richest ones and include most aspects of teaching, including the classroom management, basic teaching skills, content knowledge, and mathematical knowledge of teaching. During the interview, many teachers repeatedly referred to the effect of the teaching, including the extent of the students' understanding of the topic and the achievement of the teaching aims.

It was found from the interviews that the new teachers' main purpose in class observation is to study, so these teachers are concerned more about teaching-related items for their own learning and reference. The experienced teachers' main purpose is both to study and to critique, depending on the teaching level of teacher being observed. In particular, the leader of the lesson plan group is responsible for guiding the team members, improving their teaching levels and promoting their professional development. If the observed and observing teachers are of similar teaching levels, then the main purpose of class observing is to study, communicate, and reflect. In this case, the observer will focus mostly on student-related items and classroom environment items. They will reflect on their own teaching according to their observation, with the aims of improving their teaching and accumulating experience.

In general, the observers will communicate with the observed teacher, sharing their feelings and thinking, pointing out the strengths and weaknesses of the lesson, and giving suggestions. They will focus on typical teaching links, teaching structures or classroom events. Generally speaking, the exchange between peers is one of encouragement and praise. *More positive praise and less criticism.*

16.4.3.1 The Role of Observing and Assessing Classes

Classroom teaching is a clear indicator of a teachers' professional level. Both new teachers and excellent teachers agreed in the interview that observing and assessing classes helps to promote their professional development, especially in the case of new teachers. They can use other teachers like mirrors to reflect their own strengths and weaknesses, hence gaining a deeper understanding of their teaching and encouraging self-reflection. The lesson plan group members observe and assess each other's classes; this enhances their friendship and promotes the effective development of lesson plan group activities.

Many new mathematics teachers said in the interviews that their main purpose of class observation and assessment is to learn from other teachers, including about classroom management, teaching organization, blackboard design, teaching language and even teaching gestures, lesson introduction, the selection of example exercises and practice exercises, ways of guiding students and engaging them in activities, and anything concerned with basic teaching skills. Two of the teachers commented on their learning through this process:

> I learn a lot through class observation and assessment, such as how to manage the classroom, how to interact with students effectively, and how to deal with classroom emergencies etc.
>
> I particularly like to observe the classes implemented by the leader of our lesson plan group. His language is very concise, with no redundant word,which is an important skill that I'm trying to possess.

Experienced teachers can also improve through observing and assessing classes. It can inspire them, widen their teaching horizons, and change their teaching perspectives. One experienced teacher said that he was always inspired by the enthusiasm of new teachers:

> Observing classes can help break through teaching routine, looking for new ideas and inspiration.

16.5 Master–Apprentice Relationship

The master–apprentice relationship is used widely in training new teachers in primary and middle schools in China. When a new mathematics teacher arrives, the school will usually assign an experienced teacher to him as his *master*, aiming to help the new teacher to adapt to more quickly.

The master will direct most of the new teacher's work, including the lesson plan, classroom teaching, checking homework, and class management. The master and apprentice help each other and develop professional skills together (Fan & Liao, 2012). The excellent teachers, who have rich teaching experiences and outstanding teaching achievements, direct the new teachers, helping them to meet the school's requirements as soon as possible. The backbone teachers and subject leader teachers should also play an active role in this process. This peer cooperation reduces the new teacher's feelings of isolation or helplessness (Yu, 2003). Generally, those who are qualified to be master teachers are mathematics teachers with senior professional titles, leaders of lesson plan groups, leaders of teaching research groups, or principals. Most of the mathematics teachers can gradually form their own teaching styles and develop into fully-qualified teachers in three to five years with the direction of a master and self-reflection on their teaching practice.

16.5.1 Formats of Master–Apprentice Relationships

16.5.1.1 One-to-One

In this format, a new teacher is assigned to one master teacher. This is the most common format and has the advantage that the master's responsibility is clear. The disadvantage is that the new teacher has only one model, which may cause him/her to become a copy of the master.

16.5.1.2 Many-to-One

Here, several experienced teachers are assigned as masters to a new teacher. The advantage is to avoid the new teacher becoming a copy of a master. The new teacher is able to integrate the strengths of all the masters, thus forming a unique teaching style. However, it is a high requirement for a school's teaching resources, and many schools are unable to make this type of provision.

16.5.1.3 One-to-Many

An experienced teacher serves as a master for several new teachers at a time. The advantage is that the experiences are similar for new teachers, so they can discuss and communicate with each other when they encounter challenges. This makes more efficient use of the master's time and guidance, especially in schools that lack senior teacher resources. As well, the new teachers not only obtain the guidance of the master, but also develop their professional skills through the mutual exchanges and cooperation with their partners. The main shortcoming is that the master may be spread to thinly.

16.5.1.4 Many-to-Many

In this case, several experienced teachers are assigned to several new teachers as their co-masters. This can be said to be the most ideal one as it integrates the advantages of the other formats. Each new mathematics teachers can learn from a number of masters with different characteristics. Furthermore, the apprentices can discuss and communicate with each other and improve together (Liao, 2010).

16.5.2 The Role of the Master

16.5.2.1 Directing Teaching Skills

Mathematics teaching is not only a science but also an art. New teachers lack teaching experience and have limited teaching arts. When they enter the classroom, they meet so many challenges, including how to maintain students' attention, how to manage the classroom, and how to deal with emergencies. All of these experiences can be learned from the master.

The apprentice teacher can learn a lot from observing the master teacher's classes, including how to maintain order, how to set out the blackboard writing, the use of body language, teaching tone and words, questioning skills, dealing with students' questions, and evaluation (Ma, 2008).

16.5.2.2 Directing Teaching Content

For a long time, we have implemented a unified mathematics curriculum standard and compiled textbooks based on this standard. The textbook is an important reference material for primary and middle school teachers when designing lesson plans. The master will guide the apprentice to analyze the textbook, understand the intention of the textbook writer, and draw fully on the textbook but not be bound by it.

The selection of examples and practice exercises is an important aspect of mathematics teaching. The master will guide the newcomer to choose typical examples according to the students' existing knowledge, cognitive and psychological characteristics in the limited teaching time.

16.5.2.3 Directing Classroom Management

The master is also responsible for teaching classroom management skills, including student management, building the class atmosphere, and communicating and coordinating with other subject teachers and parents. Student management is the most important of these experiences. There are many aspects concerned with student management that consume energy, including how to counsel students at different levels, how to help excellent students maintain and make progress, how to help students with learning difficulties, and how to motivate and help the middle-level students.

The apprentice can often feel overwhelmed by these responsibilities on top of teaching the class. In this circumstance, the apprentice teachers will feel better if they can rely on the wisdom of the master to resolve their difficulties.

16.5.3 The Main Activities in the Master–Apprentice Relationship

16.5.3.1 Observing and Assessing Classes Together

One of the main master–apprentice relationship activities is to observe and assess classes together. The apprentice teacher's purpose is to learn from the master. The master's purpose is to check and guide. He will try his best to help his apprentice teacher develop his professional skills (Zheng, 2010).

16.5.3.2 Directing Oral Class Presentations and Open Classes

Usually, new teachers are required to participate in all kinds of teaching competitions and to have open classes regularly. At these times, the master will carefully help the apprentice teachers choose lesson topics, design lesson plans and practice, and refine them over time.

Taking part in oral class presentation competitions and open class competitions is different from normal classroom instruction, because the participant will try to design an outstanding lesson plan which is the crystallization of wisdom between the apprentice teacher and the master. The achievements of the apprentice teacher cannot be separated from the wisdom and efforts of the master.

16.5.3.3 Daily Consultation

Mathematics teaching happens every weekday, so the apprentice teacher may meet different situation every day. Daily consultation between master and apprentice plays an important role. The apprentice teacher consults with his master immediately when he encounters any kind of challenge in the classroom. Through this process, the apprentice teacher gradually accumulates the practical wisdom of mathematics teaching.

Compered with other guidance activities such as class observation and assessment, directing oral class presentation competitions, and so on, the random consultations with the master have an important role in improving the apprentice teacher's professional ability.

In addition to the above mathematics school-based professional development activities, teachers of the same subject often share an office. This enables them to have spontaneous discussions among themselves. This informal and non-institutionalized form of teacher assistance happens frequently. It is the mutual openness, trust, and supportive relationship between the teachers that influences their professional development profoundly (Guo, 2009).

The factors that affect the professional development of mathematics teachers are various, and this chapter has focused on the main school-based activities. As well, there are other formal training provisions for in-service teachers. Also, they can acquire practice wisdom through reflecting on their teaching (Twelve Key Normal Universities, 2002). However, compared with pre-service teacher education and the formal in-service training, the school-based professional development activities are the most common.

In 2012, the *Teachers' Professional Standards for Secondary Schools, Primary Schools, and Kindergartens* were produced, thus, for the first time since the publication of the *Teacher's Law of the People's Republic of China* (1994) that teachers' professional status was confirmed. This was a milestone for teachers' professional development. This standard set up thirteen areas and fifty-eight basic requirements from three dimensions—moral, knowledge, and ability—which provide the basic reference points for teachers as well as the directions for teacher improvement (Ministry of Education of the People's Republic of China, 2012). We believe that school-based professional development will play a more and more important role in the professional development of mathematics teachers with the implementation of the professional standards.

References

Fan, W., & Liao, Q. (2012). 基于教师专业发展的"师徒结对"的内涵及特征 [The connotation and characteristics of "apprenticeship partnership" based on teacher professional development]. *Journal of Educational Development, 9,* 45–47.

Guo, Z. (2009). 中学数学教师交流活动调查研究 [Investigation of mathematics teachers' exchange and cooperation in middle schools] (Master Degree Thesis) Beijing: Beijing Normal University.

Huang, R. (2004). 国际数学课堂的录像研究及其思考 [On international mathematics classroom video study and its implication]. *Comparative Education Review, 3,* 39–43.

Liang, W., Lu, L., & Huang, D. (2010). 中国特色基础教育教学研究制度的发展 [The development of teaching research institution of basic education with Chinese characteristics]. *Educational Research, 12,* 77–82.

Li, H., & Xiong, M. (2013). 西方社会学视域中的教师合作 [Teachers' cooperation in the view of western sociology]. *Studies in Foreign Education, 11,* 74–80.

Li, Y., & Huang, R. (2013). *How Chinese teach mathematics and improve teaching*. London: Routledge.

Liao, Q. (2010). *Research on "apprenticeship partnership" based on teacher professional development* (Master Degree Thesis). Chongqing: Southwest University.

Ma, X. (2008). *The study on the influence of "apprenticeship partnership" upon beginning teacher growth* (Master Degree Thesis). Lanzhou: Northwest Normal University

Ministry of Education of the People's Republic of China. (2012). 中学教师专业标准(试行) [Professional standards of middle school teachers]. Retrieved from http://www.moe.edu.cn/publicfiles/business/htmlfiles/moe/s6127/201112/127830.html.

Ministry of Education of the People's Republic of China. (2015). *2013* 年教育统计数据全国基本情况 [The national basic situation of educational statistics data in 2013]. Retrieved from http://old.moe.gov.cn/publicfiles/business/htmlfiles/moe/s8493/index.html.

Standing Committee of the National People's Congress of China. (1993). 中华人民共和国教师法 [Teacher's Law of People's Republic of China]. Retrieved from http://www.gov.cn/banshi/2005-05/25/content_937.htm.

Standing Committee of the National People's Congress of China. (2006). 中华人民共和国义务教育法 [Teacher's Law of People's Republic of China]. Retrieved from http://www.moe.edu.cn/publicfiles/business/htmlfiles/moe/moe_619/200606/15687.html.

Twelve Key Normal Universities. (2002) 教育学基础 [Foundation of education]. Beijing: Educational Science Press. (in Chinese).

Yang, Y., & Ricks, T. E. (2013). Chinese lesson study: Developing classroom instruction through collaborations in school-based teaching research group activities. In Y. Li & R. Huang (Eds.), *How Chinese teach mathematics and improve teaching* (pp. 51–65). New York: Routledge.

Yu, W. (2003). 论以校为本的教学研究 [A study of the school-based research on class teaching]. *Educational Research, 4,* 53–58.

Yu, W. (2006). 公开课再认识 [Re-study of open class]. *China Education Daily, 5*(2nd, June).

Yu, X. (2008). *A case study on the role of open class in the professional development of mathematics' teachers* (Master Degree Thesis). Beijing: Capital Normal University.

Zhang, H., & Zhang, J. (2011). 三十年来我国中小学教师合作研究的梳理与反思 [On the primary and high school teacher cooperation over the past thirty years]. *Journal of Hebei Normal University (Educational Science Edition), 4,* 22–26.

Zhang, Y., & Yu, X. (2010). 公开课在数学教师专业发展中作用的个案研究 [Case study on the function of demo class in the mathematics teachers' professional development]. *Journal of Mathematics Education, 19*(4), 19–22.

Zhao, J., Pei, X., Feng, R., Cheng, J., & Jin, Y. (2013). 我国教师的专业发展实践及其对学生成绩的影响:基于五城市调研的分析 [Chinese teacher professional development practice and its effects on students' achievement: Based on a survey for five sample cities]. *Global Education, 2*(22–33), 53.

Zheng, Y. (2010). 浅谈数学教学中的"师徒结对" [Study on "apprenticeship partnership" in mathematics teaching]. *Way of Success, 13,* 21.

Chapter 17
Pre-service Mathematics Teacher Education

Li Tong and Xinrong Yang

Abstract The professional education of teachers in China consists of three essential stages: pre-service, inductive, and in-service. These all come under the term "teacher education." Pre-service mathematics teacher education is concerned with training mathematics teachers for secondary and primary schools. With formal pre-service teacher education in China having developed over a period of more than 100 years, it has developed some exclusive patterns and features.

The professional education of teachers in China consists of three essential stages: pre-service, inductive, and in-service. These all come under the term "teacher education." Pre-service mathematics teacher education is concerned with training mathematics teachers for secondary and primary schools. With formal pre-service teacher education in China having developed over a period of more than 100 years, it has developed some exclusive patterns and features.

17.1 Course of Educational Development of Pre-service Teachers in China

China's pre-service teacher education has developed over more than 100 years. Based on the old maxim of "Great learning makes a teacher; moral integrity makes a model," we have become accustomed to calling the regular teacher education "normal education." Historically, the Nan Yang Public School Normal College was China's first normal school founded in 1897 specially for training teachers, founded by the main representative of the Westernization Group in the late Qing Dynasty, Sheng Hsuan-Huai. The establishment of this normal college was significant in the

L. Tong (✉)
Chongqing Normal University, Chongqing, China
e-mail: 569599691@qq.com

X. Yang
Southwest University, Chongqing, China

Y. Cao and F.K.S. Leung (eds.), *The 21st Century Mathematics Education in China*, New Frontiers of Educational Research,
https://doi.org/10.1007/978-3-662-55781-5_17

educational history of China since it marked the beginning of China's normal education (Cui, 2006). Later, the teacher's training faculty of the Imperial University of Peking (the predecessor of Beijing Normal University, a famous institution of higher learning) was founded in 1902 and was the first Normal University in history. The course structure of mixing professional courses and educational courses and the teacher-training model which was designed by the teacher's training faculty formed the foundation and pattern of China's higher normal education (Higher Normal School for short) for the next century.

Later, the Normal School was divided into two classes according to the *Approved School Regulation* ("*Gui-Mao Educational System*") promulgated by China's Qing Government on January 13, 1904; the Junior Normal School was concerned with the nature of secondary education and the Senior Normal School with the nature of higher education, and the study period was eight years. The Junior Normal School, in which the study period was five years, focused on the training of primary school teachers, recruiting students from higher primary school level. From this beginning, secondary normal education (Secondary Normal School) developed. The task of the Secondary Normal School is to train primary school and kindergarten teachers in the areas of socialist consciousness, the world outlook of dialectical materialism, communist ethics, and the cultural and professional knowledge and skills necessary to work in elementary or infant education, along with having a love for children, for a desire to serve socialist education wholeheartedly, and promote good health.

17.2 Development of Secondary Normal Schools

After the establishment of the People's Republic of China, the national government conducted a planned construction of normal education. This included reconstructing the Normal Schools of old China, in 1952, holding lots of short-term teacher-training classes, and adding Junior Normal Schools. During the first five-year plan to develop the national economy, the government stressed the development of Secondary Normal Schools, reduced Junior Normal Schools gradually, and suspended short-term training classes. Before and after 1956, the Ministry of Education of the Central Government issued *Regulations for Normal Schools*, *Regulations for Primary Schools Attached to Normal Schools* and *Educational Practice Methods for Normal Schools* for trial implementation; issued *Teaching Programs for Normal Schools* and *Teaching Programs for Kindergarten Training Schools*; and compiled and published the teaching programs and teaching materials for each subject in the normal school. In the early 1960s, the majority of Junior Normal Schools had upgraded to Senior Normal Schools and stopped recruiting students from higher primary schools. However, due to the influence of "the Great Cultural Revolution," the majority of Normal Schools had been closed; the Senior Normal Schools in various regions were restored, reorganized, and enriched until 1976. The Ministry of Education convened a national normal education working conference in June 1980, summarizing the experience of establishing secondary normal education in the 30 years since the founding of new China, and discussing the issue of how to improve secondary normal education. After the conference, the Ministry of Education issued the *Opinion of the Ministry of Education on Improving Secondary Normal Education, Regulations for Secondary Normal Schools (Draft for Trial Implementation), Teaching Programs for Secondary Normal Schools (Draft for Trial Implementation),* and *Teaching Programs for Kindergarten Training Schools (Draft for Trial Implementation).* These documents made specific provisions for the nature, tasks, educational systems, and courses of Secondary Normal Schools.

According to these provisions:

1. Essentially, the Secondary Normal School is a part of the Secondary Specialized School. The task of the Secondary Normal School is to train primary school teachers and kindergarten teachers in the ways described earlier in this paper. If possible, the Secondary Normal School also undertakes the task of training in-service primary school teachers as well as teachers and child-care workers for kindergartens, as needed.
2. The Secondary Normal School has an educational system divided into 3 years and 4 years and recruits graduates of junior middle schools or young people with the equivalent educational backgrounds.
3. The Secondary Normal School offers courses including politics, Chinese language and primary school Chinese language teaching methodology, mathematics and primary school mathematics teaching methodology, physics, chemistry, biology, primary school science teaching methodology, foreign languages, geography, history, psychology, education, physical education, music, art, and educational practice. In addition, Normal Schools for ethnic minorities offer local ethnic language courses.

According to statistics, the numbers of Secondary Normal Schools grew from 610 in the year of 1949 to 1353 in the year of 2000 throughout the country; the numbers of students increased from 152,000 to more than 500,000. Since the founding of the People's Republic of China, the Secondary Normal Schools have trained about 5 million graduates and, at the same time, have trained a large number of in-service teachers (Li, 2010).

17.3 The Development of Higher Normal Universities

Since the establishment of the People's Republic of China, Higher Normal Universities were set up independently from other comprehensive universities. The Ministry of Education issued *Provisions for Higher Normal Schools (Draft)* in 1952, specifying the study period of Normal Colleges and Universities, mainly for training senior secondary school teachers, to be four years and the study period of Normal Specialized Postsecondary Colleges, for training junior secondary school teachers, to be two years. The conference of National Higher Normal Schools was convened twice, in 1953 and 1956; before and after these conferences, many documents relating to reform Higher Normal Universities education were issued, and many professional teaching plans (or teaching schemes) were developed, as well as teaching programs for many subjects. Hence, Chinese higher normal education was standardized, and teaching quality improved continuously. The development of the higher normal education was quick; the number of Higher Normal Universities was only 12 in 1949, with 12,000 students, but the number had increased to 194 by the end of 1982 (66 Normal Universities and Normal Colleges, and 128 Normal Specialized Postsecondary Colleges), with 281,800 students.

There are 17 general majors set by the Higher Normal Universities, corresponding to middle school subjects such as Chinese language, mathematics, and English. In addition, in a few Higher Normal Universities, some other majors, such as library science and educational technology, have been introduced according to local needs for scientific and technological development. Within each major, there are five courses: political courses, including the History of the Chinese Communist Party, Political Economy, Philosophy, and Communist Moral Education; foreign language courses (mainly English); educational courses, including psychology, education, methodology for teaching a certain subject, and teaching practice; physical education; and compulsory and optional subject courses. In addition, conducting scientific research and particularly research on educational science is an important task for the Higher Normal Universities. Various research institutions on specific subjects and general education science have been set up in most normal universities to carry out deep investigations of the teaching and learning of certain subjects (Table 17.1).

In addition, in order to train teachers for higher institutions and train researchers for scientific research institutes, Postgraduate Schools have been established in some Higher Normal Universities to recruit postgraduate students and award master's and doctoral degrees. They also undertake the task of training teachers from secondary schools in various formats including correspondence schools, evening universities, training classes, and classes for advanced studies. In addition, Colleges of Education or Educational Administration Institutes specifically undertake the task of training in-service teachers and educational administrative staff from secondary schools.

Since the 1970s, the idea of lifelong education has become widespread internationally. As well, in the later period of the 1990s, with structural adjustments made to higher education in China, teacher training has undergone many reforms, so the system of independently established normal universities or schools specializing in teacher training has broken down; now, comprehensive universities and other colleges and universities also participate in training teachers. The main challenge to teacher education in China has emerged to be a mismatch between the requirements and the capacity to improve the quality (Yuan, 2004). "Normal education" has been unable to meet the demands for the development of teacher education. Therefore, the concept came to be updated from "normal education" to "teacher education" as the result of the document *Decisions of the State Council of the People's Republic of China on the Reform and Development of Elementary Education*. The policies and systems of "teacher education" have also replaced the

Table 17.1 Course setting of pre-service primary school teacher education

Learning fields	Suggestion modules	Credit requirements		
		Three-year junior college education	Five-year junior college education	Four-year college
1. Development and learning of children 2. Education basis of primary schools 3. Subject education and activity guidance of primary schools 4. Mental health and moral education 5. Professional ethics and professional development	Child development; cognition and learning of elementary school students; educational philosophy; course design and evaluation; effective teaching; the development of school education; class management; school organization and management; education policies and regulations. Subject course standard and teaching material study of primary schools; the teaching design of subjects of primary schools; interdisciplinary education of primary schools; integrative practical activities of primary schools. Psychological guidance of elementary school students; morality development and moral education of elementary school students. Professional ethics of teachers; educational research method; teacher's professional development; application of modern educational technology; teacher language; writing skill, etc.	Minimum compulsory credit 20 credits	Minimum compulsory credit 26 credits	Minimum compulsory credit 24 credits
6. Educational practice	Educational probation; educational practice	18 weeks	18 weeks	18 weeks
Minimum total credit of teacher education course (including elective courses)		28 credits + 18 weeks	35 credits + 18 weeks	32 credits + 18 weeks

Note (1) 1 credit is equivalent to 18 class hours under a teacher's guidance, which is acceptable via assessment; (2) the learning field is compulsory for every learner; the suggested modules can be selected or combined by teacher education institutions or learners and can be compulsory or elective; the credit for each learning field or module is determined by the teacher educational institutions in accordance with relevant provisions

Table 17.2 Course setting of pre-service secondary school teacher education

Learning fields	Suggestion modules	Credit requirements	
		Three-year junior college education	Four-year college
1. Development and learning of children 2. Education basis of middle schools 3. Subject education and activity guidance of middle schools 4. Mental health and moral education 5. Professional ethics and professional development	Child development; cognition and learning of secondary school students; educational philosophy; course design and evaluation; effective teaching; the development of school education; class management. Subject course standard and teaching material study of secondary schools; the teaching design of subjects of secondary schools; integrative practical activities of secondary schools. Psychological guidance of secondary school students; morality development and moral education of secondary school students. Professional ethics of teachers; teacher's professional development; educational research method; teacher language; application of modern educational technology, etc.	Minimum compulsory credit 8 credits	Minimum compulsory credit 10 credits
6. Educational practice	Educational probation; educational practice	18 weeks	18 weeks
Minimum total credit of teacher education course (including elective courses)		12 credits + 18 weeks	14 credits + 18 weeks

Note (1) 1 credit is equivalent to 18 class hours of learning under a teacher's guidance, which is acceptable via assessment; (2) the learning field is compulsory for every learner; the suggested modules are selected or combined by teacher educational institutions or learners and can be compulsory or elective; the credit of each learning field or module is determined by the teacher education institutions in accordance with relevant provisions original "normal education" gradually, embodying changes in the concepts and ways of developing teacher education (Table 17.2).

So far, China's pre-service teacher education has been developing for 100 years, initially forming a pre-service teacher education system serving various types, levels, and formats of individually established Normal Schools. In their various formats, consisting of Secondary Normal Schools, Normal Training Colleges, and Teachers' Colleges (Normal Universities), the Normal Schools have trained tens of thousands of teachers for secondary and primary schools. In addition, teacher-training majors set by Colleges of Education, Schools for Teachers' Advanced Studies, and other schools are also important providers of teacher training. Therefore, there are six types of school running various forms of teacher education in China: Secondary Normal Schools, Normal Training Colleges, Teachers' Colleges (Normal Universities), Colleges of Education, Schools for Teachers' Advanced Studies, and other types of schools. According to statistics published in 1995, China had 236 Higher Normal Schools, of which 13 could offer PhD degrees and 38 could offer master's degrees; 76 Normal Bachelor Degree Colleges with about 320,000 students; 160 Higher Normal Specialized Postsecondary Colleges with about 260,000 students; 897 Secondary Normal Schools with about 850,000 students; 242 Colleges of Education with about 214,000 students; and 2031 Schools for Teachers' Advanced Studies with about 516,000 students. In addition to all forms of individually established Normal Schools at different levels, teacher-training majors or Secondary Colleges initiated by Open Universities, Open Normal Colleges, and about 180 Comprehensive Universities have trained a large quantity of teachers for secondary and primary schools as an auxiliary path.

17.4 New Development of China's Pre-service Teacher Education

Entering the twenty-first century, due to the demands of the high-level development of normal education in China, Secondary Normal Schools were closed successively and were upgraded to Normal Training Colleges or Teachers' Colleges. However, the education given by the Secondary Normal Schools is still praised for training teachers comprehensively and with high levels of teaching skills. Generally, the Higher Normal Universities are the main providers of undergraduate courses, and these are reformed continuously over time.

17.4.1 The Making of Standards for Teacher Educational Courses

In order to implement the national educational planning outline, strengthen the reform of teacher education, specify and guide teacher education curricula and teaching, and train high-quality teacher teams, the Ministry of Education issued *Opinion of the Ministry of Education on Vigorous Promotion of Reform in Teacher Educational Courses* on October 8, 2011. The introduction of this policy has had an inevitable impact on China's normal education. The document's attachment, *Standards for Teacher Educational Courses (Trial Implementation),* documents the state's requirements for the reform of China's teacher education curriculum and is an important basis for developing teacher education curricula, developing teaching materials and curriculum resources, carrying out teaching and evaluation, and identifying teaching qualifications. *Standards for Teacher Educational Courses* clearly proposes the fundamental philosophy of "people-oriented education, practical orientation, and lifelong learning" and offers suggestions regarding education course objectives and content for pre-service teachers in kindergartens, primary schools, and secondary schools.

There are three main curriculum objectives: (1) belief in education and responsibility, which mainly refers to teachers developing correct views of students and corresponding behaviors, teachers and corresponding behaviors, and education and corresponding behaviors; (2) educational knowledge and ability, which refers to the knowledge and skills of understanding students, educating students, and implementing self-development; and (3) educational practice and experience, which refers to the experience of observing and emulating, participating in and researching educational practice.

There are six compulsory learning fields in the curriculum, and several other fields which can be used as optional models are suggested in the standard. Detailed information is as follows.

17.4.1.1 Educational Courses for Pre-service Teachers in Primary Schools

Course objective: The educational courses for pre-service primary school teachers aim to guide future teachers to understand the characteristics and differences during the growth of primary school students and to create supporting, challenging learning environments that will address primary school students' thirst for knowledge and desire to perform. Another aim is to guide future teachers to understand the life experiences of primary school students and the significance of on-site resources, and to design and organize appropriate activities that will guide children to make their own decisions, collaborate and explore learning, thereby forming good learning habits. Future teachers are also guided to understand the importance and nature of communication in the development of primary school students and to organize various collective and partner activities, thereby supporting children to grow happily in a meaningful school life.

17.4.1.2 Educational Courses for Pre-service Teachers in Middle Schools

Course objective: The educational courses for pre-service secondary school teachers aim to facilitate future teachers to understand the characteristics of adolescence and the effect of adolescence on the lives of secondary school students, and to guide secondary school students to spend adolescence smoothly. As well, these courses guide pre-service teachers to understand the cognitive characteristics and learning styles of secondary school students and to create learning environments that encourage students to think independently and explore subject knowledge in multiple ways. A third objective is to guide pre-service teachers to understand the personality and cultural characteristics of secondary school students, and to respect their self-awareness and guide them to plan their lives, thereby developing social skills in diversified activities.

17.4.2 Reform in the Authentication of Teacher Qualifications

Owing to a large demand for teachers at the basic education stage in China, and the subsequent need to encourage university graduates to select teaching as their future profession, graduates from normal universities and schools were allocated by the state and enjoyed the treatment of state public institutions before 1993. Article 10 of the *Teacher Law of the People's Republic of China,* promulgated by China in 1993, declared that the state institutes a system of qualifications for teachers. All Chinese citizens, who abide by the Constitution and Laws, take a keen interest in education, have sound ideological and moral characters, possess a record of formal schooling as stipulated in this Law or have passed the national teachers' qualification examinations, and have educational and teaching abilities, after being evaluated as qualified, may obtain qualifications to be teachers (Ministry of Education of the People's Republic of China, 1993). The State Council of the People's Republic of China promulgated the *Regulations for Teacher Qualifications* in 1995 in accordance with the provisions of *Teacher Law*, and the new regulations stated that graduates from normal schools could apply for teacher qualification certificates only after having passed educational science and educational psychology courses and achieving higher than Level 2 Rank B (Level 2 Rank A for Chinese language major) in a common provincial Mandarin examination. Non-normal students and other social personnel can apply for teacher qualification certificates only by passing the certification examination.

The teacher qualification examination is divided into provincial and national examinations, and the certification is universal throughout the country. The provincial examination covers two subjects, educational science and educational psychology, and the national examination also covers two, comprehensive attainment and educational knowledge and ability. After 2000, provinces and cities gradually phased out the provincial examination. In 2013, the Ministry of Education published *Sequential Methods for Qualifying Examinations for Primary and Middle School Teachers*, giving its examination center the responsibility of formulating uniform examination standards and examination outlines, organizing written examinations and interview questions, and establishing a question pool. The teacher qualifications of graduates from Normal Schools are not recognized directly any longer, but are incorporated into the examination scope. The teacher qualification examination certificate is valid for three years; this provides a strong incentive for Normal Schools to adjust their training models and course settings. The main aspect of this reform is the emphasis on training in educational practice for Normal School students.

17.4.3 Reform of No-Fee Pre-service Teacher Education

Teachers are the foundation of a great education policy; the overall quality and levels of teacher teams in primary and secondary schools are critical to educational development, and therefore, the training and development of outstanding teachers is an issue of concern for masses of people. This is particularly important to strengthen even further China's tradition of respecting teachers and valuing education. To achieve this goal and encourage more outstanding young people to choose education as their lifelong careers, the State Council of the People's Republic of China introduced no-fee pre-service teacher education for students in Normal Universities, directly under the Ministry of Education, in May 2007. This provision was made in six Normal Universities under the Ministry of Education, Beijing Normal University, East China Normal University, Northeast Normal University, Central China Normal University, Shaanxi Normal University, and Southwest University, for students entering the school in autumn of 2007 (the education for normal students was free once before, in the 1990s).

Before entering the school, no-fee pre-service teachers are required to sign an agreement with the administrative department for education in the province in which they are studying, making a commitment to work in primary or secondary schools for more than 10 years after graduation. No-fee graduates can work in schools in cities and towns but are required to teach in rural schools for two years. The state encourages the no-fee normal graduates to work in basic education for a long time or for life. Correspondingly, the no-fee pre-service teachers enjoy preferential policies offered by the state: (1) The central finance is responsible for arranging their tuition and accommodation at the university and granting living allowances of RMB 600 Yuan per month (granted for 10 months every year); (2) the provincial-level administrative department for education administers the graduates' teaching posts; (3) the no-fee pre-service teachers can move between schools, having the opportunity to work in educational management positions; (4) the state provides a guarantee, with good conditions, for the no-fee pre-service teachers to continue with further education, enabling those who are eligible to enroll in university as master's degree candidates of education to learn specialized courses while working.

17.5 Methods of Training Pre-service Mathematics Teachers in China

Along with the overall development of pre-service teacher education, the model for pre-service mathematics education in China has formed gradually. The Colleges of Education, or Colleges of Elementary Education of Normal Schools (Normal Training Colleges, Teacher Colleges, and Normal Universities), mainly undertake the training of primary school mathematics teachers; these include three-year junior college, five-year junior college, and four-year and six-year colleges, but the four-year college is the most common. The Schools of Mathematics of Normal Universities mainly undertake the training of secondary school mathematics teachers, and most offer four-year undergraduate programs. In addition, some primary and secondary school mathematics teachers are trained at Teachers' Colleges of Education in some Comprehensive Universities.

17.5.1 The Training of Pre-service Primary School Mathematics Teachers

In China, primary school mathematics teachers do not only teach mathematics. Sometimes, they need to teach several other subjects as well, for example, art or calligraphy; thus, there is no specific mathematics major in pre-service teacher-training institutions to train primary school mathematics teachers. Generally, in each institution, a primary school education major is set and, under this major, there will be several orientations (such as mathematics, Chinese language, English, music education, social science orientation, or science). The duration of the primary school education major for those who have only completed junior secondary school is either five years, for a college degree, or six for a bachelor's degree. Those who have completed senior secondary school take three years to earn a college degree or four for a bachelor's degree.

17.5.1.1 Training Objective

Each institution formulates its own training objectives by combining its own characteristics and development directions, and the training objectives set for pre-service primary school teachers, as stated in the *Standards for Teacher*

Educational Courses. The training objectives of three representative institutions, Northeast Normal University, Chongqing Normal University, and Xi'an University, are provided in Table 17.3. These three were selected because they represent different academic levels and training modes. Northeast Normal University is the key university directly subordinate to the Ministry of Education of China, and its primary school education major is characterized by comprehensive training with a single major, primary education. Chongqing Normal University is directly subordinate to the province, and its primary school education major is characterized by training in several major specializations. Xi'an University is also directly subordinate to the province, and its primary school education major is characterized by general training in social science and science.

As shown in Table 17.3, different institutions base their training objectives on different viewpoints. However, there are some common characteristics:

1. Orientation of training: Determine the "nature" of education, the "direction" in primary schools, and the undergraduate "standard." In relation to determining the "nature" of education, the three universities all focus on occupational features, educational concepts, specialized knowledge, and teaching skills. Determining the "direction" in primary schools refers to training specialized primary school teachers or professional personnel working in primary school education and scientific research management. Determining the undergraduate "standard" reflects the reality that the majority of primary school education majors are now at undergraduate level in China. Even though the overall academic levels of the three selected institutions are different, their primary school education majors are all at undergraduate level, indicating that, with the development of society, the requirements regarding the quality of primary school teachers in China have been raised.
2. Professional quality: From the training objectives listed in Table 17.3, it can be seen that primary school teachers are required to develop morally, intellectually, and aesthetically and have a deep understanding of educational theory, subject matter, teaching skills, and scientific research. This also reflects that a main goal of Chinese pre-service mathematics teacher training is to train pre-service teachers with high overall quality, a solid theoretical basis, and high levels of practical ability.

17.5.1.2 Training Format

Currently, primary school mathematics teachers are mainly trained in Colleges of Education or Colleges of Primary School Education in Normal Universities. In addition, some are trained in Normal Colleges, which were upgraded from secondary Normal Schools several years ago (e.g., Nantong Normal College and Hunan First Normal College). Generally speaking, there are three training formats in these institutions: an integrated-subjects mode, a subject-orientation mode, and an intermediate mode.

Table 17.3 Training objectives of the selected institutions

Schools	Training objective
Northeast Normal University (primary education major)	Implement the Party's educational policy comprehensively, guide and promote students to become autonomous learners having insight, ability, and sense of responsibility, and train high-quality and specialized teachers with loyalty to educational business, modern educational concepts, deep educational and theoretical attainment, broad educational views, and high ability levels in education, teaching, scientific research, and management as well as a spirit of innovation and ability to adapt to the needs of the twenty-first century
Chongqing Normal University (primary mathematics education major)	Based on the developing tendency for specialization by primary school teachers, students become high-quality primary school mathematics teachers and professionals adapting to the requirements of the times and ongoing reforms in primary school mathematics education. Students are also encouraged to develop morally, intellectually, physically, and aesthetically, to be able to integrate subject attainment and mathematics literacy, to have a mastery of basic theoretical knowledge professional skills of mathematics education in primary schools, teaching and scientific research, etc. They become emotionally equipped to teach primary school mathematics education, develop innovative teaching skills, and being capable to develop further to higher levels
Xi'an University (science component of primary education major)	This major equips students with the technical skills and a love of primary school education, an understanding of the basics of education and the ability to adapt to reforms, solid mastery of knowledge, theory and skills related to the primary school education major and modern education concepts, practical abilities, and the ability and attitudes to sustain their future professional growth

Note The contents of this table have been taken from the colleges' recent training programs

In the integrated-subjects mode, the primary school subjects are not separated. The focus is on the comprehensive development of all subjects, so that graduates are able to teach several subjects. For example, as described above, this mode is used by the Northeast Normal University. In the subject-orientation mode, teachers are trained in particular primary school subjects, based on the notion that, in practice, many primary school teachers only teach one subject. Therefore, a professional primary school teacher should be equipped with specialized subject knowledge and teaching skills. The mode developed in Chongqing Normal University is one example of this. The intermediate mode lies between the above two, based on the idea that primary school teachers should have comprehensive preparation in various primary school subjects even though they may eventually only teach one specific subjects (or, sometimes, one main subject plus one or two other subjects, such as drawing and physical education). In this mode, pre-service teachers are trained with either a science or a social science orientation. Zhejiang Normal University and Xi'an University are two examples of institutions that have adopted this training mode.

17.5.1.3 Course Settings

The primary school education course structures of each of the selected universities are shown in Figs. 17.1, 17.2, and 17.3. Even though these courses are different in name and classification, there are some common features that include general courses, theoretical educational courses (some are included in subject foundation courses or teacher educational courses), subject teaching courses, professional ethics courses, skill training courses, and subject specialist courses.

1. General courses: These are courses that are studied by all undergraduates in China, with the aims of expanding their horizons, increasing the breadth of their knowledge, and improving their quality. Generally, the general courses include Ideological and Moral Training and Fundamentals of Law, Outline of Modern and Contemporary Times of China, Situation and Policy, Fundamental Tenets of Marxism, Introduction to Mao Zedong Thought and Theoretical Systems of Socialist Theory with Chinese Characteristics, College English, College Sports, College Chinese, Fundamentals of College Computing, and Advanced Mathematics. The credit for these courses accounts for about 30% of the total.
2. Educational theory courses: These courses require pre-service teachers to understand and master fundamental education principles and rules and to establish the solid basic knowledge of educational theory. Generally, the

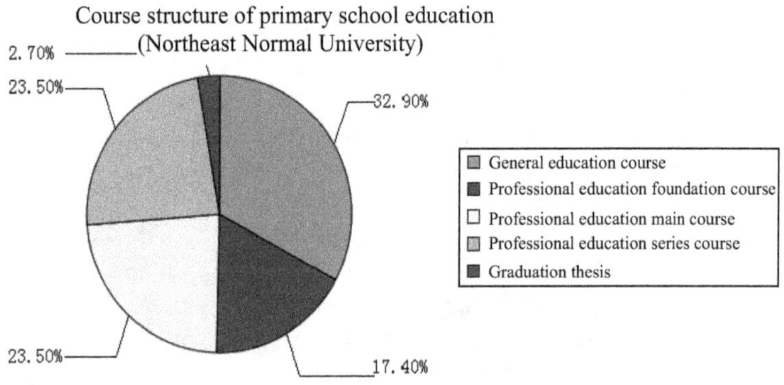

Fig. 17.1 Course structure of primary school education (Northeast Normal University)

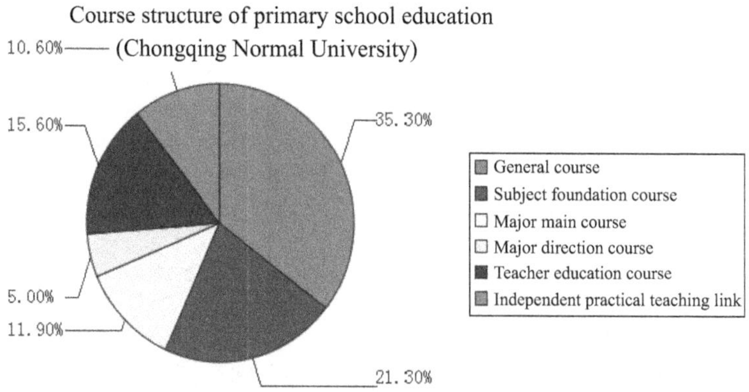

Fig. 17.2 Course structure of primary school education (Chongqing Normal University)

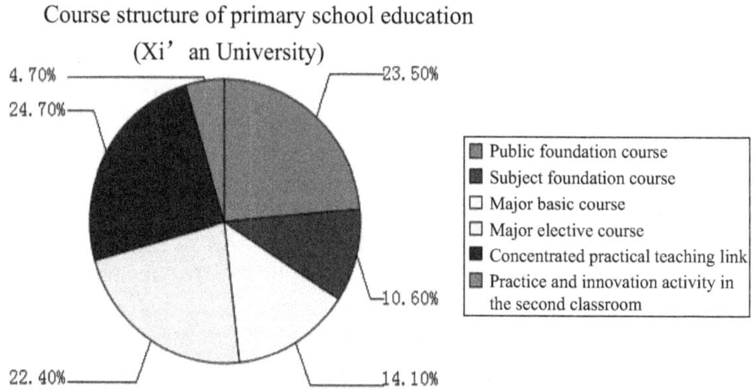

Fig. 17.3 Course structure of primary school education (Xi'an University)

educational theory courses include Psychology Principles, Educational Principles, History of Chinese Education, History of Foreign Education, Educational Statistics, Developmental Psychology, Educational Psychology, Educational Research Methods, Curriculum and Teaching Methodology, Educational Philosophy, and Educational Management. Such courses contribute to a high proportion of the integrated-subjects mode. For example, the credits of such courses of Northeast Normal University account for about 19%, while they account for about 10% in the subject-orientation and intermediate modes.

3. Subject teaching courses: These courses focus on how to teach certain subjects. Generally, the primary school mathematics teaching courses include Curriculum and Teaching Methodology for Primary School Mathematics, Teaching Material Analysis and Teaching Design for Primary School Mathematics, Problem-Solving Study for Primary School Mathematics, Competition and Counselling for Primary School Mathematics, Elementary Mathematics Study,

Curriculum Standards and Teaching Material Analysis for Primary School Mathematics, and Mathematics Learning Psychology. The credits for these courses account for about 5% of the total.

4. Teacher skill training courses: These focus on developing teaching skills. Generally, the teacher skill training courses include Mandarin, Three Calligraphy, Physique and Dance, Writing and Drawing, Vocal Music, Oral Training, Teaching Skills, and Design of Training Aids and Learning Materials. These account for about 15% of the total courses.
5. Mathematics courses: These courses are used for improving pre-service mathematics teachers' understanding of mathematical knowledge. Usually, they include Mathematical Analysis, Higher Algebra, Probability and Mathematical Statistics, Elementary Number Theory, Mathematics Thoughts and Methods, Advanced Mathematics, Modern Algebra, and Outline of Primary School Mathematical Knowledge. The credits for such courses are high in proportion to the subject-orientation mode, accounting for about 20%, but only account for about 7% in the other two modes.

17.5.1.4 Practical Teaching

Practical teaching for pre-service primary school mathematics teachers focuses mainly on Educational Probation, Educational Practice, and the Graduation Thesis. Detailed information is shown in Table 17.4.

Each training institution has established cooperative partnerships with many primary schools, some of which are used regularly as practice bases for the institutions. In the practical teaching, the educational probation mainly includes attending lectures and observing and emulating in the primary schools for about two weeks. The educational practice includes activities of trial teaching and educational investigations in primary schools. The educational practice is normally carried out

Table 17.4 Information about practical teaching component of primary school pre-service teacher training

	Educational probation	Educational practice	Graduation thesis
Northeast Normal University	1 credit, flexible	5 credits, in the 7th semester	4 credits, in the 8th semester
Chongqing Normal University	2 credits, in the 4th and 5th semesters, respectively	6 credits, in the 7th semester	6 credits, in the 8th semester
Xi'an University	2 credits, in the 4th semester	4 credits, in the 6th semester	10 credits, in the 8th semester

in the second half-semester of grade 3 or the first half-semester of grade 4 and is intensive for about 3 months. The graduation thesis is started in the second half-semester of grade 4 and takes about 8 weeks.

17.5.2 Training of Pre-service Secondary School Mathematics Teachers

17.5.2.1 Training Objective

The training object for pre-service secondary school mathematics education is to prepare junior and senior secondary school mathematics teachers. Different training institutions formulate their own training objectives by combining their own characteristics and developmental directions, as stated in *Standards for Teacher Educational Courses*. The training objectives of the pre-service secondary school mathematics teachers in six normal universities are described here. These six are representative of a range of provisions offered in China. Beijing Normal University and Southwest University are universities directly subordinate to the Ministry of Education; Zhejiang Normal University and Chongqing Normal University are normal universities directly subordinate to the province; and Xi'an University and Chongqing University of Arts and Sciences are universities which are directly subordinate to the province. Not only do these six universities represent China's universities in different regions and academic levels, but they also represent universities at different levels in the same region (Southwest University, Chongqing Normal University, and Chongqing University of Arts and Sciences are all located in Chongqing).

Even though these training objectives are different in expression, some common characteristics can be identified:

1. Training orientation: The training objective is to train excellent mathematics teachers at basic education level or high-level secondary school level. Beyond that, teaching researchers, personnel in teaching management organization, and educators for teacher-training institutions are included.
2. Quality requirements: There is a very high expectation of quality for excellent teachers competent to teach secondary school mathematics education. From the knowledge perspective, graduates are expected to have solid mathematical knowledge, knowledge of general pedagogy, and knowledge of mathematics education; and with respect to competences, they need to have mathematical thinking ability, practical teaching skills, self-development attitudes and abilities, and problem-solving skills.

17.5.2.2 Training Form

At present, pre-service secondary school mathematics teachers are mostly trained by the Schools of Mathematics in normal universities. Generally, the training period is four years, and after the completion of the training, a bachelor's degree will be awarded (Table 17.5).

Table 17.5 Training objective of the selected universities

Universities	Training objective
Beijing Normal University	Students learning for 4 years can master the basic theory, basic knowledge, and basic approaches of the science of mathematics, master the basic principles of mathematics education, are subjected to strict mathematical thinking training, can solve problems by using mathematical knowledge and computers, and have strong practical abilities for education and teaching and knowledge renewal capacities. Graduates can work on teaching, scientific research and management work in key secondary schools, and departments of teaching research and educational management
Southwest University	The mathematics and applied mathematics (normal) major mainly trains high-level mathematics teachers of middle schools, who can tightly combine mathematics theory and mathematics classes and have "perfect personality, solid foundation of mathematics, prominent teacher quality, and strong comprehensive abilities." It requires that the mathematics teachers master the basic theory and basic methods of mathematics, accept basic training in mathematical modeling, computing, and mathematical software, have high scientific literacy and strong consciousness of innovation, and can solve practical problems by means of computers via mathematical knowledge Mathematics teachers master the basic theory and skills of modern education, can think and interpret primary and secondary school mathematics education by comprehensively applying the learned mathematics, mathematics education, and knowledge from other fields, and are qualified for mathematics education tasks at basic education level
Zhejiang Normal University	The mathematics teachers are trained to work as secondary school mathematics teachers, educational scientific researchers, and other educators. They develop high-quality scientific knowledge and skills, mastery of the basic theory and basic methods of mathematics, solid foundations of mathematics and good mathematical thinking abilities, mastery of modern educational technology, adaptation to the need for reform and development of basic education, and a spirit of innovation and practical ability

(continued)

Table 17.5 (continued)

Universities	Training objective
Chongqing Normal University	The mathematics teachers are trained to have good political qualities, scientific and cultural literacy, strong learning and practical abilities, mastery of the basic theory, basic knowledge and basic methods of mathematics, the ability to solve actual problems by using mathematical knowledge and computers, and modern educational concepts and solid professional basic knowledge
Xi'an University	The major aims are to train graduates to master the basic theory and basic methods of mathematics, have good mathematical thinking abilities and mathematical literacy, solve problems occurring in engineering and actual life by using mathematical knowledge and relevant education and teaching theories and working on teaching jobs of basic education, and provide applied talents with solid foundation, strong ability, and high quality for local social and economic development and local basic education. Graduates can work in the fields of education and teaching, engineering mathematics, data analysis and processing, algorithm research, software development, etc. in the education industry, scientific research and educational training institutions, the software industry, and enterprises
Chongqing University of Arts and Sciences	The aim is to train excellent teachers for mathematics education at basic education level, adapting to the needs of economic and social development and educational reform and development, developing morally, intellectually, physically, and aesthetically, having a solid professional basis in mathematics subjects, prominent education and teaching and self-development abilities and good comprehensive qualities, "noble teachers' ethics, consummate teachers' abilities, intelligent teachers' wisdom, and the ability to teach happily and suitably"

17.5.2.3 Course Setting

According to the basic standard requirements for teacher educational courses, normal universities have quite different courses for mathematics and applied mathematics (teacher education orientation). Generally speaking, however, the courses are divided into compulsory and elective. Most courses are compulsory, and the elective courses account for a smaller proportion. Detailed information about the distributions of compulsory and elective courses in the selected universities is shown in Fig. 17.4.

The courses can also be divided into theory and practical courses; the theory courses are mainly those taught by teachers in class, accounting for about 77% of the credit proportion, while the practical courses mainly include educational probation, educational practice, and practices carried out by students themselves as a supplement to the theory courses. Examples include a computer operations

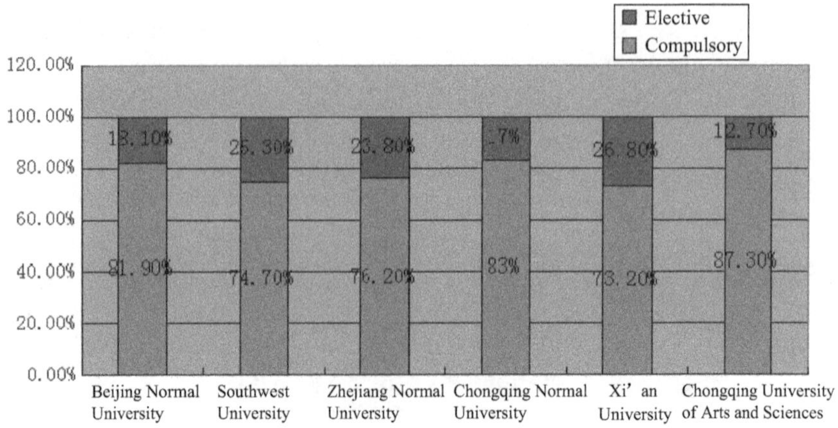

Fig. 17.4 Distribution situation of compulsory and elective courses of the selected universities

component of the computing course and experiment component of the college physics course, accounting for about 23%. Detailed information about the distribution of theory courses and practical courses is shown in Fig. 17.5.

The course structure for each university is different. However, the courses can be classified roughly into the following five categories according to the course content: general courses, educational theory courses, mathematics courses, mathematical educational courses, and teacher skill training courses. The distribution of course types in the selected universities is summarized in Table 17.6.

1. General courses, accounting for about 27.8%, mainly include Ideological and Moral Training and Fundamentals of Law, History of Modern and Contemporary China, Fundamental Tenets of Marxism, Introduction to Mao

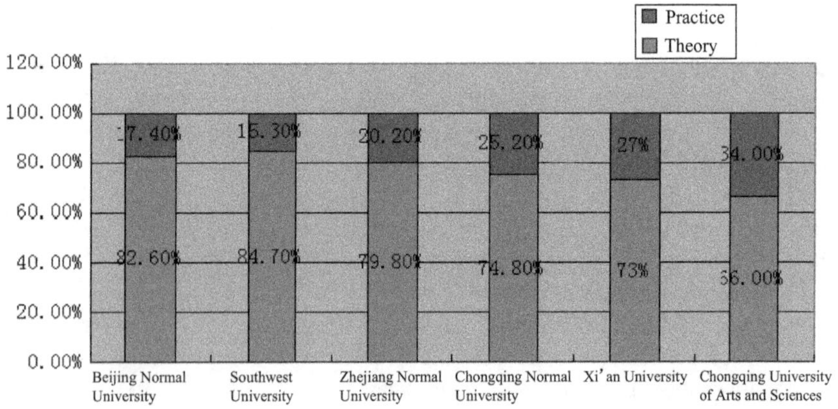

Fig. 17.5 Distribution situation of theory courses and practical courses of the selected universities

Table 17.6 Course structure in the selected universities

	General courses (%)	Educational theory courses (%)	Mathematics courses (compulsory) (%)	Mathematics educational courses (%)	Teacher skill training courses (%)
Beijing Normal University	23.9	3.9	38.7	8.4	2
Southwest University	41	2.9	31.2	5.9	7.1
Zhejiang Normal University	23.8	6	35.1	6.5	1.8
Chongqing Normal University	27.5	3.9	34.8	10.3	1
Xi'an University	25.5	7.1	35.3	5.3	1.8
Chongqing University of Arts and Sciences	25	4.2	46.6	6.9	2.6

Zedong Thought and the Theoretical System of Socialist Theory with Chinese Characteristics, Practice of Ideology and Politics, Situation and Policy, College English, Fundamentals of Computer Culture, College Sports, Military Course, and Career Planning and Occupational Guidance.

2. Educational theory courses account for about 4.7% and mainly include Educational Psychology, General Pedagogy, Educational Research Methods, and Application of Modern Education Technology.
3. Mathematics courses, of which compulsory courses account for about 40%, mainly include Mathematical Analysis, Higher Algebra, Analytical Geometry, Differential Equation, Probability Theory, Higher Geometry, Modern Algebra, College Physics, Complex Variables Functions, Numerical Analysis, Functions of Real Variable, and Operational Research. Elective mathematics courses are generally selected from the following courses: Functional Analysis, Topology, Graph Theory, Combinatorial Mathematics, Mathematical Modeling, Mathematical Statistics, Differential Geometry, Fuzzy Mathematics, and Mathematical Physical Equations.
4. Mathematical educational courses, accounting for about 7.2%, are calculated by combining compulsory and elective courses. Normally, such courses include Mathematical Education Theory, Mathematical Classroom Teaching Case Analysis, Teaching Design for Secondary School Mathematics, Course Standard and Study for Secondary School Mathematics, Competition Mathematics, Elementary Algebra Study, and Elementary Geometry Study.

5. Teacher skill training courses, accounting for about 2.7%, only refer to the training courses listed independently in the training program and generally include Teaching Trial Lecture, Practical Training of Teaching Skill for Middle School Mathematics, Three Calligraphy Training, Mandarin Training, Writing Skill, Geometry Software and Courseware Making, Music Basic Ability Training, Art Basic Ability Training, and Oral Competence Training.

17.5.2.4 Practical Teaching

Practical teaching, mainly including educational probation, educational practice, and graduation thesis (design), accounts for about 10% of credits. Detailed information about practical teaching in the selected universities is listed in Table 17.7.

Each normal university has established cooperative partnerships with many local secondary schools, and some secondary schools are regularly turned into the practice bases for some normal universities, in which the pre-service mathematics teachers' practical teaching is done. Teaching observing and emulation mainly happen in secondary schools in the first or second half-semester of grade 3 before teaching practice for educational probation, which is for about two intensive weeks; educational practice is carried out in secondary schools for trial teaching, study, and educational survey in the second half-semester of grade 3 or the first of grade 4, followed by three months. Generally speaking, the "two-mentor system" is implemented: On the one hand, secondary school mathematics teachers and class teachers are responsible for mentoring the pre-service teachers, and as well, professional teachers studying mathematics education in the same universities are responsible for mentoring. The Graduation Thesis starts from the second semester of grade 4 and takes about 9 weeks.

Table 17.7 Information about practical teaching in the selected universities

	Educational probation time	Educational practice time	Graduation thesis or design
Beijing Normal University	1 credit, the 6th semester	10 credits, the 7th semester	4 credits, the 6th–7th semester
Southwest University	Not listed individually	8 credits, the 6th semester	4 credits, the 8th semester
Zhejiang Normal University	2 credits, the 6th semester	8 credits, the 7th semester	8 credits, the 7th–8th semester
Chongqing Normal University	2 credits, the 5th semester	6 credits, the 6th semester	6 credits, the 8th semester
Xi'an University	1 credit, the 5th semester	10 credits, the 7th semester	10 credits, the 8th semester
Chongqing University of Arts and Sciences	2 credits, the 3rd semester	9 credits, the 7th semester	8 credits, the 8th semester

17.6 Characteristics of Pre-service Mathematics Teacher Education in China

After around one hundred years of development, pre-service mathematics teacher education in China has developed some unique characteristics. Generally, the following characteristics can be identified.

17.6.1 Adaptation to the Mathematics Curriculum Reform at Basic Education Level

The main task of pre-service mathematics teacher training is to prepare qualified mathematics teachers for basic education in China. Thus, the training of the pre-service mathematics teachers needs to meet the new requirements for the teaching of mathematics at the basic education level. In the initial period of the twenty-first century, basic education underwent its eighth educational reform. The *Mathematics Curriculum Standard for Compulsory Education* was issued in 2001, and the *Mathematics Curriculum Standard for Senior Secondary Schools* was issued in 2003. The objectives, content, implementation, and assessment of the mathematics curriculum were reformed to a large degree. Of course, the mathematics curriculum reform also influenced the training of pre-service mathematics teachers at basic education level. Correspondingly, the structure, content, and assessment of the pre-service mathematics teacher education curriculum were changed systematically to meet the needs of the new mathematics curriculum in China. Many courses adapting the mathematical teaching practices of primary and secondary schools have been added, including *Teaching Design and Case Analysis for Mathematics of Middle (Primary) Schools*, *Course Standard and Textbook Research for Mathematics of Middle (Primary) Schools*, *Mathematical Culture*, *Mathematical Methodology*, and *Geometry Software and Courseware Making*. The content of some courses, such as *Elementary Number Theory*, *History of Mathematics*, and *Graph Theory,* focuses on the primary and secondary mathematics content. To assess the students' mastery of the curriculum, in addition to quantitative assessment, some qualitative assessment has been added, for example, assessment of the teaching design of some curriculum content at primary and secondary school levels, and the assessment of teaching practices.

17.6.2 Curriculum Setting Stressing General Education and Promoting Pre-service Teachers' Accomplishments

In both primary and secondary school pre-service mathematics teacher education, the general education components account for large proportions. General educational courses include ideology and politics and history as well as literature, art, and natural sciences. The learning of the general educational courses aims to equip pre-service mathematics teachers with wide and comprehensive knowledge bases in various fields and to improve their overall qualities as teachers.

However, we need to discuss the rationality of the content included in the general educational courses. At present, there are too many courses on ideology and politics. Pre-service teachers do not realize the necessity for learning such courses. The approach to general educational courses used by Beijing Normal University might be used in other universities. In Beijing Normal University, the general educational courses are classified in the form of modules and the reasons for learning them are established. The general educational courses are classified into six modules: patriotism and value ideals, international views and dialogue among civilizations, classic study and cultural inheritance, mathematical basis and scientific literacy, artistic creation and aesthetic experience, and development and civic responsibility. These courses are arranged in such a way as to enable teachers and students to understand their purpose, thereby achieving positive effects and promoting students' accomplishments.

17.6.3 Graduates Are Willing to Work as Primary or Secondary School Mathematics Teachers

In China, the traditional philosophy of "respecting teachers and valuing education" means that teachers enjoy a high social status. Their incomes are gradually and stably increasing with the development of China's economy, and mathematics teachers receive particular attention due to the fact that mathematics is a compulsory and important subject in the College Entrance Examination. The vast majority of graduates trained as mathematics teachers are willing to work as mathematics teachers in primary or secondary schools. More than 80% of graduates are employed every year. In addition, for the reasons mentioned above, many graduates from other comprehensive universities who are not trained as mathematics teachers are also willing to be primary and secondary school mathematics teachers. Since the reform to the qualification authentication system, those who are not trained as mathematics teachers can be primary or secondary school mathematics teachers after graduation as long as they acquire the appropriate teaching qualifications. The real situation in China is that primary and secondary schools still like to select graduates from mathematics majors who are trained in Schools of Mathematics in normal universities. Here, the time that primary and secondary school mathematics teachers spend learning mathematics is longer than is the case for teachers from other countries. Before they begin work as primary or secondary school mathematics teachers, they have already studied mathematics for 16 years. This, to a certain degree, ensures that most primary and secondary school mathematics teachers have deep mathematical knowledge foundations and profound understanding of primary and secondary school mathematical knowledge.

17.6.4 Attaching Importance to Mathematical Knowledge, But Neglecting the Mathematical Teaching Knowledge

The heavy emphasis on mathematics teachers' mathematics knowledge base is influenced by the traditional education concept in China. It is also influenced by traditional social and cultural valuing of teachers' knowledge in China. A popular and famous Chinese saying, "To give students a bowl of water, the teacher must have a bucket of water," is often used to tell pre-service and in-service teachers that a solid professional knowledge is the basis of the growth and development of teachers. From the information about required courses, it can be seen that both primary and secondary school pre-service mathematics teachers need to learn a wide range of mathematics-related subjects. Besides mathematics knowledge at basic education level, pre-service mathematics teachers in China are required to learn quite a lot of difficult advanced mathematics knowledge. Learning experience like this gives primary and secondary school mathematics teachers a profound understanding of mathematics knowledge.

However, on the other hand, it should be noted that not much attention is paid to mathematics educational courses in the training of pre-service mathematics teachers in China. Generally speaking, during the period of pre-service mathematics teacher-training, mathematics educational courses are elective in nature and do not account for many credits. As well, in practice, theoretical learning is heavily emphasized in mathematics educational courses, which may lead to the consequence that students cannot integrate knowledge of education theory and mathematical knowledge. This will further influence the accumulation of knowledge about how to teach specific primary and secondary school mathematical topics, the mathematical teaching content knowledge. Therefore, after pre-service mathematics teachers' graduate, they need to spend quite a long period of time adapting to the real situation of primary and secondary school mathematics teaching and cannot teach mathematics effectively.

17.6.5 Emphasizing the Training of Teaching Practice Ability and Ignoring the Training of Teaching Research Ability

Teachers are reflective practitioners. The training of pre-service teachers is inseparable from teaching practice, and theoretical knowledge of education and teaching can really only develop in the teaching practice. Pre-service teachers grow gradually through reflecting on their own teaching practices and behaviors continuously. With time, they will form their own teaching styles. Therefore, in China, during pre-service teacher-training periods, a lot of attention is paid to the training of the pre-service teachers' teaching abilities. Most normal universities will work closely with primary and secondary schools to develop every aspect of pre-service mathematics teaching ability. Activities in schools, such as educational probation and educational practice, can facilitate the development of these abilities.

However, on the other hand, high-quality primary and secondary school mathematics teachers should also be able to conduct research on primary and secondary school mathematics teaching practice. During the pre-service mathematics teacher-training period, research skills are developed through a course called Educational Research Ability and the writing of the Graduation Thesis. It has been observed that educational research, particularly mathematics educational research courses, is insufficient. As well, the study of educational research courses is not connected tightly with teaching practice. Particularly, pre-service mathematics teachers do not have the experience of participating in educational research projects.

17.6.6 Limited Self-selecting Courses During Pre-service Teacher Training

Both compulsory and elective courses are offered in the training of pre-service mathematics teachers, which conforms to the characteristics of the ages and personalities of undergraduates. The compulsory courses mainly consist of general educational courses, foundation courses in mathematics, educational foundation courses, and practical teaching. Elective courses mainly comprise a small proportion of general educational courses (about 4%), mathematics courses (about 7%), and mathematics educational courses (about 6%). It has been observed that the credits for the elective courses are small in proportion; most of the general educational courses and the mathematics courses are compulsory. However, only a few choices of mathematics teaching courses are available. Therefore, pre-service mathematics teachers have quite limited choices of courses in the field of mathematics learning and mathematics education, which is not good for the overall development of their professional competence.

References

Cui, Y. (2006). *History of teacher education in China*. Taiyuan: Shanxi Educational Press.

Li, C. (2010). *The change and development of secondary normal schools in china (Unpublished master thesis)*. Nanjing: Nanjing Normal University.

Ministry of Education of the People's Republic of China. (1993, October 31). 中华人民共和国教师法 *[The teachers law of the People's Republic of China]*. Retrieved January 22, 2016, from the MOE Web Site: http://www.moe.edu.cn/publicfiles/business/htmlfiles/moe/moe_619/200407/1314.html.

Yuan, Z. (2004). 从"师范教育"向"教师教育"的转变 [The change from "normal education" to "teacher education" in China]. *China Higher Education, 5,* 29–31.

Chapter 18
Post-service Education of Mathematical Teachers

Xiaoya He

Abstract The post-service education of mathematics teachers in China addresses three levels, novice, common, and backbone teachers. The education of a novice teacher is overseen by a tutor who is a member of the mathematics team in the teacher's school. The education of common teachers has two main categories; one is regular teaching and research activities (public lectures, discussion lectures on different ways of teaching certain content, teaching philosophy contests, problem-solving contests, etc.) that are held by schools or teaching and research groups at city or county level. The other category is online in-service training that is organized and implemented by municipal governments' trusted training institutions. The training of backbone teachers is divided into city and national levels. It usually consists of short-term, focused training projects combined with subsequent follow-up seminars. The main focus of training is the implementation of new basic education courses, as well as to meet personalized needs relating to teachers' professional development, and to guide teachers in their professional growth. The training content can be divided into six categories, development of teachers' professional ethics, revision and expansion of knowledge, mathematics education theory and its applications, research methods in mathematical education, modern education skills, and discussions about teaching practice problems. The administration and assessment processes are concerned mainly with the state's construction of information and administrative systems for national teacher training, the establishment and monitoring of training quality standards, and the regular implementation of training quality assessment.

X. He (✉)
South China Normal University, Guangdong, China
e-mail: hexy@scnu.edu.cn

Y. Cao and F.K.S. Leung (eds.), *The 21st Century Mathematics Education in China*, New Frontiers of Educational Research,
https://doi.org/10.1007/978-3-662-55781-5_18

The post-service education of mathematics teachers in China addresses three levels, novice, common, and backbone teachers. The following sections will describe each of these according to training objectives, training methods, and training content.

18.1 Training for Novice Teachers

Novice teachers are those who have been in service for one to two years. Most of them have graduated from normal (teacher education) universities or colleges, and only a few from non-normal universities. In the case of certain senior middle schools, teachers are hired for their high academic achievements even if they have not received specific education training.

18.1.1 Training Objective

The training objectives for novice teachers include to explicate professional ethics code for teachers, to become familiar with a school's administrative regulations for teaching, teachers in charge, and teachers, as well as to become familiar with the content and requirements of mathematics teaching, master basic mathematics teaching methods, and become familiar with daily administrative tasks for teachers in charge.

18.1.2 Training Method

18.1.2.1 Collective System

Usually organized by the school which engages them, training for all novice teachers in all subjects is carried out under the supervision of the principal, vice-principal, dean, and grade leaders, mainly by means of thematic reporting, consultations about studies, and sharing with outstanding peers.

18.1.2.2 Tutorial System

Generally, the school will assign every novice teacher to a senior teacher who has been teaching for more than ten years and is experienced as a tutor. New teachers observe their tutors' classes and consult with them, and the tutors observe the new teachers' classes to give feedback and guidance. This tutorial system, under which the experienced are leading the inexperienced, has greatly advanced the development of novice teachers' professional skills.

18.1.2.3 Post-system

Unless it is necessary for a particular school or teacher to have a high ability level, novice teachers are generally not required to undertake the tasks of a teacher in charge of a class. However, as long as the new teacher's performance is reasonably satisfactory, he/she will be assigned some tasks of a teacher in charge from the second or third year. The purpose of this process is to familiarize the novice teacher with daily administrative tasks and gradually develop class administration skills.

18.1.3 Content of Training

The training content for novice teachers includes professional ethics; education laws and regulations; teaching administrative requirements of the school; administrative regulations for teachers in charge; administration rules and regulations for teachers; establishing and understanding standards for teaching courses; outlines of entrance examinations; content and requirements of collective preparation; skills for teaching mathematics; and daily administrative tasks for teachers in charge.

18.2 Training for Common Teachers

Common teachers are those who have been in service for more than two years, and the training is led by clusters of schools or teaching and research groups at city or county level, or municipal governments' trusted training institutions.

18.2.1 Training Objectives

The objectives of common teacher training are focused on general enhancement of their abilities to teach mathematics and reviews of their teaching skills. The professional development aims at developing teachers' understanding of newly added content, mastery of corresponding teaching strategies, familiarity with challenges and obstacles to students' learning of this content, and strategies for expanding and optimizing mathematics teachers' professional knowledge structures.

Professional development at this level is also aimed at defining the standards of mathematics teaching design, attending and evaluating demonstration classes, discussing mathematics teaching problems, rethinking their own class teaching, and learning from the experiences of outstanding mathematics teachers and outstanding teachers in charge.

18.2.2 Training Methods

18.2.2.1 School-Based Training

School-based training refers to training that is organized and implemented by the school, including novice teachers' public lectures, internal and external outstanding teachers' public lectures, and the introduction of internal and external outstanding in-charge teachers' experiences.

18.2.2.2 Region-Based Training

Region-based training is organized and implemented by regional teaching and research groups or educational institutions. The formats are usually demonstration lessons that are attended by related schools in the local region, discussion forums about different approaches to teaching certain topics, teaching philosophy contests, and problem-solving contests.

18.2.2.3 Distance Training

Distance training refers to online in-service training that is organized and implemented by municipal governments' trusted training institutions. For example, the South China Normal University Cyber Institute (www.gdou.com) undertakes distance training for junior middle school mathematics teachers in some regions; in-service training for Guangdong senior middle school mathematics teachers is offered through www.teacher.com.cn and www.teacheredu.cn.

18.2.3 Training Content

Here, only the content of region-based training and distance training is introduced.

18.2.3.1 Content of Region-Based Training

1. **Demonstration lessons**

Demonstration lessons are given by outstanding backbone teachers or senior teachers who are selected in local regions. After a lesson, the demonstration teacher describes the lesson design and reflects on the strengths and weaknesses; the observers describe what they have learned from the demonstration and critique the teaching. This model of demonstration lessons for clusters of mathematics teachers in a region is effective in enhancing the teachers' professional growth.

2. **Discussions about different approaches to teaching particular content**

In this approach, the same topic is taught in different classes by different teachers, after which they meet to review and discuss their experiences. Because of the differences in demonstrators' abilities, years of teaching experience, teaching standards and teaching designs, the teaching processes, and teaching effects will differ a great deal, thus providing teaching cases for different effects, styles, and value standards to the observers of these activities, which will improve their reflections on teaching.

3. **Teaching philosophy contests**

In teaching philosophy contests, 10–20 min lectures are given to expound the design of a lesson or a unit of work, based on theories of education and teaching, and making use of multimedia. The audience is a group of peers, administrators, or teaching and research professionals, and the topics can include lesson objectives and planning, discussion of difficult or key points, and the assessment of teaching effects and quality (He & Yao, 2012).

Other content of teaching philosophy contests includes the analysis of teaching materials (status, vertical and horizontal connections, functions), the analysis of students' cognitive development and learning difficulties, the designing of teaching objectives (knowledge and skill, process and method, affective factors), the selection of appropriate teaching methods, and examples of teaching processes.

The participating competitors are selected by their schools, and the experts in the audience give feedback and select winners.

4. **Problem-solving contests**

Problem-solving contests involve teachers discussing how to solve a mathematical problem and the teaching strategies for doing so, based on theories of mathematics education. Again, the audience consists of peers, experts, teaching, and research professionals.

The content of problem-solving competitions mostly includes analysis of the problem context, analysis of the thinking involved in solving the problem, and the problem's value as a teaching tool.

Again the audience of experts gives feedback and selects the best presentations.

As problem-solving contests have favorable effects, some normal colleges and universities have also launched this kind of contest. One such contest was conducted for Guangdong Normal School students majoring in mathematics in the South China Normal University Cyber Institute in August 2013. The problems selected are described in Fig. 18.1.

Problem 1. As shown in the diagram, the edge BC of the rectangle $ABCD$ is on the x-axis of the rectangular coordinate system. Fold the edge AD, so that the point D joins with the point F on the x-axis. The crease is AE, and it is known that $AB = 8$, and $AD = 10$. Set the coordinates of point B as $(m, 0)$, $m > 0$.

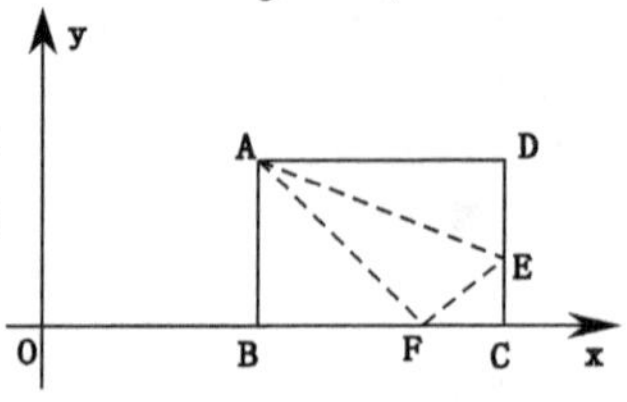

(i) Figure out the coordinates of the points E and F (to be expressed in a formula that involves m);

(ii) Connect OA. If ΔOAF is an isosceles triangle, try to figure out the value of m.

Problem 2. Try to prove the basic theorem of the plane vector.

Problem3. An express company has n $(n \geq 2)$ delivery sites on a straight path, and it needs to build a goods distribution site on this straight path. Please figure out where the distribution site should be built to obtain a minimum sum of the distances from it to each delivery site.

Fig. 18.1 Sample problems

18.2.3.2 Remote Training Content

The remote training curriculum for the South China Normal University Cyber Institute's Network Institute for Mathematics Teachers of Junior Middle Schools is shown in Table 18.1.

The 2014 remote training curriculum for high school mathematics teachers in Guangdong Province is described below and summarized in Table 18.2.

The first subject is the professional qualities of mathematics teachers, followed by dissemination of research into teaching methods, then a series of commentaries on mathematics teaching. Each of these subjects is presented as a brief introductory text, specialist lecture videos, text case descriptions and comments on the cases on videos and in readings, readings about mathematical thinking and activities, reference materials, and readings on homework design.

The training time is 2–4 weeks with 60 h of class time, combining the remote training with school-based educational research. The participants are required to select at least one module from each of subjects one, two, and three, and 10 modules should be selected in total.

The curriculum specialist team implements the program. The chief specialist (author of this chapter) allocates the responsibilities to team members as follows:

Table 18.1 The remote training curriculum for the Mathematics Teachers of Junior Middle Schools

Training module	Training subject (45 min for each subject)
Hot issues for research and study	Basic ideas and strategies for implementing the *Mathematics Curriculum Standard for Full-time Compulsory Education*
	"Numbers and Algebra" and "Geometry and Graphics" content analysis
	"Statistics and Probability" and "Synthesis and Practice" content analysis
	Basic qualities and professional standards of junior middle school and middle school mathematics teachers
	An educational war on mathematics
	The future of mathematics education
	Misunderstandings in mathematics education research
	Mathematical education research case studies
	New ideas for mathematics teaching design
	Setting mathematics objectives
	The essence of learning mathematical concepts
	The essence of learning mathematical principle
	Teaching mathematical problem solving
Professional qualities	Topics on mathematical competence (I)
	Topics on mathematical competence (II)
	Exploration of numbers
	Depiction of graphics
	Solving equations
	Angle trisection
	Discussing probability problems
	Non-euclidean geometry
Number and algebra	Main knowledge, methods, and ideas of numbers and algebra
	Teaching and learning algebraic expressions in the middle school
	Teaching and learning equations and inequalities in the middle school
	Teaching and learning functions in the middle school
	"Number and Algebra" Features of questions and analysis of questions in the senior high school entrance examination
	"Number and Algebra" teaching materials and class case analysis (first)
	"Number and Algebra" teaching materials and class case analysis (second)
	"Number and Algebra" teaching materials and class case analysis (third)
	"Number and Algebra" teaching materials and class case analysis (fourth)

(continued)

Table 18.1 (continued)

Training module	Training subject (45 min for each subject)
Graphics and geometry	How to understand "intuitive geometry" in middle school mathematics
	How to understand the "geometric transformation" in middle school mathematics
	How to understand the "coordinate geometry" in middle school mathematics
	How to understand the "reasoning and argumentation" in middle school mathematics
	"Geometry and graphics" Features of questions and analysis of questions in the senior high school entrance examination
	"Geometry and graphics" teaching materials and class case analysis (I)
	"Geometry and graphics" teaching materials and class cases analysis (II)
	"Geometry and graphics" teaching materials and class case analysis (III)
	"Geometry and graphics" teaching materials and class cases analysis (IV)
Statistics and probability	Understanding data analysis concepts
	How to teach statistics
	Understanding the relationship between statistics and probability
	Understanding probability theory and its implementation
	"Statistics and Probability" Features of questions and analysis of questions in the senior high school entrance examination
	"Statistics and Probability" teaching materials and class case analysis (I)
	"Statistics and Probability" teaching materials and class case analysis (II)
	"Statistics and Probability" teaching materials and class case analysis (III)
	"Statistics and Probability" teaching materials and class case analysis (IV)
Synthesis and practice	Reasons for setting up synthesis and practice
	Basis of setting up synthesis and practice
	Cases of synthesis and practice classes

1. Developing the training curriculum
   Developing the content of the training curriculum includes demands analysis, scheme formulation, curriculum design, making videos, text materials, reference materials, homework design, and marking criteria.

Table 18.2 The 2014 remote training curriculum for high school mathematics teachers in Guangdong Province

	Category	Curriculum name	Source of curriculum
1	Subject III	Linear inequalities in two variables and planar domain (course video and comments) (1)	Guangdong curriculum
2	Subject III	Linear inequalities in two variables and planar domain (course video and comments) (2)	GC
3	Subject III	Linear inequalities in two variables and planar domain (course video and comments) (3)	GC
4	Subject III	Cases and reflection in teaching the same mathematical subject by different teachers	GC
5	Subject III	The world of mathematical modeling	GC
6	Subject III	New design of mathematical induction teaching	GC
7	Subject I	High school teaching research	Continuing education website
8	Subject I	High school mathematics test question setting: theories and techniques	CEW
9	Subject II	High school mathematics module teaching research	CEW
10	Subject II	High school mathematics "Set and Logic" teaching research	CEW
11	Subject II	High school mathematics "Concepts and Properties of Functions" teaching research	CEW
12	Subject II	High school mathematics "Derivative and its Applications" teaching research	CEW
13	Subject II	High school mathematics "Transformation of Triangle Functions and Solving Triangles" teaching research	CEW
14	Subject II	High school mathematics Plane Vectors teaching research	CEW
15	Subject II	High school mathematics "Spatial Vectors and Solid Geometry" teaching research	CEW
16	Subject II	High school mathematics "Counting Principles" teaching research	CEW
17	Subject II	High school mathematics "Probability" teaching research	CEW
18	Subject II	High school mathematics "Statistics" teaching research	CEW
19	Subject II	High school mathematics "Algorithms and Block Diagram" teaching research	CEW
20	Subject II	High school mathematics "Applications of Functions" teaching research	CEW
21	Subject II	High school mathematics "Conic Section" teaching research	CEW

(continued)

Table 18.2 (continued)

	Category	Curriculum name	Source of curriculum
22	Subject II	High school mathematics "Preliminary Analytic Geometry" teaching research	CEW
23	Subject II	High school mathematics "Preliminary Solid Geometry" teaching research	CEW
24	Subject II	High school compulsory mathematics 1 Module introduction	CEW
25	Subject II	High school compulsory mathematics 1 "Function Monotonicity" teaching research	CEW
26	Subject II	High school compulsory mathematics 1 "Logarithm Operations" teaching research	CEW
27	Subject II	High school compulsory mathematics 1 "Applications of Functions" teaching research	CEW
28	Subject II	High school compulsory mathematics 1 High-end lesson preparation	CEW
29	Subject II	High school compulsory mathematics 2 Module introduction	CEW
30	Subject II	High school compulsory mathematics 2 "Preliminary Solid Geometry introductory course" teaching research	CEW
31	Subject II	High school compulsory mathematics 2 "Parallel and Vertical" teaching research	CEW
32	Subject II	High school compulsory mathematics 2 "Plane Analytic Geometry review lesson" teaching research	CEW
33	Subject II	High school compulsory mathematics 2 High-end lesson preparation	CEW
34	Subject II	High school mathematics 1 Module introduction	Teachers' website
35	Subject II	High school mathematics "Function Monotonicity" teaching discussion	TW
36	Subject II	High school mathematics "Logarithm Operations" teaching discussion	TW
37	Subject II	High school mathematics "Applications of Functions" teaching discussion	TW
38	Subject II	High school mathematics 1 High-end lesson preparation	TW
39	Subject II	High school mathematics 2 Module overall introduction	TW
40	Subject II	High school mathematics "Preliminary Solid Geometry" teaching discussion	TW
41	Subject II	High school mathematics "Parallel and Vertical" teaching discussion	TW

(continued)

Table 18.2 (continued)

	Category	Curriculum name	Source of curriculum
42	Subject II	High school mathematics "Plane Analytic Geometry" teaching discussion	TW
43	Subject II	High school mathematics 2 High-end lesson preparation	TW
44	Subject II	High school mathematics III Module overall introduction	TW
45	Subject II	High school mathematics "Emphasis and Functions of Algorithm Lessons" teaching discussion	TW
46	Subject II	High school mathematics "How to Have a Good Activity Lesson" teaching discussion	TW
47	Subject II	High school mathematics "Probability" teaching discussion	TW
48	Subject II	High school mathematics III High-end lesson preparation	TW
49	Subject II	High school mathematics IV Module overall introduction	TW
50	Subject II	High school mathematics "Functions of Unit Circle in Trigonometric Function Study" teaching discussion	TW
51	Subject II	High school mathematics "$y = A \sin(\omega x + \varphi) + b$" teaching discussion	TW
52	Subject II	High school mathematics "Improve Operational Capability" teaching discussion	TW
53	Subject II	High school mathematics IV high-end lesson preparation	TW

2. Guiding the training process

   - These duties include sampling and commenting on the homework of the trainees and the expanded resources recommended by the coaching teachers. The number of pieces sampled each day for each subject should not be less than six on average.
   - Interactive online Q&A
     This involves organizing at least one online Q&A discussion per week during the whole training period, collecting the Q&A achievements once a week to make a Q&A Summary and publishing it in the Subject Announcement column.

3. During the training period, the specialist team members need to take it in turn to be on duty, in groups of no less than three. They need to assist trainees with problems and answer their questions when they experience difficulties. The

platform has a Specialist Homepage for each specialist participating in the training. Each specialist is required to enrich his/her own homepage so as to provide convenient learning modes and resources for the trainees. The content to be added includes expanded learning resources that the specialists consider worth recommending; specialist subject posts; samples and comments on the trainees' homework and expanded resources; and recommendations for school-based activity achievements that the trainees have evaluated as excellent.

4. Composing the subject briefing
   One- or two-person teams are formed to compose subject briefings. After being checked and confirmed by the specialist group leader, these are uploaded to the platform homepage.
5. Summarizing the training
   After the training, the chief specialist needs to write a summary report. The content should include a basic evaluation of the curriculum training, the experiences and achievements occurring in the training process, problems arising, and some suggestions for carrying out the next training.

18.2.3.3 Assessment Scheme

1. Purpose and principles of the assessment

The purpose is to guide the trainees to participate in the study and to communication actively. The assessment process is intended to guarantee the training quality and effectiveness and to conduct an overall, systematic evaluation of the training. The evaluation focuses separately on the remote study (accounting for 80%) and the school-based training (20%). The minimum score to pass the course is 60%, and no makeup examinations are provided. The evaluation scoring is done according to the following formula, which is the sum of:

- the score for the objective questions (30%);
- the score for the subjective questions (30%);
- participation in the online mentoring (10%);
- posting and replying (5%);
- commenting on the subject and the brief reports (5%);
- score for School-based Activity Achievements (20%).

2. Assessment criteria for the remote study component

The assessment is composed of four parts: project work, participation in online discussion, posting and replying as well as contributing brief reports and School-based Activity Achievements project work. This training applies the "5 + 1" elective system, which means trainees must randomly pick five curriculum modules in the subject curricula (of which there must be at least one from the special topic III) and one general education compulsory module.

The assessment requirements of the project work are carried out as follows:

- Each set of objective questions is worth a total score of 100, and the scores are calculated automatically by the system. The final score of objective questions is equivalent to 30% of the total score.
- Trainees must and can only submit one subjective item marked by tutors.
- Participation in online discussion: Every trainee must take part in the online Q&A discussion organized by the experts. They get 1 point for every follow-up discussion, 3 points if their post is set as a sticky post by the experts, and no points if their discussion has no practical meaning and is deleted.
- Posting and replying as well as contributing brief reports: Trainees are required to participate actively in their class forum discussions and communications, and to read and comment on brief subject reports. Points are awarded as follows:
  - 0.5 point for every post and 2 points for every sticky post (sticky post should be recommended by tutors);
  - 1 point for every reply to another trainee's post or comment on a brief report;
  - No points if the post has no practical meaning and gets deleted;
- School-based Activity Achievements: Trainees are expected to take part actively in the school-based activities which are based on this project, and the class president needs to submit a complete list of attendance at the school-based activities. The counting rules are:
  - Trainees must and can only submit one School-based Activity Achievement.
  - Trainees must read and appraise at least three students' School-based Activity Achievements; for these, they will receive three review scores.
  - If one is judged by three students to have cheated, he/she gets no points.
- Criteria for excellent trainees
  - Excellent trainees need to show that they have studied for no less than 3 h every day, as evidenced by the online course learning records showing no less than 900 min. As well, they are expected to participate actively in online discussion.
  - The assessment scores must all be over 85, and all examinations must show even scores, with no zeros recorded.
  - Contributions such as resources have been selected for uploading as brief reports.
  - The students take part actively in the school-based activities that are based on high school teachers post-training projects; the original School-based Activity Achievements are recommended.

18.3 Training of Backbone Teachers

The backbone teachers include those chosen to be specially trained (divided into provincial-level and national-level) from the common and the professional teachers (the future trainers). In order to learn more comprehensively and deeply about China's backbone teacher training, we will first introduce the national-level ideology, policies, and measures, and then give details about the training of the national-level backbone teachers and trainers.

18.3.1 State Policies for Teachers' Professional Training

In order to adapt to the economy's high-speed development, at the beginning of the twenty-first century, a big reform was introduced to China's basic education. In July, 2001, the Chinese Ministry of Education released the *Mathematics Curriculum Standards (Experiment Draft) for the Full-Time Compulsory Education Stage* and, in April, 2003, and launched the *Mathematics Curriculum Standards (Experiment) for Regular High Schools*. The success of this reform depends mainly on the quality of mathematics teacher teams. For this purpose, in September, 2004, the Chinese Ministry of Education launched the new set of *2003–2007 Plan for Training All Workers and Staff for Primary and Middle School Teachers*. This plan established the policy of "everyone is included; the backbones are highlighted; attention is given to rural areas," and focused on "new ideas, new courses, new technologies" as well as promoting the development of teachers' ethics, implementation of a new training program for all primary and middle school teachers, improving teachers' ethical levels and business qualities, and guaranteeing human resources to promote quality education and the reform and development of rural education.

In August, 2008, China started to prepare and, on July 29, 2010, the *State Medium and Long-Term Educational Reform and Development Planning Outline (2010–2020)* was launched. The major content included 10 aspects: promoting selected units for experiments in quality education, balanced development of compulsory education, vocational education, selecting units for experiments in the construction of lifelong education mechanisms, selecting units for experiments in developing top innovative talents, the entrance examination system, the modern university system, in-depth schooling mechanisms, regional education input safeguard mechanisms, and the provincial government's education integration.

After that, the Ministry of Education and Ministry of Finance implemented *The National-level Training Plan for Primary and Middle School Teachers* (in short *The National Training Plan*), starting from 2010. It mainly includes primary and middle school backbone teacher training and distance training, training for class advisers, training for primary and middle school subjects that were in short supply or suffering from weaknesses, and other pilot demonstration projects. Based on this document, backbone teachers can be trained, demonstrations can be made, and a set of excellent training courses and teaching resources can be developed and provided, offering powerful support, to teacher professional development. The process included establishing the management of the national teacher training information system, designing the project, making provisions to bid for projects, establishing training standards, building expert databases and a resource library, and developing assessment processes. The outcomes were standardized management and improved quality. By the end of 2013, the national training plan had trained 3500 thousand backbone teachers, of whom 3350 thousand were rural teachers, which accounted for 96% of the total number.

In the two files *Opinions of the Ministry of Education about Enhancing Primary and Middle School Teacher Training (Teacher [2011] No. 1) and Opinions of the Ministry of Education's National Development and Reform Commission's Financial Department about Strengthening the Educational Reform* (Teacher [2012] No. 13) launched by the Ministry of Education, the basic ideology and overall mission of the teacher training are specified, and the general requirements are proposed.

In 2012, the Ministry of Education released the *Kindergarten Teachers' Professional Standards, Primary School Teachers' Professional Standards and Middle School Teachers' Professional Standards*, which outline the standards that backbone teachers' training must comply with.

In another document, the *Guidelines and Opinions of the Ministry of Education about Deepening Primary and Middle School Teacher Training and Improving Training Quality* (Teacher [2013] No. 6), the following are specifically required:

- First, carry out training in accordance with needs; this includes the process of training planning, project design, organization implementation, and quality monitoring.
- Second, enhance practical training, which should take up no less than 50% of teacher training courses; first-class teachers should account for no less than 50% of the trainers' team.
- Third, motivate teachers to take part in the training. By promoting the teachers' selections of topics and training credit management system, teachers can be encouraged and motivated to take part in the program.
- Fourth, utilize information technology, build teachers' online research and study communication, promote school-based research and innovative training modes; use informalization management platforms to enhance the management and supervision of the training process.

18.3.2 National-Level Backbone Teachers' Training

This section illustrates the national-level backbone teachers' training with the example of the National Training Plan (2011) developed by the South China Normal University.

18.3.2.1 Guidance

The national-level backbone teachers' training program aims to encourage high schools to experiment with ideas pertaining to the mathematics curriculum reform and based on modern theories of mathematics education. The aims included the application of new training modes that are targeted, practical and feasible; improve the professional levels of backbone mathematics teachers, along with their teaching ability and the capacity to guide young teachers, and the training of excellent backbone high school mathematics teachers who are ethical, modern in their educational outlooks, and have innovative spirits for the nation.

18.3.2.2 Target of Training

There are several targets that need to be taken into account:

1. Insist on the correct political direction; love the teaching career; establish modern education ideas; establish scientific views about teachers, teaching, quality, and talents; be ethical teachers and uphold vocational ideals.
2. Understand the new curriculum and new textbooks of high school mathematics thoroughly, learn new knowledge, theories, and methods of mathematical education, and expand and optimize the backbone mathematical teachers' professional knowledge structures, so as to promote their professional development.
3. Develop the ability to apply new theories, methods, and measures to the teaching practice, gradually form the distinct teaching style or specialty, and have a strong educational reform consciousness and innovation ability.
4. Through educational scientific research practice, cultivate scientific research ability to undertake important teaching and research and educational reform tasks independently; be able to put forward some significant research topics and practice schemes according to the requirements of the mathematics education reform; and develop the ability to organize, guide, and carry out scientific research and educational reform.
5. Understand the concepts of innovation and reform; be able to play an exemplary role in the field of middle school mathematics teaching, and develop into an educational expert and academic leader of middle school mathematics, step by step.

18.3.2.3 Training Objectives and Methods

The training focuses on excellent senior high school backbone mathematics teachers from across China. The training includes three stages and lasts for about one year. The first stage is mass learning, and this lasts for 15 days; the second stage is post-practice and action study, which lasts for 11 months; and the third stage is achievement exhibition and training summary. The training methods include:

1. Specialist lectures, group communication, collaborative discussions, case teaching, on-the-spot teaching, and problem solving;
2. Emphasis on communication and interaction between the specialists and trainees and among the trainees, to achieve the goals of emotion touching, thought provoking, question generating, and consensus reaching;
3. Linking of specific and general aspects, with deep analysis of specific problems;
4. Working in groups with mathematics education specialists from the School of Mathematical Science of South China Normal University as a community of learning.

18.3.2.4 Training Curriculum (He, 2005)

Module 1: General training

The Basic Education Training and Research Institute of the South China Normal University Cyber Institute is responsible for this module, including comparisons of Chinese and foreign basic education reforms, the professional intelligence of teachers, professional ethics and affective factors for teachers, teaching problems and their solutions, and action research studies.

Module 2: Professional training

Professional training design is based mainly on the following four principles:

1. Learn the new concepts, knowledge, and skills underpinning the senior high school mathematics course reform, and update the educational knowledge and structure.
2. Master the concepts, goals, structure, content, and teaching requirements for implementing the new senior high school mathematics course, and address teachers' problems and questions.
3. Master professional educational scientific research, conduct research in the real context of mathematics teaching, and improve the educational scientific research level.
4. Exchange experiences of education and teaching, improve the level of interaction and cooperation among teachers, promote trainees' teaching reflections, and enhance the professional level of backbone teachers.

The specific courses include the following six categories.

1. Special topics for improving mathematics professional attainment:

   - Introduction to the mathematics discipline is taught by three highly regarded professors, Jingxue Yin, Daochun Sun, and Bolian Liu. They mainly introduce the history, thoughts, and methods of the branches of mathematics as well as their own understanding of mathematics, and they expand the trainees' horizons and improve their teaching attainment.
   - Mathematical modeling in middle schools: aims at trainees mastering the basic methods of mathematical modeling and improving their ability to use mathematics to solve practical problems.
   - Probability and statistics problems: aims to expand trainees' knowledge about probability and statistics and improve their ability to utilize the new teaching materials.
   - Research on mathematical problem solving: aims to discuss approaches to middle school mathematical problem solving and develop trainees' abilities to solve mathematical problems.

2. Special topics in mathematics education

   - Some common problems domestically and overseas: aims to help trainees to understand the concepts, goals, and content of the mathematics courses of different countries and master the latest news about the development of international mathematics courses.
   - Case study of mathematics teaching: disseminates information about the problems, countermeasures, and trend in teaching mathematics from the perspective of international mathematics education and provides the trainees with cases about teaching mathematics under the global vision.
   - "Double-base" teaching in Chinese mathematics education: aims to let the trainees master the connotation, contents, changes, and experience of the double-base teaching of mathematics.
   - Philosophy and culture of mathematics aims to help the trainees to understand the philosophy of mathematics, comprehend all kinds of mathematics education concepts, understand the essence of mathematics education, and improve their understanding of the culture of mathematics.

3. Special topics in mathematics teaching
   The purpose of the special topics is to improve the trainees' skills in mathematics teaching design and enable them to design scientific and reasonable teaching according to the features of the objects, teaching objectives, and environmental conditions.

   - Learning and teaching of mathematics concepts and principles: aims to help the trainees master the laws of mathematics concepts and solve standard problems relating to mathematics teaching design (He, 2003, 2004, 2011).

- Examples of mathematics teaching design: aims to solve the inherent problems of algebra teaching and teaching problems of mathematical modeling through the discussion of two cases of national teaching design champions (He & Yao, 2008, 2012).
- Research on mathematics teaching cases: aims to help trainees to master the basic requirements and implementation processes of case teaching and learn to create mathematics teaching cases step by step.

4. Special topics in mathematics education research
The purpose of these topics is to improve the teachers' mathematics ability and develop positive attitudes toward educational research and its practical implications.

    - How to do mathematics education research involves demonstrating the process of teaching, research, and the growth of secondary school mathematics teachers from the angle of a special-grade middle school mathematics teacher.
    - Research on assessment issues: involves studying and discussing the questions, methods, and experiences of middle school mathematics learning.

5. Special topic on modern education skills
The goal of Modern Mathematical Teaching Technology is to enable teachers to use modern education skills in their teaching and in developing course materials.
6. Special topic on teaching practice

    - Discussion on Issues in Teaching of New Mathematical Courses: The goal is to provide trainees with a platform for interaction between teachers and students and interaction, exchange, and discussion among students, to solve trainees' problems of trainees relating to teaching and learning.
    - Mathematics Teaching Research: The goal is to share experiences, cooperate, exchange, and make common progress.
    - Implementation of Middle School Mathematical Exploration Activity: The goal is to study and discuss the connotation and meaning, characteristics, basic methods, teaching requirements, and cases of mathematical exploration.
    - On-the-Spot Teaching: The goal is to view and emulate the teaching of typical lessons on the spot, develop standard teaching plans, teach and evaluate classroom teaching, and discuss strategies for improvement.

7. Module 3: Studies in the mathematical practice. This module encourages teachers to conduct action research relating to various topics and write papers. These papers are exhibited and reviewed by peers.

18.3.2.5 Examination and Evaluation

1. The project expert team is responsible for evaluating the teaching of the whole training.
2. The research topics are evaluated by the tutors and defense committee.
3. Process evaluation is the main method used to examine the trainees' learning, self-assessment, and peer assessment; evaluation is conducted by the experts and the teacher who is charge of the class based on individuals' learning performances, summaries, and papers. Trainees who pass the examination are awarded the certificate of training completion from the Ministry of Education.

18.3.3 Training the Trainers

The example discussed here is the *Skills Upgrading of Outstanding Front-Line Teachers-Middle School Mathematics* (South China Normal University Cyber Institute), focusing specifically on the *State Training Plan (2014)*.

18.3.3.1 Training Goal

This project is dedicated to the improvement of ideas, methods, and practices of middle school mathematics teachers and their professional quality. The goals are:

1. To help teachers to improve their abilities of training, teaching, organization, and implementation;
2. To understand new developments in domestic and foreign teacher training, learn and understand the theories and methods of modern teacher training, and master the methods of design and planning of teacher training projects and school-based teaching research and training;
3. To expand participants' mathematical horizons, gain insights into the essence of mathematics learning and teaching, thoroughly understand the new curriculum standards and new teaching materials at the Compulsory Education Stage, and expand and optimize the trainees' professional knowledge structure;
4. To improve the trainees' abilities to solve the problems associated with implementing middle school experiments in the mathematics curriculum reform;
5. To exchange, assess, and summarize the experiences of middle school backbone mathematical teachers to prepare them for teacher training according to the national training plan.

18.3.3.2 Training Courses

Module A introduces trainees to the status quo, features, and trends of domestic and international teacher training, the essence of teacher professional development, the *Middle School Teachers' Professional Standards Trial*, and directions in teacher training from a macroscopic view. It also guides trainees to master the design and planning of teacher training projects and school-based teaching research and training. The content is as follows:

A1. Interpretation of the *Middle School Teachers' Professional Standards Trial*;
A2. The experiences of international basic education teacher training;
A3. The modes and methods of teacher training;
A4. Interpretation of *Guidelines for Strengthening the Reform* and design and organization of teacher training projects;
A5. Design and organization of school-based teaching research and training;
A6. The micro-lecture as a teacher training innovation;
A7. Teachers' moral education.

Module B expands the mathematical horizons of trainees, to develop a thorough understanding of the essence of mathematical teaching, master the theory and technological standard of mathematics teaching design, and improve the professional quality of mathematics educators.

B1. Analysis of basic mathematical ideas;
B2. Overall understanding and analysis of content of secondary school mathematical curriculum;
B3. Mathematics teaching practice;
B4. Mathematics education methods;
B5. Research and discussion on improving the effectiveness of middle school lesson observation and evaluation.

Module C introduces trainees to the latest developments in the middle school mathematics course reform, and to solve common problems in the implementation process so as to improve the pertinence, importance, and practice of middle school mathematical teachers' training scheme design.

C1. In-depth analysis of *Standards for Compulsory Education Stage (2011)*;
C2. Problems and case analysis (Huang, Li, & He, 2010);
C3. International teacher education training and case analysis.

Module D addresses the basic methods of teacher training scheme design and prepares middle school backbone mathematics teachers through special reports, case research and discussion, query and reflection, interaction and exchange, and other research and training methods.

D1. Selection and optimization of content of middle school mathematics teachers training;
D2. Research on middle school algebraic problems;
D3. Research on middle school geometric problems.

Module E introduces trainees to methods of effective teacher training and organization and management, through teaching practice, observation, inspection, query and reflection, interaction and exchange, and other research and training methods.

E1. Research and discussion on middle school mathematics teaching practices;
E2. Organization and management experience exchanges.

18.3.3.3 Training Methods

1. Expert lectures
   Experts introduce relevant advanced ideas, leading theories, and practical experiences in order to expand the trainees' professional perspectives, establish theoretical frameworks, and improve professional quality.
2. Participative training
   In this format, trainees participate in training, teaching, and discussions, thus learning and making progress together with other trainees. This training mode is based on five principles: equal participation and mutual cooperation; showing respect for diversity as well as flexible and diverse forms; taking advantage of existing experience to construct knowledge; paying attention to training process; and emphasizing linking theory with practice and combining the concrete with the abstract. The methods commonly used are group discussion, case analysis, watching videos, and interviews.
3. Task-driven approach
   The task-driven training mode involves presenting complex and meaningful problem situations. Learners cooperate with each other to solve a series of relevant real teaching problems. This mode mainly consists of the following sections: creating situations; determining problems; independent learning; cooperative learning; and performance evaluation.
4. Case study
   In this mode, trainees discuss and exchange new problems and typical cases arising in the experiment of new mathematics curriculum reform in the past ten years, and based on relevant advanced theories.
5. Class observations and evaluations
   Trainees visit and evaluate classes in selected schools.

These visits are beneficial as an exchange of ideas and experiences. Evaluations focus on such aspects as teaching objectives, content handling, teaching methods, basic teaching skills, and teaching effects.

6. Field trips
   Field trips providing opportunities for trainees to exchange experience with peers and learn from each other, as well as to gain ideas about effective teacher training.
7. Mixed style training
   This combines centralized face-to-face instruction and Web-based advanced studies, using online learning platforms. It includes on-site diagnoses and case studies involving practice and observation, situational experiencing to help the trainees to solve problems occurring in mathematics education and teaching practice. A Web-based community has been established in which the trainees can carry out personalized learning, exchange experiences, share results, receive follow-up tutoring, and communicate with each other at the post-training stage.
8. Follow-up and guide
   A training service team has been set up to oversee the post-training follow-up exchange between trainers and trainees as well as learners through Web-based platforms for advanced studies. The details are as follows:

   - The training organization provides the trainees with the latest information on the plans, schemes, implementation, etc., for training middle school backbone mathematical teachers of all provinces and cities.
   - The trainees share their training experiences via network platforms.
   - Mentor groups provide the trainees with some latest training resources, solve the problems they encounter, and interact with them;.
   - During the stage of centralized learning, the trainees are divided into several learning groups, and a South China Normal University Cyber Institute specialist will be assigned to each group as a mentor to build a learning community, so as to provide an institutional guarantee for continuous learning.

18.3.3.4 Assessment and Evaluation

Assessment is based on five aspects: the disciplinary performance, learning results, interactions, design of training scheme and an overall conclusion of the trainee during advanced studies, and by ways of self-evaluation, trainees mutual evaluations, experts' evaluations, and evaluation by the teacher in charge of the class. Those who complete the course requirements receive a Certificate of Completion issued by the Ministry of Education.

References

He, X. (2003, 2004, 2011).数学学与教的心理学 [Psychology for math learning and teaching]. Guangzhou: South China University of Technology Press.

He, X. (2005). Construction of further education curricula system for math teachers of senior high school. *Research in Mathematics Education, 9*(2), 135–151.

He, X., & Yao, J. (2008, 2012). 中学数学教学设计 [Instructional design for mathematics in middle school]. Beijing: Science Press.

Huang, R., Li, Y., & He, X. (2010). What constitutes effective mathematics instruction: A comparison of Chinese expert and novice teachers' views. *Canadian Journal of Science, Mathematics and Technology Education, 10*(4), 293–306.

Chapter 19
A Study of the Status of Teacher's Professional Knowledge

Jiwei Han, Qiong Wu, Xinxin Ji and Lijuan Wu

Abstract Teacher knowledge is an important factor of teaching level and also is an important indicator to evaluate the quality of education. In order to understand the state of Chinese teachers' knowledge, this study investigates the state and type of in-service mathematics teachers and pre-service mathematics teachers in the capital of northeast three provinces. Then by means of comparing the difference of all kinds of teachers' knowledge between different types of teachers and analyzing the influence of different sources on teachers' knowledge development, this study explores the development of teachers' knowledge. Results of the study were as follows: In-service junior high school mathematics teachers behave well in teacher knowledge, and there are three types of their knowledge structure: balanced, experiential, and branch, and then the order from big to small of the importance of all kinds of knowledge in professional development is PCK, math subject knowledge, pedagogical content knowledge, and math curriculum knowledge; Pre-service teachers got lower scores in mathematics curriculum knowledge and PCK, and there are three types of their knowledge structure: balanced, weak, and branch; There are significant differences between in-service and pre-service teachers in math curriculum knowledge, math subject knowledge, and PCK, and in-service teachers scored higher on this three knowledge; The main sources of teacher knowledge for in-service teachers are teaching experience and reflection, and daily communication with colleagues; however, the main sources for pre-service teachers are educational probation and micro-teaching. Implication for theory is discussed in order to provide reference for teacher education.

J. Han (✉) · Q. Wu · X. Ji · L. Wu
Northeast Normal University, Changchun, China
e-mail: hanjw617@nenu.edu.cn

Y. Cao and F.K.S. Leung (eds.), *The 21st Century Mathematics Education in China*, New Frontiers of Educational Research,
https://doi.org/10.1007/978-3-662-55781-5_19

19.1 Introduction

In the past ten years, many countries have implemented basic education curriculum reforms. China has also been doing so, since the beginning of the twenty-first century. With its implementation, lots of problems have appeared. There is a need for these problems to be addressed by educational policy makers, scholars, and society. The teacher is an important executor of curriculum reform, so teacher quality is critical. As a part of this, teachers' professional knowledge is considered as an important and basic factor. It has a significant impact on classroom teaching and students' development (Ball, 2000; Li, 2009; Leinhardt & Smith, 1985; Ma, 1999). Various empirical studies have found that teachers' professional knowledge and beliefs have a crucial influence on the teacher's teaching behavior, decisions, and other aspects. First, teachers' professional knowledge influences classroom teaching (Ball, 1990; Brown & Borko, 1992; Frykholm, 1996). Second, it influences the teacher's decisions, for example, how to choose and arrange teaching content, activities, tasks, and teaching materials (Sanchez & Linares, 2003; Shulman & Grossman, 1988). Third, it influences the teacher's understanding of students' cognition (Dooren, Verschaffel, & Onghena, 2002; Schmidt, 1996; Schmidt & Bednarz, 1997). Last, it also influences the implementation of curriculum (Belfort & Guimaraes, 2002).

In the past ten years, many international comparative research studies have been carried out on mathematics teachers' professional knowledge. COACTIV, as an example, focuses on the relationship between mathematics teachers' professional knowledge and their students' PISA mathematics scores. Other international comparative studies have included MT21 (Mathematics Teaching in the twenty-first century, cited in Schmidt, Blömeke, & Tatto, 2011), TEDS-M (Teacher Education and Development Study, cited in Blömeke, Suhl, Kaiser, & Döhrmann, 2012), LMT (Learning Mathematics for Teaching, Hill, Ball, & Schilling, 2008). All of these have concurred that teachers' professional knowledge plays a significant role in mathematics education. For example, TEDS-M, the largest scale international comparative study of mathematics-major pre-service teachers' professional knowledge, found that the rankings of mathematics-major pre-service teachers' professional knowledge levels correlated with the ranking of students' scores on PISA and TIMSS in each country (Blömeke & Delaney, 2014).

Taking into consideration the importance of teachers' professional knowledge to, there has been a need to address a number of questions. What is the actual status of in-service mathematics teachers' professional knowledge? What are the characteristics of teachers' professional knowledge structures? Furthermore, a lot of time, energy, and finances have been devoted to training excellent teachers, but how well is teachers' professional development promoted? What are the alternative methods teachers use to develop their professional knowledge? To address these questions, the purposes of this study were the following:

1. To examine the status of in-service teachers' professional knowledge, the characteristics of this professional knowledge, and the function of professional knowledge in teachers' professional development.
2. To investigate the source of in-service teachers' professional knowledge.

It is hoped that the answers will be helpful in providing an understanding of mathematics teacher education in China as well as giving some constructive suggestions about the training of mathematics teachers.

19.2 Methodology

19.2.1 Research Subjects

There are 31 provinces in China (excluding Hong Kong, Macao, and Taiwan). According to their development levels, they are divided into three zones. The first zone includes the provinces with high levels of development, such as Beijing, Shanghai, Tianjin, Zhejiang, Jiangsu, and Guangdong. The second zone refers to the provinces with secondary levels of development, mainly in the middle or east part of China, such as Liaoning, Jilin, Heilongjiang, Shanxi, Hunan, Hubei, Hebei, and Henan. The development level of the third zone is low, mainly in the western part of China. The education levels are also different in the three zones. In the present study, the research subjects were in-service teachers from three provincial capitals in the second zone.

The sample of this study is selected by stratified sampling and cluster sampling combined. First, all junior middle schools in the three provincial capitals were stratified. Second, several junior middle schools at each level were chosen, and all mathematics teachers in these schools were considered as the research subjects. The in-service mathematics teachers were from Shenyang, Changchun, and Harbin. The total number of in-service mathematics teachers is 150; of the 150 questionnaires administered, 123 valid responses were received. It is noted that the 123 teachers account for 4% of all mathematics teachers in junior middle schools in these three provinces. In this case, it met the accuracy requirement of the research.

19.2.2 Questionnaire

The questionnaire consisted of three parts. The first part aimed to investigate the background information, such as mathematics teachers' education certification, professional titles, and teaching experience. The second part aimed to investigate the sources of teachers' professional knowledge, and the last part examined the status of their professional knowledge.

19.2.2.1 Sources of Teachers' Professional Knowledge

There were two steps involved in developing the questionnaire on the sources of teachers' professional knowledge. Step 1 was an open questionnaire used to investigate some effective ways of developing teachers' professional knowledge. There were two questions in this questionnaire. The first asked about courses in the respondents' universities that were helpful to develop their professional qualities. The second question asked about methods that were helpful to developing respondents' professional qualities in their work. Here, professional qualities included knowledge of mathematics, knowledge of education theory, knowledge of mathematics teaching, and knowledge of the mathematics curriculum. Actually, the professional quality means the teachers' professional knowledge, but this term was not used in the questionnaire, because the concept of professional quality is easier to understand. In Step 2, by analyzing data from this questionnaire, several variables relevant to the development of teachers' professional knowledge were identified. These variables are the foundation for the main questionnaire. At the same time, by summarizing related studies, other variables are also identified (Fan, 2003). Finally, all of these variables were defined as items relating to developing teachers' professional knowledge in the formal questionnaire. There was a total of 49 items, including nine about educational theoretical knowledge, 10 about knowledge of the mathematics curriculum, 10 about subject matter knowledge, and 10 about pedagogical content knowledge. This questionnaire had a four-point score scale. For example, on the item "To what degree does the educational internship and probationary period enrich your knowledge of educational theoretical?" the responses were 4 very useful, 3 useful, 2 somewhat useful, and 1 useless; a further option of "no experience" was included and scored as 0.

19.2.2.2 Teachers' Professional Knowledge

There are various types of teachers' professional knowledge, and the range of teachers' professional knowledge is wide. The first step in designing the questionnaire, therefore, was to identify the range and sources of this knowledge. This was done in two ways: first, by summarizing related research, and second, by analyzing the status of teacher education in China. From this, four dimensions of teachers' professional knowledge were defined. Before compiling the questions, an item specification was formed, which described the range of content in the test, its level, the number of questions, and the percentage of each category of questions. The item specification was a guide for the compilation and evaluation of the test. Then, different researchers were responsible for different questions. After a pilot testing, some questions were deleted or added. This was followed by a second pilot testing and further revisions.

In the final version of the formal, there were four dimensions of teachers' professional knowledge: educational theoretical knowledge, knowledge of the mathematics curriculum, subject matter knowledge, and pedagogical content knowledge. The test of educational theoretical knowledge was compiled by university teachers. There were 11 multiple-choice questions, for an example, "In Hainan province, when a teacher introduces the snowflake, he shows students a video and also throws scraps of paper into the air. The purpose is to let students experience the snow. Which approach has the teacher used? A. Entity B. Language C. Virtual D. Model". For the test of knowledge of mathematics curriculum, there were both multiple-choice and essay questions. The total number was seven. These questions were compiled by three people: a teaching researcher, a university teacher, and an expert mathematics teacher. The university teacher's field of study was mathematics education in junior middle schools. An example of a question is: "Which is not the key point of the chapter on integral expression? A. To strengthen students' sense of symbol, B. To analyze, summarize, and conclude, C. Basic calculation ability, D. To explore the law and regularity of calculation." For the test of subject matter knowledge, there were three parts: test of concepts, test of open questions, and test of reasoning questions. They were compiled by an expert mathematics teacher. For the test of pedagogical content knowledge, there are totally four questions, compiled by a teaching researcher and an expert mathematics teacher. The test of pedagogical content knowledge aimed to investigate two aspects: first, how teachers deal with students' difficulties in class, and, second, how to design a better mathematical representation for promoting students' learning. There were four questions, for example: "When teaching the concept of the system of binary linear equations, students think that $\begin{cases} x+y=7 \\ y-3=1 \end{cases}$ is not a system of binary linear equations. If you do not force students to accept the concept, how do you explain it?"

The arrangement of questions and their scores is presented in Table 19.1.

Table 19.1 Arrangement of the test questions of teachers' professional knowledge

	Number	Score per examination question	Total score
Educational theoretical knowledge	11	2	22
Knowledge of mathematics curriculum	7		16
Multiple-choice question	6	2	12
Essay question	1	4	4
Subject matter knowledge	13		50
Multiple-choice question	7	2	14
Essay question	4	5 + 5 + 3 + 3	16
Reasoning question	2	10 + 10	20
Pedagogical content knowledge	4		36
Question 1	1	10	10
Question 2	1	10	10
Question 3	1	6	6
Question 4	1	10	10

19.2.3 Data Collection and Analysis

The examination was conducted by a team, including university teachers and postgraduates. There was no time limit. In general, it took about 1 h to finish the examination. For the essay questions and pedagogical content knowledge, two junior middle school teachers and one university teacher worked together to grade the score. First, the three teachers worked together to set a standard for the grading. Then, the two junior middle school teachers graded each paper simultaneously. If there was a large gap between the two scores, the third teacher would grade the paper again. In most cases, the two teachers doing the initial grading concurred; the percent of coherence was 95%.

19.2.4 The Degree of Difficulty and the Degree of Distinction

The classical test theory is used in this study to measure the test's degree of difficulty and the degree of distinction. For the test of educational theoretical knowledge, the degree of difficulty of every item ranges from 0.16 to 0.85, with the mean result being 0.42. For the test of knowledge of mathematics curriculum, the degree of difficulty of every item ranged from 0.28 to 0.83, with the mean result being 0.50. For the test of concepts in subject matter knowledge, the degree of difficulty of every item ranged from 0.41 to 0.68, with the mean result being 0.54. For the open questions on subject matter knowledge, the degree of difficulty of every item ranged from 0.48 to 0.68, with the mean result being 0.58. For the test of reasoning in subject matter knowledge, the degree of difficulty of every item ranged from 0.17 to 0.68, with the mean result being 0.43. The average degree of difficulty of the test of subject matter knowledge was 0.54. For the test of pedagogical content knowledge, the degree of difficulty of every item ranged from 0.47 to 0.52, with the mean result being 0.49. In summary, for the test of teachers' professional knowledge, except for three items of educational theoretical knowledge and one item of subject matter having slightly higher degrees of difficulty, all items ranged from 0.2 to 0.8, with the mean result being about 0.5, thus meeting the requirements for this test.

The degree of distinction is measured by the degree of relationship between total scores and every item. For educational theoretical knowledge, the degrees of distinction ranged from 0.30 to 0.54, with the mean result being 0.43. For the knowledge of mathematics curriculum, the degrees of distinction ranged from 0.20 to 0.47, with the mean result being 0.37. For subject matter knowledge, the range was 0.25–0.91, with the mean result being 0.51. For pedagogical content knowledge, it was 0.57–0.74, with the mean result being 0.65. According to Ebel, a degree of distinction above 0.2 is acceptable; those ranging from 0.30 to 0.39 are good; and above 0.4 is even better. So, in the present study, the degree of distinction met the requirement.

19.3 Research Result

19.3.1 In-Service Mathematics Teachers' Professional Knowledge Status

19.3.1.1 Professional Knowledge Status and Its Features

The in-service mathematics teachers' professional knowledge status is shown in Table 19.2. Since the test was aimed at mathematics teachers from junior middle school, to ensure it was sufficiently discriminatory, the topic of foundations was not allocated a large proportion, and most of the problems were above medium difficulty.

In order to investigate the impact of diversity in the teachers' backgrounds\their teaching ages (the length of time spent teaching), the time spent teaching graduate grades, their education backgrounds, locations and teacher categories were analyzed. The result showed that there were no significant differences between teaching education backgrounds and locations. There were significant differences in teachers' teaching ages, the times spent teaching graduate grades, education backgrounds, and teacher categories. To be specific, there were significant differences between teachers' subject matter knowledge ($F = 5.06$, $P = 0.00$) and pedagogical content knowledge ($F = 5.06$, $P = 0.00$). With increasing teaching age, teachers' subject knowledge and pedagogical knowledge increased gradually. There were significant differences between subject matter knowledge ($F = 3.32$, $P = 0.01$) and pedagogical content knowledge ($F = 4.88$, $P = 0.001$) for teachers with different teaching experiences of graduate grades (see Fig. 19.1). Teachers' understandings of subject matter knowledge and mathematics curriculum knowledge had a positive correlation with their teaching experiences of graduate grades (see Fig. 19.2).

It can be seen that the development of teachers' professional knowledge was related closely to their teaching practices, teaching ages, and experiences of teaching graduate grades, but it was not highly related to their teaching backgrounds, locations, or other factors.

Table 19.2 Status of mathematics teachers' professional knowledge in junior middle school

	Full marks	Average score	Standard deviation	Maximum value	Minimum value
Educational theoretical knowledge	22	9.429	4.265	20.00	0.00
Curriculum knowledge	16	6.774	2.475	13.00	2.00
Subject matter knowledge	50	25.988	8.439	46.00	11.00
Pedagogical content knowledge	36	18.113	9.008	36.00	0.00

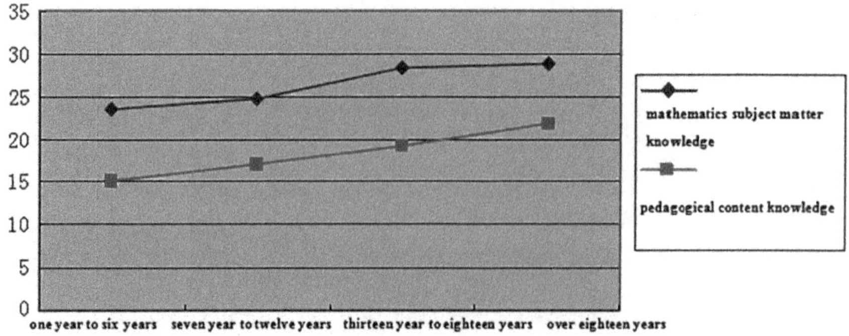

Fig. 19.1 Differences in teachers' professional knowledge by teaching age

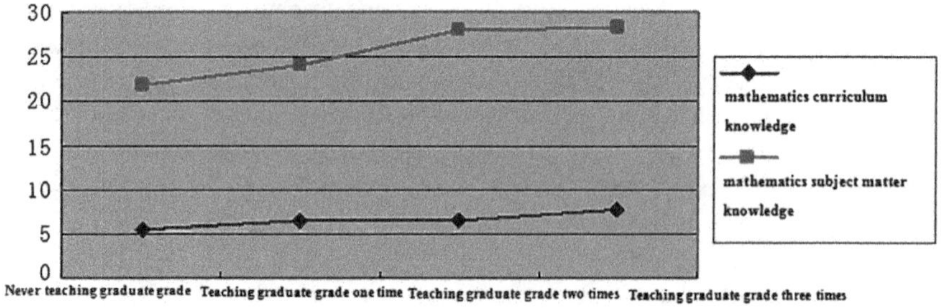

Fig. 19.2 Differences in teachers' professional knowledge by experience of teaching graduate grades

19.3.1.2 Different Knowledge Structure Types of in-Service Mathematics Teachers

Educational theoretical knowledge, subject matter knowledge, curriculum knowledge, and pedagogical content knowledge are different kinds of knowledge. Different teachers have different professional knowledge structures. In order to investigate these structures, the scores of mathematics teachers' professional knowledge were classified into different types by using clustering analysis. First, teachers' professional knowledge was classified into five types, but there were no significant differences between two of these, so these categories were reduced to four. The internal characteristics of one of these were found to be inconspicuous. Finally, we found that three types were reasonable for the cluster analysis (Table 19.3; Fig. 19.3).

The scores of teachers belong to first type were not only much higher than the other two groups in pedagogical content knowledge but also the highest in curriculum knowledge and educational theoretical knowledge. Due to the balanced nature of their professional knowledge, this category was named "balanced teachers." Balanced teachers accounted for 36% of the total number, and most were

Table 19.3 Professional knowledge structure of mathematics teachers in junior middle school

	Teachers' professional knowledge types			F value	Significance
	Balanced	Experienced	Scattered		
Educational theoretical knowledge	11.85	6.73	8.77	9.826	0.000
Knowledge of curriculum	7.81	5.41	6.19	7.169	0.001
Subject matter knowledge	32.78	15.82	28.73	82.336	0.000
Pedagogical content knowledge	29.07	13.27	12.62	87.678	0.000
Proportion (%)	36	29	35		

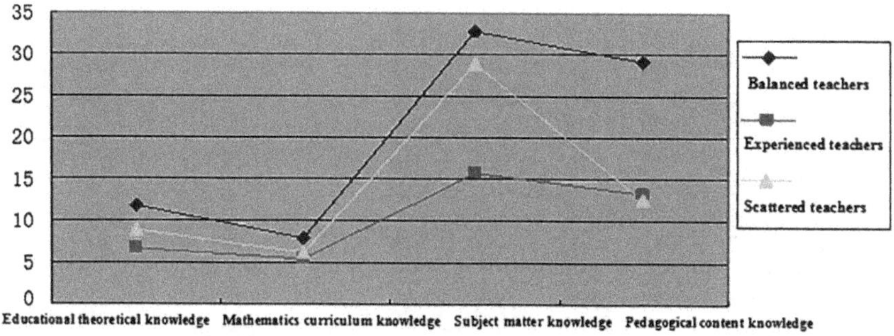

Fig. 19.3 Teachers' professional knowledge types in junior middle school

key teachers. The second type of teachers got the lowest scores in the dimensions of educational theoretical knowledge and professional knowledge of mathematics. Because they had accumulated some practical knowledge in teaching practice, they did not lack pedagogical content knowledge, so this group was called "experienced teachers." Experienced teachers accounted for 29% of the total number, and most were middle-aged. The third group's scores for educational knowledge and mathematics knowledge were lower than those of the balanced teachers, but higher than the experienced teachers; the scattered knowledge was moderate, and their pedagogical content knowledge in teaching practice was not good, so this group was named "scattered teachers." Scattered teachers accounted for 35% of the total number and were mostly young.

19.3.1.3 The Impact of Different Teachers' Knowledge on Teachers' Professional Development

During the process of teachers' professional development, which kind of teachers' professional knowledge is more important? In order to examine this question, a discriminant analysis was carried out to compare prominent and ordinary teachers. This research was not aimed at distinguishing prominent teachers or ordinary teachers; its aim was to explore the importance of different kinds of teachers' professional knowledge in their professional development. Prominent teachers are those who have over ten years' teaching experience and win the honorary title of "key teacher." There are two kinds of ordinary teachers, young and middle-aged. The young ordinary teachers are those who have teaching experience lesser than three years. The middle-aged ordinary teachers are those with teaching experience of more than ten years but who have not become key teachers. Over 20 observations were included in the discriminant analysis; in order to have a commensurate classification, it was important that there was little variation in each group. Twenty prominent teachers were included in this research; 20 young ordinary teachers and 20 middle-aged teachers were also selected randomly for the study.

Two discriminatory functions were produced from the discriminant analysis. The eigenvalue of the first was 1.66, Wilk's Lambda = 0.4, χ^2 (8, 60) = 50.898, and $p < 0.05$. The eigenvalue of the second was 0.155, Wilk's Lambda = 0.866, χ^2 (3, 60) = 7.999, and $p < 0.05$. The standardized criterion weights, also called discriminant coefficients, are shown in Table 19.4; the bigger the discriminant coefficient is, the more discrimination power the variable has. Table 19.5 shows the structure matrix; it is the correlation index between variable quantity and discrimination function. A bigger absolute value indicates a greater impact on the discrimination function. As the structure matrix was not affected by the collinearity of the variable, the result of the analysis was reliable.

When two or more discrimination functions were kept, the index of comprehensive abilities was needed to describe the impact of the variables on significant functions. The calculation of the index of comprehensive abilities was carried out as follows: First, calculate the index of comprehensive abilities in every function, the ability value of variable in function = (identifying load)^2 × values of function character. The relative characteristic value of function = values of identifying function/sum of all the values of significant functions. Then, calculate all the indexes of comprehensive ability of significant functions. A comprehensive index indicates the sum number of the comprehensive ability value of the variable in every significant function. The comprehensive indexes of four types of teachers' professional knowledge were as follows: pedagogical content knowledge −0.4747, mathematics knowledge −0.3166, mathematics curriculum knowledge −0.2354, educational theoretical knowledge −0.3022. The simple scatter plot is shown as Fig. 19.4.

Table 19.4 Discriminant coefficients

	Function	
	1	2
Educational theoretical knowledge	0.487	0.843
Subject matter knowledge	0.316	−0.145
Pedagogical content knowledge	0.561	−0.474
Knowledge of curriculum	0.341	0.001

Table 19.5 Structure matrix

	Function	
	1	2
Pedagogical content knowledge	0.704 (*)	−0.563
Subject matter knowledge	0.597 (*)	−0.129
Knowledge of curriculum	0.513 (*)	−0.163
Educational theoretical knowledge	0.497	0.847 (*)

*The numbers which are quite bigger which show greater influences on the discrimination functions

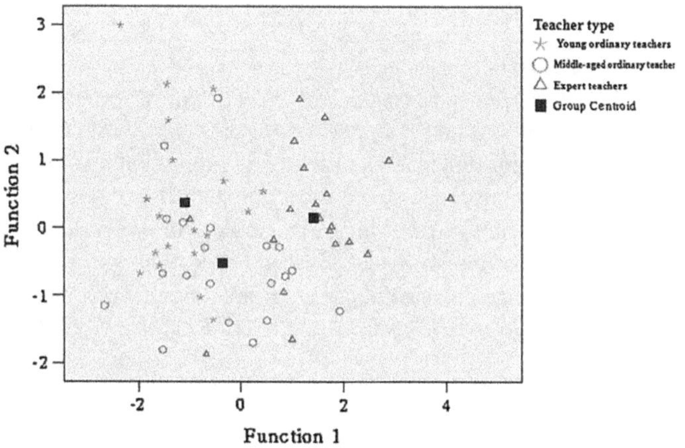

Fig. 19.4 Distribution diagram of three kinds of teacher

It is shown in Fig. 19.4 that the prominent teachers had significantly higher scores on the first function; pedagogical content knowledge, mathematics subject matter knowledge, and mathematics curriculum knowledge were more highly correlated with the first function. The young teachers had higher scores on the second function; however, educational theoretical knowledge was more highly correlated with the second function.

19.3.2 Source of in-Service Teachers' Professional Knowledge

There is no doubt that teachers' professional knowledge has an important influence on teaching practice. So, where do teachers learn and develop this knowledge? On the basis of Fan Lianghuo's study, the present study has listed many sources of teachers' professional knowledge. Some are courses in normal universities for pre-service teachers, including mathematics subject courses, teaching methods courses, solving mathematics problems in junior middle schools, educational internship, and probation and micro-teaching. Others are sources of teachers' professional knowledge for in-service teachers, including education with records of formal schooling, training for in-service teachers, teaching practice and reflection, demonstration lectures, daily communication with colleagues, and reading professional books. The teachers in this study were asked to evaluate the importance of each source item on the four-point scale. For each dimension, descriptive statistics and ordinal polytomous logistic regression were used to investigate the importance of each source. The model of ordinal polytomous logistic regression is presented in the Appendix.

The result for the descriptive statistics was the same as the result for the logistic regression. Due to limitations of space, the descriptive statistics are not presented here.

From the logistic regression, on the whole, there were three levels of importance assigned by teachers to their professional knowledge: most important, secondary importance, and least important. Table 19.6 presents the importance of various sources for each of the four dimensions of teachers' professional knowledge. It can be seen clearly that teaching practice and reflection and daily communication with colleagues were seen as the two most important sources of professional knowledge for in-service teachers. For the in-service teachers, education with record of formal schooling was considered as the least important source. For the pre-service teachers, the educational internship and probation and micro-teaching were the two most important sources.

19.3.2.1 Source of Educational Theoretical Knowledge

1. Logistic regression model

To see whether the cumulative logistic regression model was suitable or not, a parallel hypothesis test was done. The value of L.R. Chi-square was not obvious (Chi-square = 16, df. = 18, $p = 0.593$), which supported the cumulative logistic regression model as suitable for the present study.

The significance testing of the logistic regression model involves doing a likelihood ratio test, while all the partial regression coefficients are zero. The result shows that the model was significant (Chi-square = 126.965, df. = 9, $p = 0.000$). According to the results of the Chi-square goodness of fit test, there was no significant deviation for two variables of goodness of fit: Pearson Chi-square (Chi-square = 12.979, df. = 18, $p = 0.793$) and D-statistic (Chi-square = 16, df. = 18, $p = 0.593$). That is to say, there was a perfect goodness of fit between the model and data.

2. Parameter estimation of the logistic regression model

To evaluate the importance of each source of teachers' professional knowledge on the whole, the logistic regression model was established in the present study. The importance of each source of theoretical knowledge was ranked from the most

Table 19.6 In-service teachers' evaluation of importance of each source of professional knowledge

Source of Teachers' professional knowledge	Educational theoretical knowledge	Knowledge of mathematics curriculum	Subject matter knowledge	Pedagogical content knowledge
Most important	Teaching practice and reflection Daily communication with colleagues Reading professional books	Teaching practice and reflection Daily communication with colleagues Demonstration lecture	Teaching practice and reflection Daily communication with colleagues Reading professional books Demonstration lecture	Teaching practice and reflection Daily communication with colleagues
Secondary important	Demonstration lecture Teaching methods courses	Educational intern and probation Reading professional books Training for in-service teachers Micro-teaching	Educational intern and probation Teaching methods courses Micro-teaching Training for in-service teachers Mathematics subject courses	Reading professional books Demonstration lecture Educational intern and probation Micro-teaching
Least important	Educational courses Micro-teaching Training for in-service teachers Education with record of formal schooling	Teaching methods courses Educational courses The education with record of formal schooling	The education with record of formal schooling	Training for in-service teachers Teaching methods courses Educational courses The education with record of formal schooling

important to the least. The results are presented in Table 19.7. It can be seen that the most important source of educational theoretical knowledge was teaching practice and reflection. The least important source of educational theoretical knowledge was education with record of formal schooling.

Based on the results of the logistic regression, the importance of each source of educational theoretical knowledge was divided into three levels. Compared to the educational intern and probation, three sources are considered as the more important, at the level of 0.05: teaching practice and reflection, daily communication with

Table 19.7 Logistic regression for the importance of each source of educational theoretical knowledge

Variable	Parameter estimator	SE	Wald value	df	Significance	95% CI	
						Lower	Upper
Intercept 1	−3.524	0.277	161.853	1	0.000	−4.067	−2.981
Intercept 2	−2.275	0.241	89.398	1	0.000	−2.747	−1.803
Intercept 3	−0.352	0.222	2.515	1	0.113	−0.787	0.083
Teaching practice and reflection	0.961	0.351	7.487	1	0.006	0.237	1.649
Daily communication with colleagues	0.906	0.348	6.788	1	0.009	0.225	1.588
Reading professional books	0.761	0.340	5.016	1	0.025	0.095	1.427
Demonstration lecture	0.492	0.328	2.256	1	0.133	−0.15	1.134
Educational intern and probation[a]	0	–	–	0	–	–	–
Teaching methods courses	−0.554	0.306	3.273	1	0.070	−1.153	0.046
Educational courses	−0.852	0.305	7.815	1	0.005	−1.449	−0.255
Micro-teaching	−0.937	0.305	9.465	1	0.002	−1.535	−0.34
Training for in-service teachers	−1.117	0.305	13.435	1	0.000	−1.715	−0.520
The education with record of formal schooling	−1.485	0.306	23.566	1	0.000	−2.085	−0.886

[a]The parameter value of educational intern and probation is defined as o, because the model of UNSATURATED is used here

colleagues, and reading professional books. The demonstration lecture, educational internship and probation, and teaching methods courses were of equal importance. Still, compared to the educational internship and probation, four sources were marked as less important sources, namely educational courses, micro-teaching, training for in-service teachers, and education with record of formal schooling.

19.3.2.2 Sources of Mathematics Curriculum Knowledge

1. Logistic regression model

The same method of testing the source of educational theoretical knowledge was used here. To analyze the source of mathematics curriculum knowledge, the parallel hypothesis test of cumulative logistic regression model was used. The value of L.R. Chi-square was not obvious (Chi-square = 21.407, df. = 18, $p = 0.259$), which supported the cumulative logistic regression model as suitable for the present study. The result of the significance testing showed that the model was significant

(Chi-square = 165.280, df. = 9, $p = 0.000$). According to the results of the Chi-square goodness of fit test, there was no significant deviation for two variables of goodness of fit: Pearson Chi-square (Chi-square = 16.354, df. = 18, $p = 0.568$) and D-statistic (Chi-square = 21.407, df. = 18, $p = 0.259$). In other words, there was a perfect goodness of fit between the model and data.

2. Parameter estimation of the logistic regression model

By analyzing the result of the logistic regression, the importance of the sources of mathematics curriculum knowledge was ranked from most to least important. The results are presented in Table 19.8. It can be seen that the most important source of mathematics curriculum knowledge is teaching practice and reflection. The least important source of mathematics curriculum knowledge was education with record of formal schooling.

Based on the result of the logistic regression, three levels of importance for the sources of mathematics curriculum knowledge were identified. Three sources were considered as more important than the educational internship and probation, at the level of 0.05: teaching practice and reflection, daily communication with colleagues, and demonstration lectures. The reading of professional books, training for in-service teachers, micro-teaching, and educational internship and probation were of equal importance. Still, compared to the educational internship and probation, three sources were marked with less important sources, namely education with record of formal schooling, teaching methods courses, and educational courses.

19.3.2.3 Sources of Subject Matter Knowledge

1. Logistic regression model

For sources of subject matter knowledge, the value of L.R. Chi-square was not obvious (Chi-square = 39.042, df. = 27, $p = 0.063$); hence, the cumulative logistic regression model was accepted as suitable for the present study. The model was found to be significant (Chi-square = 84.97, df. = 9, $p = 0.000$). According to the results of the Chi-square goodness of fit test, there was no significant deviation for two variables of goodness of fit: Pearson Chi-square (Chi-square = 32.971, df. = 27, $p = 0.198$) and D-statistic (Chi-square = 39.042, df. = 27, $p = 0.063$), that is, the goodness of fit between the model and data was perfect.

2. Parameter estimation

From the logistic regression, the sources of subject matter knowledge were ranked from most to least important (Table 19.9). The most important source was teaching practice and reflection. The least important was education with record of formal schooling.

Three levels of importance for the sources of subject matter knowledge were identified. Four sources were considered more important than the educational internship and probation, at the 0.05 level: teaching practice and reflection, daily

Table 19.8 Logistic regression of the importance of each source of mathematics curriculum knowledge

Variable	Parameter estimator	SE	Wald value	Df	Significance	95% CI	
						Lower	Upper
Intercept 1	−2.908	0.253	132.447	1	0.000	−3.403	−2.413
Intercept 2	−1.886	0.231	66.612	1	0.000	−2.338	−1.433
Intercept 3	−0.207	0.217	0.903	1	0.342	−0.633	0.220
Teaching practice and reflection	1.220	0.356	11.765	1	0.001	0.523	1.917
Daily communication with colleagues	1.220	0.356	11.765	1	0.001	0.523	1.917
Demonstration lecture	0.867	0.334	6.724	1	0.010	0.212	1.523
Reading professional books	0.543	0.320	2.866	1	0.090	−0.086	1.171
Educational internship and probation[a]	0	–	–	0	–	–	–
Training for in-service teachers	−0.252	0.303	0.693	1	0.405	−0.845	0.341
Micro-teaching	−0.508	0.300	2.861	1	0.091	−1.097	0.081
Teaching methods courses	−1.107	0.299	13.684	1	0.000	−1.693	−0.520
Educational courses	−1.181	0.299	15.561	1	0.000	−1.767	−0.594
The education with record of formal schooling	−1.343	0.300	20.076	1	0.000	−1.931	−0.756

[a]The parameter value of educational intern and probation is defined as o, because the model of UNSATURATED is used here

communication with colleagues, reading professional books, and demonstration lectures. The teaching methods courses, micro-teaching, and training for in-service teachers were of equal importance. The mathematics subject courses and education with record of formal schooling were considered as less important sources.

19.3.2.4 The Source of Pedagogical Content Knowledge

1. Statistical test of the logistic regression model

A L.R. Chi-square value of 26.890 (df. = 18, $p = 0.081$) shows that the model is reasonable. It was also found to significantly support the logistic regression model (Chi-square = 116.303, df. = 9, $p = 0.000$). According to the results of the Chi-square goodness of fit test, there was a perfect goodness of fit between the model and data: Pearson Chi-square (Chi-square = 21.770, df. = 18, $p = 0.242$) and D-statistic (Chi-square = 26.890, df. = 18, $p = 0.081$).

Table 19.9 Logistic regression for importance of each source of subject matter knowledge

Variable	Parameter estimator	SE	Wald value	Df	Significance	95% CI	
						Lower	Upper
Intercept 1	−2.792	0.250	124.482	1	0.000	−3.282	−2.302
Intercept 2	−1.645	0.223	54.449	1	0.000	−2.082	−1.208
Intercept 3	0.134	0.212	0.398	1	0.528	−0.282	0.550
Teaching practice and reflection	1.987	0.389	26.106	1	0.000	1.225	2.749
Daily communication with colleagues	1.645	0.359	21.031	1	0.000	0.942	2.348
Reading professional books	1.111	0.326	11.589	1	0.001	0.471	1.750
Demonstration lecture	1.042	0.323	10.391	1	0.001	0.408	1.675
Educational internship and probation[a]	0	–	–	0	–	–	–
Teaching methods courses	−0.198	0.296	0.449	1	0.503	−0.779	0.382
Micro-teaching	−0.526	0.295	3.180	1	0.075	−1.104	0.052
Training for in-service teachers	−0.542	0.295	3.375	1	0.066	−1.120	0.036
Mathematics subject courses	−0.651	0.295	4.879	1	0.027	−1.229	−0.073
The education with record of formal schooling	−0.768	0.295	6.792	1	0.009	−1.346	−0.190

[a]The parameter value of educational intern and probation was defined as o, because the model of UNSATURATED was used here

2. Parameter estimation

The results are presented in Table 19.10. The most important source of pedagogical content knowledge was teaching practice and reflection, and the least important was education with record of formal schooling.

Two sources were considered as the more important sources at 0.05 levels: teaching practice and reflection and daily communication with colleagues. Reading professional books, demonstration lectures, micro-teaching, and educational internship and probation were of equal importance. Four sources were considered as the less important: training for in-service teachers, teaching methods courses, educational courses, and education with record of formal schooling.

Table 19.10 Logistic regression for importance of each source of pedagogical content knowledge

Variable	Parameter estimator	SE	Wald value	Df	Significance	95% CI	
						Lower	Upper
Intercept 1	−3.475	0.256	184.820	1	0.000	−3.976	−2.974
Intercept 2	−2.317	0.222	108.680	1	0.000	−2.752	−1.881
Intercept 3	0.949	0.204	21.553	1	0.000	0.548	1.349
Teaching practice and reflection	1.054	0.283	13.891	1	0.000	0.500	1.609
Daily communication with colleagues	0.663	0.282	5.546	1	0.019	0.111	1.215
Reading professional books	0.480	0.282	2.889	1	0.089	−0.073	1.033
Demonstration lecture	0.351	0.283	1.542	1	0.214	−0.203	0.906
Educational internship and probation[a]	0	–	–	0	–	–	–
Micro-teaching	−0.291	0.289	1.015	1	0.314	−0.857	0.275
Training for in-service teachers	−0.590	0.291	4.116	1	0.042	−1.161	−0.020
Teaching methods courses	−0.778	0.292	7.109	1	0.008	−1.350	−0.206
Educational courses	−0.942	0.292	10.382	1	0.001	−1.514	−0.369
The education with record of formal schooling	−1.170	0.292	16.049	1	0.000	−1.743	−0.598

[a]The parameter value of educational intern and probation is defined as o, because the model of UNSATURATED is used here

19.4 Major Findings and Suggestions

19.4.1 Major Findings

19.4.1.1 In-Service Mathematics Teachers' Professional Knowledge Still Needs Improvement

Some comparative studies of teachers' professional knowledge of mathematics in different countries, for example, Ma (1999), Leung and Park (2002) have found that teachers in China are better than those in other countries, such as America and Korea. What's more, in some international mathematics tests, such as the IMO (International Mathematical Olympiad), Chinese students have outstanding performances (e.g., Stevenson et al., 1990). In other words, Chinese students have scored higher on almost every part of the test, whether it is about algebra or spatial concepts. These findings suggest that mathematics education in China is very successful and that the teachers' professional knowledge of mathematics teachers in China is good, especially the subject matter knowledge. The present study also supports these findings. In the present study, on the whole, the teachers' professional knowledge of in-service mathematics teachers was found to be good, but can still be better. In this case, considering the in-service mathematics teachers' ignorance, antipathy, and inability to adapt to the *New Curriculum Reform* ten years ago, it seems that they have now adapted much better. They have mastered the professional knowledge that is required in the *New Curriculum Reform.* The progress of these mathematics teachers has encouraging implications for the implementation of the *New Curriculum Reform.* It is noted that the present study was carried out in the second zone in China, so there is still a need to see whether the findings are the same in other zones, especially the third. Also, the present study focused on mathematics teachers in cities, so it is necessary to compare these results to those in other areas.

A country's education has crucial influence on the comprehensive national strength and the quality of the next generation, so the modernization of education is an important thrust of education reform in each country. In China, the *New Curriculum Reform* has been underway for ten years and has reached a key point now. It is clear that the reform does not mean revising curriculum standards, revising teaching materials, or revising the courseware. What is the most important factor of this reform? The teacher. According to the results of this study, it has been difficult for mathematics teachers to accept the new content of the *New Curriculum Reform*. They need to update their professional knowledge continually, particularly subject matter knowledge and new teaching methods and approaches.

19.4.1.2 Differences in Teachers' Professional Knowledge Structures

In the present study, scores for the structure of professional knowledge varied considerably. Some were stronger on educational knowledge, some on professional knowledge, and some on practice knowledge, suggesting that different teachers have different professional knowledge structures. This implies the need to train different teachers differently. Scattered, mainly novice, teachers need more teaching practice and reflection to develop their professional knowledge. Experienced teachers, who are not expert teachers, need to update their professional knowledge.

19.4.1.3 The Influence of Teaching Practice on the Development of Teachers' Professional Knowledge

In the present study, analyses of the teachers' background information indicated that the years of teaching experience and of teaching final-year students influenced their professional knowledge, while the level of certification and the zone in which they lived had no effect. Furthermore, of the four dimensions of professional knowledge, pedagogical content knowledge was found to be the most important key factor in distinguishing the types of teachers, novice, experienced, or expert. Pedagogical content knowledge is related to teaching practice. Both subject matter knowledge and educational knowledge need to be consolidated as pedagogical content knowledge in practice. This suggests that reflection on teaching practice can be helpful for teachers at all levels of experience.

The analyses of the in-service teachers' preferred sources of professional knowledge indicated two main sources that were considered important: teaching experience and reflection; and daily communication with colleagues. However, in teachers' continuing education, there are more courses on educational theoretical knowledge. There is still a need to find answers to the question of how to prefect teachers' professional knowledge by relating it to teaching practice.

19.4.2 Suggestions

19.4.2.1 The Role of Educational Theoretical Knowledge and Practice Knowledge in Teacher Education

In China, the in-service teacher education is organized mainly by universities, provincial education institutes, city education institutes, and in-service training schools. The main form is the lecture by experts in universities, teaching researchers, and excellent teachers. The content of lectures is mainly theoretical educational knowledge. However, the present study has suggested that this may not be enough. According to the results of the present study, the training is the least important source of teachers' professional knowledge, while the teaching experience, reflection, and daily communication with colleagues are the most important. This gives rise to the question of whether teaching practice knowledge should be given more prominence and less theoretical educational knowledge.

The answer is no. In fact, it is very complex. In this study, the only conclusion that can be drawn is that the result of the training, mostly introducing educational theoretical knowledge, is less effective than teaching experience, reflection, and daily communication with colleagues in developing their teachers' professional knowledge. The finding just reflects the fact, rather than the reason. Actually, the theory and the practice are both important. The most important thing is to keep developing both the theory and the practice in cycles according to teachers developing abilities and the problems that arise. There is a need to move to a new micro-level of teacher training that includes reflection and discussion of teaching practice, and addressing teachers' beliefs. This cyclic process of training can be applied for both novice teachers and experienced teachers.

19.4.2.2 Certificate Requirement for Teachers

The present study found that education with record of formal schooling was the least important source of teachers' professional knowledge. Recently, in China, lots of colleges are ungraded: Special secondary technical schools upgraded to junior colleges and junior colleges upgraded to colleges. With these changes, there is also a higher certification requirement of teachers. In the past ten years, less and less teachers have studied by correspondence education as undergraduates in normal universities. More and more teachers have continued education as postgraduates while working in primary schools or junior middle schools. Furthermore, various departments of education have published some policies to encourage in-service teachers to develop their certification. So it is disappointing that the result impact of education with record of formal schooling is less than what we had expected. It is important to investigate further into the various and complex reasons for this. Interestingly, Begle and Geeslin (1972) suggested that the development of teachers' professional knowledge has no relationship to the effectiveness of their teaching, as measured by the students' test scores. Begle and Geeslin (1972) put forward a suggestion that there may be a certain optimal certification standard for teachers and that once they reach this standard, it is not necessary for them to have more continuing education. This suggests the possibility that paying too much attention to seeking higher certification may not produce the value we expect, whereas reflection on teaching and communication with colleagues are more important.

References

Ball, D. L. (1990). Prospective elementary and secondary teachers' understanding of division. *Journal for Research in Mathematics Education, 21*(2), 132–144.

Ball, D. L. (2000). Bridging practices: Intertwining content and pedagogy in teaching and learning to teach. *Journal of Teacher Education, 51*(3), 241–247.

Begle, E. G., & Geeslin, W. (1972). *Teacher effectiveness in mathematics instruction (National Longitudinal Study of Mathematical Abilities Reports: No. 28).* Washington, DC: Mathematics Association of America and the National Council of Teachers of Mathematics.

Belfort, E., & Guimaraes, L. C. (2002). The influence of subject matter on teachers' practices: Two case studies. In A. D. Cockburn & E. Nardi (Eds.), *Proceedings of the 26th conference of the International Group for the Psychology of Mathematics Education* (Vol. 1, pp. 73–80). Norwich.

Blömeke, S., & Delaney, S. (2014). Assessment of Teachers' professional knowledge across Countries: A Review of the State of Research. In S. Blömeke, F. Hsieh, G. Kaiser, & W. H. Schmidt (Eds.), *International perspectives on teachers' knowledge, beliefs and opportunities to learn* (pp. 541–585). New York: Springer Science + Business Media.

Blömeke, S., Suhl, U., Kaiser, G., & Döhrmann, M. (2012). Family background, entry selectivity and opportunities to learn: what matters in primary teacher education? An International Comparison of Fifteen Countries. *Teaching and Teacher Education, 28,* 44–55.

Brown, C. A., & Borko, H. (1992). Becoming a mathematics teacher. In D. A. Grouws (Ed.), *Handbook of research on mathematics teaching and learning* (pp. 209–239). New York: Macmillian.

Dooren, W. V., Verschaffel, L., & Onghena, P. (2002). The impact of preservice teachers' content knowledge on their evaluation of students' strategies for solving arithmetic and algebra word problems. *Journal for Research in Mathematics Education, 33*(5), 319–351.

Fan, L. (2003). 教师教学知识发展研究 [A study on the development of teacher's teaching knowledge]. Shanghai: China East Normal University Press.

Frykholm, J. A. (1996). Pre-service teachers in mathematics: Struggling with the standards. *Teaching and Teacher Education, 12,* 665–681.

Hill, H. C., Ball, D. L., & Schilling, S. G. (2008). Unpacking pedagogical content knowledge: Conceptualising and measuring teachers' topic-specific knowledge of students. *Journal for Research in Mathematics Education, 39*(4), 372–400.

Leinhardt, G., & Smith, D. A. (1985). Expertise in mathematics instruction: Subject matter knowledge. *Journal of Educational Psychology, 77*(3), 247–271.

Leung, F., & Park, K. (2002). Competent students, competent teachers? *International Journal of Educational Research, 37,* 113–129.

Li, Q. (2009). 教师专业发展的知识基础——教学专长研究 [The knowledge base of teacher's professional development]. Beijing: Beijing Normal University Press.

Ma, L. (1999). *Knowing and teaching elementary mathematics: Teachers' understanding fundamental mathematics in China and the United States*. Mahwah, NJ: Lawrence Erlbaum Associates.

Sanchez, V., & Llinares, S. (2003). Four student teachers' pedagogical reasoning on functions. *Journal of Mathematics Teacher Education, 6,* 5–25.

Schmidt, S. (1996). Problem solving as a privileged context for a fruitful connection between arithmetic and algebra. *Revue de Sciences de l'Education, 22,* 277–294.

Schmidt, S., & Bednarz, N. (1997). Arithmetical and algebraic reasoning in a problem-solving context: difficulties met by future teachers. *Educational Studies in Mathematics, 32,* 127–155.

Schmidt, W. H., Blömeke, S., & Tatto, M. T. (2011). *Teacher education matters: A study of the mathematics teacher preparation from six countries*. New York: Teachers College Press.

Shulman, L. S., & Grossman, P. L. (1988). *Knowledge growth in teaching: A final report to the Spencer Foundation*. Stanford, CA: Stanford University.

Stevenson, H. W., Lee, S. Y., Chen, C. S., Lummis, M., Stigler, J., Fan, L., et al. (1990). Mathematics achievement of children in China and the United States. *Child Development, 61* (4), 1053–1066.

Chapter 20
Chinese Teachers' Beliefs About Mathematics Teaching

Shengying Xie and Jinfa Cai

Abstract Cultural beliefs about teaching do not directly dictate what teachers do, but teachers do draw upon their cultural beliefs as a normative framework of values and goals to guide their teaching. In other words, teachers' beliefs and values concerning effective mathematics teaching influence their instructional practice. Based on the findings from a number of studies, we provide a retrospective review and present a profile of Chinese teachers' beliefs about mathematics teaching. In particular, we discuss these teachers' beliefs about mathematics, the learning of mathematics, and the teaching of it. We then discuss the findings in a cross-national comparative context and point out some future directions for this line of research. This chapter not only helps us understand Chinese teachers' beliefs about mathematics teaching from a cross-national comparative perspective, but it also provides insights about the ways to nurture these beliefs. In the past fifteen years, we have seen a growing interest in affect research in mathematics education. Common to all research into affect is the idea that the categories of affect are based on mental systems, and that these systems have a crucial influence on all processes of mathematics learning and teaching. Since the launch of a new round of basic education mathematics curriculum reforms in China, teachers have been seen widely as the key to its success. It is believed that educational change will happen only if the teachers critically reflect on their own beliefs and accordingly change their teaching behaviors to meet the new reforms. In summarizing successful experiences in education reform, the United Nations Educational Scientific and Cultural Organization (UNESCO) suggested that "without the help and active participation of teachers" or "against their will," education reform will never succeed. However, just like difficult system engineering, successful teacher change relates to various complicated factors. And, of all aspects of teacher change, belief change is the most difficult—yet, it holds the core position.

S. Xie (✉)
School of Mathematical and Computer Sciences, Hunan Normal University, Changsha, China
e-mail: syxie@hunnu.edu.cn

J. Cai
Department of Mathematical Sciences, University of Delaware, Newark, USA

Y. Cao and F.K.S. Leung (eds.), *The 21st Century Mathematics Education in China*, New Frontiers of Educational Research,
https://doi.org/10.1007/978-3-662-55781-5_20

20.1 Introduction: A Historical Review of Research on Teachers' Beliefs About Teaching in China

Beliefs are basic knowledge and general understandings which a person holds. This chapter is focused on the beliefs held by mathematics teachers related to teaching and learning mathematics, and to mathematical thoughts and ideas. These beliefs are interrelated and interact with each other, forming a belief system. Although the belief system may be influenced by views on mathematics philosophy or by other people, it originates more from a person's own practical experiences. Through accumulation and participation in intellectual activities, perspectives penetrate into the holders' minds and beliefs are gradually formed by constant exposure and osmosis over a period of time (Ball, 1990).

In the early twentieth century, social psychologists were the first to study the essence of belief and its impact on people's behavior. Research on Chinese mathematics teachers' beliefs, however, has only existed for less than two decades.

Since the late 1990s, mathematics education researchers in China have begun to pay close attention to teachers' beliefs and attitudes about mathematics, teaching, and learning, as well as the relationship between these constructs. For his master's thesis, Zhou (2001) conducted the first investigation of college mathematics teachers' beliefs in China. In the same year, Xu (2001) used a case study, interviewing a high school mathematics teacher and attending one of his classes to understand how his beliefs about teaching could affect his teaching practice. After a period of time, many studies shifted to focus on more specific mathematical conceptions. Among this research, some were influential, including that of Wong (2002), who interviewed high school teachers in Mainland China about their views of mathematics, and that of Huang (2004), who took a theoretical approach to explore teachers' views of mathematics and mathematics education.

It was only after Zhou (2007) completed his doctoral dissertation on the development of preservice mathematics teachers' beliefs about teaching that the research on mathematics teachers' beliefs begins to flourish. One year later, Ding and Jia (2008) used questionnaires and interviews to investigate and study high school mathematics teachers' beliefs about teaching and their teaching behavior.

Especially during the past three or four years, the research has begun to dig deeper. Researchers have studied the beliefs of mathematics teachers at the different grade levels: elementary, junior high and senior high (Wang, 2009; Li, 2013; Tuo, 2014). Others have investigated beliefs about specific subject content, such as plane geometry and information technology (Zhao, 2014; Li, 2015). Still, others have developed a framework and research tool for studying mathematics teachers' beliefs (Yu, 2013; Xie, 2014). In addition, scholars have started to explore the relationship between mathematics teachers' beliefs and specific teaching behaviors, trying to interpret the nature of the influence by using path analysis of teachers' beliefs, students' beliefs, and learning interest (Liu & Zhang, 2014; Wang, 2014).

Through this historical review of the research on Chinese mathematics teachers' beliefs about teaching, we can see a rough map of the overall research in this area.

However, what are Chinese mathematics teachers' beliefs like as a whole? How can we describe and understand them from a cross-national comparative perspective? And, what are the future directions for research? In response to these problems, this chapter will discuss four aspects of Chinese mathematics teachers' beliefs: (1) beliefs about mathematics, (2) beliefs about mathematics teaching, (3) relationships with other factors, and (4) future research directions and insights about the ways to nurture Chinese teachers' beliefs. We can use the following model (Fig. 20.1) to describe a more comprehensive relationship between mathematics teachers' beliefs and students' mathematical learning.

20.2 Teachers' Beliefs About Mathematics

A teacher's beliefs about mathematics form his or her conception of its nature. These beliefs, concepts, views, and preferences constitute the rudiments of a philosophy of mathematics, although for some teachers, they may not be developed and articulated into a coherent philosophy (Ernest, 1989). Out of a number of possible variations, Ernest (1989) distinguished three conceptions of mathematics: the problem-solving view, the Platonist view, and the instrumentalist view. Lerman (1983, cited in Thompson (1992)) identified two alternative conceptions of the nature of mathematics, which he called absolutist and fallibilist views. According to Thompson (1992, p. 132):

> "From an absolutist perspective, all of mathematics is based on universal, absolute foundations, and, as such, it is 'the paradigm of knowledge, certain, absolute, value-free and abstract, with its connections to the real world perhaps of a platonic nature.' From a fallibilist perspective, mathematics develops through conjectures, proofs, and refutations, and uncertainty is accepted".

Some researchers have divided mathematical beliefs into traditional absolute or non-traditional constructivist views (Zhou, 2001; Barkatsas & Malone, 2005; Shahvarani & Savizi, 2007). This dualistic view not only corresponds to Lerman's absolute and constructivist points of view, but also to Ernest's Platonist and

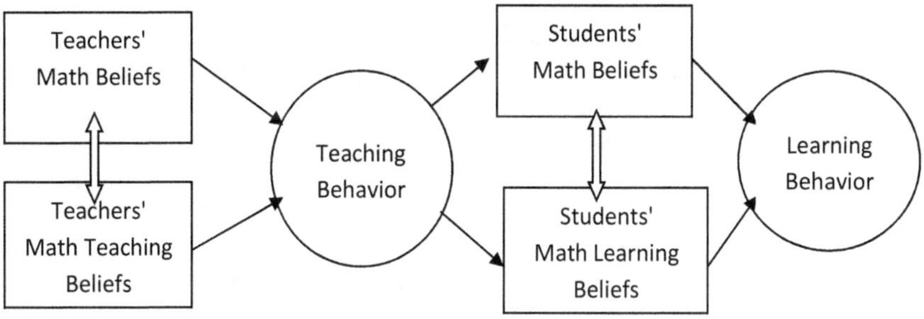

Fig. 20.1 Mathematics teachers' and students' beliefs

problem-solving views (Thompson, 1992). Both the dualistic and Lerman model were thus included in the model developed by Ernest (Jin, Guo, & Dai, 2009). Törner and Grigutsch (1994) described mathematics as a toolbox encompassing beliefs about it as a set of rules, formulae, skills, and procedures. Recently, Wang (2014) found that teachers' mathematical beliefs form a system which is a multi-dimensional 3D model with five dimensions: structure, stability, source, verification, and value. Each dimension has three orientations in theory, namely the traditional orientation, mixed orientation, and modern orientation.

Based on the mathematics belief models discussed above, a number of Chinese researchers have investigated teachers' mathematical beliefs. Using Lerman's absolute-constructivism model, the mathematical beliefs of teachers from 10 local normal universities in 7 provinces were found to be neither completely an absolute point of view nor totally constructivist. However, their beliefs tended to be a bit more absolute than constructivist (Zhou, 2001). This result accords with what Ball (1990) discovered: Teachers' mathematical beliefs may be a combination of the two views. But there are still questions unanswered about Chinese elementary school and secondary school mathematics teachers and their beliefs. Primary school teachers' beliefs about mathematics have been classified into four categories: traditional, deformed, vague and advanced (Chen, 2008). Wong (2002) used semi-structured individual interviews, designed on the basis of Ernest's model, to study Chinese Mainland high school teachers' mathematics belief systems. He found their beliefs were mainly of a Platonist point of view, and that they saw mathematics as a discipline associated with logic, rigor, graphs, and precise numerical computations. Wu (2006) also found, through a questionnaire and interviews, that elementary teachers' beliefs about mathematics were more inclined toward Ernest's Platonism. In theory, Chinese mathematics teachers' beliefs about mathematics were thought to incorporate naturalism, utilitarianism, pragmatism, and scientism (Huang, 2004). From the perspective of a multidimensional 3D model, nearly all five dimensions of Chinese secondary school teachers' mathematical beliefs reflect a combination of orientations. Only in the structural dimension does the orientation tend toward a modern one; the other four dimensions are close to the mixed orientation (Wang, 2014).

There have been only a few cross-national studies of Chinese teachers' beliefs about mathematics. Leung did some comparative research on beliefs from the perspective of different Eastern and Western cultural traditions (Leung, Graf, & Lopez-Real, 2006). When talking about comparative studies of mathematics teachers' beliefs across China and the USA, it is important to mention that Ma (1999) compared Sino-US teachers' views about mathematical knowledge. In cross-regional comparative research, both Hong Kong and Mainland Chinese mathematics teachers have tended to see mathematics as "just a school subject," but to believe that thinking as an important feature. In actual teaching practice, the Hong Kong teachers were often found to place mathematical operations and mathematical thinking in opposition to each other. Teachers from Mainland China have exhibited such opposition as well, but there have been some successful reform experiences integrating regular skills training and logical thinking ability development (Wong, 2002).

Some scholars have made cultural comparisons between ancient Chinese and ancient Greek mathematical beliefs. They found that the ancient Chinese beliefs were based on empiricism and pragmatism. Mathematics was regarded as a tool to solve practical problems, a kind of skill. Moreover, the recognition of mathematical laws and conclusions was in accordance with empirical validation, without any emphasis on logical deductive argument. On the contrary, ancient Greece had absolute and humanistic mathematical beliefs. Mathematics could be divorced from the facts, and it was a science which could be developed independently through axioms and logical deductive development. Moreover, mathematics was an important characteristic of a cultured person (Liu, 2005). Hong Kong scholar Leung (2001) compared and analyzed the content of *Nine Chapters on Mathematical Procedures* and *Euclid's Elements of Geometry*, proposing that algorithms and an emphasis on application are the two major features of traditional Chinese mathematics. Precision in farming practice and culture influenced the field of Chinese mathematical beliefs. These beliefs attach great importance to the "double-base" education in China—basic knowledge and basic skills. The kingship culture of ancient China advocated using mathematics in life, especially in agriculture and trade. That was quite different from the Western culture of emphasizing the "reason" of rational thinking. The patriarchal system of the Confucian culture resulted in contempt for mathematics education; mathematics could exist as folk knowledge, but did not have a place in the mainstream culture. The unique examination culture, which aimed to pursue high achievement, helped to clear the direction of mathematics education, and the long history of advocating rigorous argumentation led to greater importance being attached to skills training, to the detriment of creative thinking (Zhang, 2007).

According to a Chinese saying, "There can be no identification without contrast." Mathematics can no longer be seen solely as a subject required for learning other disciplines, but as a culture of its own, a kind of basic quality that a contemporary social citizen needs. In this context, the cross-cultural comparison is particularly important, especially within the context of the curriculum reform currently ongoing across the world. In terms of teachers' mathematical beliefs, there is a wide variety of ways to conduct comparative studies. Instructional design structure and content, teaching practice and its assessment, can all be associated with teachers' mathematical beliefs. Even in different regions, domestic comparisons between teachers from different types of schools have been found to have particular significance. For example, Ma, Lam, and Wong (2006) did a comparative study about mathematics teachers in urban and rural areas in the northeast part of China. In the past, there have been some disparities in Eastern and Western mathematics education theory.

In short, Chinese teachers' mathematical beliefs have three main characteristics. First, as a system, they have complexity, diversity, and compatibility. Our research mainly focused on models of mathematics beliefs. These models have evolved from two or three dimensions to pluralism, and from pluralism to a multiple-layered orientation. Second, the "correctness" of mathematical beliefs can be interpreted as a relative concept. Researchers have struggled to find the "right" and "appropriate" mathematical beliefs for Chinese teachers, attempting to change their original "unreasonable" mathematical beliefs. In fact, we should understand "correctness" relatively (Zhang, Wong, & Lim, 2009). Under a certain mathematics curriculum reform background and the corresponding curriculum design framework, the mathematical beliefs which accord with the connotations and ideas might easily be considered "correct" and "reasonable." Otherwise, they would probably be labeled "unreasonable." Without a specific educational situation to ground a discussion about beliefs and their so-called correctness, the discussion will be biased, at least, and might be carried too far. Third, Chinese mathematics teachers' mathematical beliefs have been shaped and polished mainly by Confucian culture. Ancient Chinese mathematical beliefs placed emphasis on mathematics knowledge and indicated the pragmatic idea of "To be lasting and practical" (经世致用). Mathematics was, and still is, regarded as a tool and skill to solve practical problems. The recognition of mathematical laws and conclusions is in accordance with empirical validation but the logic deductive argument is not recognized.

20.3 Teachers' Beliefs About the Teaching of Mathematics

Teachers' mathematics teaching beliefs are also called mathematics teaching views. What a teacher considers to be the desirable goals of the mathematics program, his or her own role in teaching, the students' role, appropriate classroom activities, desirable instructional approaches and emphases, legitimate mathematical procedures, and acceptable outcomes of instruction are all part of the teacher's conception of mathematics teaching (Thompson, 1992). The research in this field has produced multiple models or perspectives to describe teachers' teaching beliefs.

20.3.1 Oriented Theory

Kuhs and Ball identified at least four dominant and distinctive views of how mathematics should be taught:learner-focused, content-focused, content-focused with an emphasis on performance, and classroom-focused (Thompson, 1992).These four models are useful in describing major differences among current views of mathematics teaching. As with conceptions of mathematics, a given teacher's conception of mathematics teaching is more likely to include various aspects of several models than it is to fit perfectly into the description of a single model.

For example, a behaviorist and constructivist binary mathematics teaching belief was proposed by Yu (2013). According to a behaviorist teaching belief, teachers are the transferors of knowledge and students are the recipients. Teachers have the authority. They manipulate the whole teaching process. Generally speaking, the stimuli which are supplied for students do not allow them to have free imaginations. The teaching goal is refinement and emphasizing operational practice, and teaching evaluation is based on observations of students' behavior changes (Haser, 2006). A constructivist teaching belief regards teaching as a bilateral activity between teachers and students; students should play a dynamic role and active function in the teaching. As well, teachers need not only to design problems according to the curriculum content, but also to prepare meaningful and valuable questions. During this teaching process, students are expected to take the lead in solving the problems. The teacher's role is to provide support and guidance, organizing the students to discuss and cooperate.

Some researchers have put forward the notion of "teacher-centered" and "student-centered" teaching beliefs (Yu, 2004; Tsailexthim, 2007; Correa, Perry, Sims, Miller, & Fang, 2008). In essence, the "teacher-centered" view corresponds to what Kuhs and Ball described as "content-focused," "with an emphasis on performance" and "classroom-focused." The idea of "student-centered" teaching is in line with the category of "learner-focused" teaching beliefs.

Based on the models above, the discrepancies in mathematics teachers' beliefs across different cultures and traditions have been investigated closely. For instance, Correa et al. (2008) made a comparison between the teaching beliefs of teachers in Beijing and Illinois. It was found that those from Beijing paid more attention to arouse their students' interest in learning mathematics, and attempted to connect the mathematical content with actual life. In contrast, their American counterparts valued their students' learning styles more and adopted a more "hands-on" teaching method. This prompted the question of whether their teaching beliefs would be more consistent if the mathematics teachers came from similar cultural traditions. Wong, Lam, Huang, Ma, & Han, (2002) investigated mathematics teachers from Taiwan, Hong Kong, and the Chinese Mainland. He found that Taiwanese teachers had the least tendency to hold a transfer-of-knowledge belief and were the most student-focused. Hong Kong high school mathematics teachers were most likely to see teaching as delivering knowledge. The least student-focused mathematics teachers were from Changchun. At the same time, Wong pointed out that the more senior the teachers were, the less student-focused were the beliefs they espoused.

A growing number of high school mathematics teachers have begun to focus on teaching how to apply mathematics in real life, and not just on the students' mathematics test scores. Nevertheless, in 2007, a survey showed that there were still a few teachers who did not exhibit this change; they always gave priority to the problem sets and mathematics exams (Peng, 2007). By means of a questionnaire and scale, Zhou (2007) studied the teaching beliefs of students in a normal (i.e., teacher training) university. He discovered that the student teachers' teaching behaviors were not consistent with the new curriculum requirements for independent inquiry or the construction of mathematics teaching beliefs. During their college studies, the student teachers mainly formed teacher-focused and traditional teaching beliefs.

20.3.2 Effective Teaching

From the perspective of effective teaching, there are several representative teaching beliefs. Effective teaching has three meanings: effect, efficiency and benefit; economically speaking, these are input and output. For example, effective teaching means "the teacher uses as little time, energy and material inputs as possible to obtain as good as possible a teaching effect" (Zhang & Yu, 2010). Considering students' progress and development, effective teaching refers mainly to their specific progress and development after a certain time period of teaching. Effective teaching is defined as when the "teacher, through teaching, causes the student to obtain specific progress and development" (Chen & Jiang, 2009). Xu (2010) argued that effective teaching "should promote the accumulation and development of the students' mathematics accomplishment." Effective teaching has three levels: the surface layer is a kind of teaching form; the middle is a kind of teaching thought; and the upper is a kind of teaching ideal realm (Long & Chen, 2005). Effective teaching is happy and efficient instruction (Cai & Che, 2013). Some scholars have also reflected further on effective teaching. For example, Wu (2011) put forward that teaching cannot "get rich" quickly, and educators should not just consider their immediate interests. It is necessary to have a long-term and sustainable development horizon, pluralized values, and flexible views of effective teaching.

Some researchers (Chen & Jiang, 2009; Xu, 2010; Zhang and Yu, ibid) supported their argument by invoking more cases in mathematics teaching. In order to explore in-service secondary school mathematics teachers' beliefs about effective teaching more directly and with larger sample sizes, Wu (2014) designed a questionnaire based on an effective teaching research framework and surveyed 316 high school mathematics teachers' understanding of effective teaching in Jilin province. The results showed that there were significant demographic differences related to understand effective teaching. The largest differences were discovered between male and female teachers. There were also significant differences in nine other variables—feedback, evaluation process, learning style, interaction between teachers and students, love of the content, teaching enthusiasm, dedication, teaching reflection, and ethics. However, there were minimal differences in teachers' understanding of effective teaching with respect to different grades and different titles. The results indicated that gender may be one of the most important variables of effective mathematics teaching in future studies.

Cai and Wang (2010) adopted an approach of semi-structured interviews to survey mathematics teachers' beliefs about effective teaching in China and the USA. They found that the sample of US teachers put more emphasis on student understanding with concrete examples, and the sample of Chinese teachers put more emphasis on abstract reasoning after using concrete examples. The US teachers highlighted a teacher's abilities to facilitate student participation, manage the classroom, and have a sense of humor, while the Chinese teachers emphasized a teacher's solid mathematical knowledge and careful study of textbooks. Both groups of teachers agreed that memorization and understanding cannot be separated. However, for the US teachers, memorization came after understanding, while these were reversed by the Chinese teachers. The findings of this study showed the value of examining teachers' beliefs about effective teaching from a cultural context. Historically, the conceptions of learning and teaching in China and the USA can be traced back to their original philosophical frameworks raised by Confucius and Socrates, respectively. For Confucius, knowledge and truth should be acquired by learning from authority figures/masters (e.g., a teacher) rather than being generated by the learners themselves. In contrast, Socrates believed that knowledge and truth should largely be self-generated individually by learners through questioning other people, including authority figures. In teaching and learning, the Confucian tradition emphasizes the teacher's authority and students' hard work, whereas the Socratic tradition emphasizes the students generating knowledge by questioning themselves and others.

20.3.3 Instructional Coherence

Coherence has long been a construct in research related to discourse, and recently, it has become a construct in educational research (Cai, Ding, & Wang, 2014). Schmidt, Wang, and McKnight (2005) defined coherence as a sequence of topics and performances consistent with the logical and, if appropriate, hierarchical nature of the disciplinary content from which the subject matter is derived. There are both micro and macro levels of coherence. At the micro level, "connections between propositions in composite sentences and successive sentences" are present (Van Dijk, 1997, p. 4). At the macro level, coherence refers to the global meaning or theme of the discourse.

Many researchers have adopted frameworks and research methods from other disciplines to study instructional coherence in mathematics, with a focus on lesson structures and events within and across lessons. For example, international studies have revealed that instructional coherence is a distinguishing characteristic of mathematics teaching in China when compared with the USA (Leung, 1995; Cai, 2005). These studies have revealed some important differences in classroom enactment of instructional coherence, which reflect distinctive cultural beliefs about teaching (Cai, Kaiser, Perry, & Wong, 2009; Cai & Wang, 2010). Behind these observed instructional features, what are the underlying views that guide teachers' designing of coherent lessons? What is the essence of instructional coherence in teachers' eyes? How do teachers actually view the possible ways to achieve instructional coherence?

In order to understand teachers' views about the meaning of instructional coherence and the ways to achieve it, Cai, Ding and Wang (2014) surveyed 20 exemplary Chinese teachers and 16 exemplary US teachers. With respect to the meaning of instructional coherence, while the majority of US teachers paid attention to connections between teaching activities, lessons, or topics, the majority of Chinese teachers emphasized the interconnected nature of mathematical knowledge beyond the teaching flow. The US teachers expressed their views about ways to achieve instructional coherence through managing a complete lesson structure. In contrast, the Chinese teachers emphasized the pre-design of teaching sequences, transitional language, and questioning based on the study of textbooks and students beforehand. Moreover, they emphasized addressing student thinking and dealing with emerging events in order to achieve "real" coherence. The findings of this study contribute to our understanding of the meaning of instructional coherence and ways to achieve it in different cultural contexts.

20.3.4 Representations

One of the main goals for research on beliefs about teaching is to improve the learning opportunities for all students. A number of studies have revealed remarkable differences between US and Chinese students' mathematical thinking and reasoning in problem solving and problem posing. Although the findings about the differences between US and Chinese students across studies seem to be relatively consistent, we are just beginning to uncover why there are differences between US and Chinese students' mathematical problem solving.

Through analyzing US and Chinese teachers' scoring of 28 student responses, Cai (2004) examined their views of solution representations and strategies. A group of 59 Chinese elementary mathematics teachers from Guizhou and a group of 52 US middle school mathematics teachers from Delaware, North Carolina, Pennsylvania, and Wisconsin participated in the study. The findings clearly showed that the US and Chinese teachers viewed students' responses involving concrete strategies and visual representations differently. In particular, the US teachers valued responses involving concrete strategies and visual representations much more than the Chinese teachers did. Moreover, although both the US and Chinese teachers valued responses involving more generalized strategies and symbolic representations, the Chinese teachers expected 6th graders to use the generalized strategies to solve problems, whereas the US teachers did not. The research reported in this paper contributed to our understanding of the differences between US and Chinese students' mathematical thinking. This research also established the feasibility of using teachers' scoring of student responses as an alternative and effective way of examining their beliefs.

20.4 Influence of Mathematics Teachers' Beliefs on Other Factors

The purpose of studying mathematics teachers' beliefs is to analyze their influence on teaching behavior and student development, namely the relationships between mathematics teachers' beliefs and other factors.

The first question in such research is, therefore, "Will mathematics teachers' beliefs influence the students' development?" There have not been many studies of this issue in China. Through scale measurements and a questionnaire survey, the Chinese researcher Zhang (2006) explored teachers' mathematical beliefs and the relationship between these beliefs and students' mathematics anxiety. The study uncovered that there were significant correlations between teachers' mathematical beliefs and student-learning behavior, and also between teachers' mathematical beliefs and students' mathematics anxiety. If the teachers held a fallibilist perspective, then they would treat students' faults with more tolerance. If the mathematics teachers held absolutist mathematical beliefs, they would have tough attitudes toward the problems which occurred while students were learning mathematics. This tough attitude might lead to students doubting their mathematical ability and even become anxiety-producing.

So, is the influence of mathematics teachers' beliefs on students direct or indirect? It would seem that the influence is indirect, operating via the teachers' teaching concepts, teaching designs, teaching behaviors, teaching organization, and teaching evaluations (Yu, 2007).Empirical research appears to have bolstered this view. Li, Ni, & Xiao, (2007) found that teachers' mathematical beliefs did not have a direct influence on elementary students' mathematics studies; only their teaching beliefs and actions had a significant impact on the formation of students' mathematics learning views. The reasoning behind this is that teachers' understanding of mathematical knowledge itself is more implicit, and it does not transform directly into external learning results. Only when the teachers expose their own subject knowledge as effective teacher–student interactions, does the teachers' knowledge have direct influence on students' learning, especially when they present their subject knowledge based on the students' way of thinking or understanding.

No doubt, teachers' teaching behaviors or teaching methods are always influenced by their beliefs. For example, Wang, Deng, and Wei (2008) described four ways of teaching under the influence of different mathematical beliefs: "stuffing" exercises, mechanically fixed pattern strategies, constructivism, and mathematical thinking. Teaching beliefs also have effective predictive power for teaching behaviors, with the most predictive power for mathematics teaching and evaluation of the teaching. However, the influence of beliefs on teaching behavior can be restricted by external environmental factors, and beliefs are not fully reflected in actual teaching practice. In addition to beliefs about the nature of mathematics, factors of mathematics teaching beliefs have direct or indirect effects on the teaching behavior. Furthermore, teachers' mathematical beliefs, especially about understanding the nature of mathematics, do not just impact directly on teaching behaviors. The beliefs about mathematics teaching play an intermediate role between beliefs about mathematics and teaching practice (Ding, 2008).

20.5 Discussion and Future Directions

It is clear that there has been an active body of research on mathematics teachers' beliefs. These have laid a solid foundation for theory and practice for subsequent research. Nevertheless, there is much room for further work, particularly with respect to the content, methods, and integral connection of the research.

First, with respect to the content of the research, more attention needs to be paid to social and cultural factors. Belief is the by-product of an individual engaging in social activities, and it is inseparable from the social culture in which the individual is immersed. In the process of teaching, the classroom culture is especially important. It has a great influence on teachers and students. Only by returning to the specific educational situation can we reveal the truth and get a better understanding of teachers' beliefs. When judging teachers' beliefs apart from the social culture and class culture, the results may be biased. To understand the relationship between teachers' mathematical beliefs, learning beliefs and teaching beliefs, the analysis depends on returning to the mathematics classroom context. Moreover, if we wish to change the existing system of beliefs we need to start from a cultural perspective to analyze the core beliefs. Unfortunately, most recent studies in China have been limited to theoretical discussions, and the field lacks empirical research. Many questions remain unaddressed. For example, in the process of new curriculum implementation, is there a difference in the selection and utilization of new teaching materials when teachers hold different beliefs? Would these teachers' beliefs affect their feelings about the new mathematics textbooks?

Second, with respect to research methods, we should adopt the combination of quantitative and qualitative methods. Most existing studies have used quantitative research methods with questionnaires. Admittedly, using a scale to assess the stability of the belief system is highly effective. However, to reveal further the slowly changing process of mathematics teachers' belief systems and to evaluate their impact on the teachers' practice, quantitative research methods such as belief scales have some limitations. We could consider adopting more ethnographic research methods, or other qualitative research methods such as video, tracking observation, and interview methods. All of these could help us to uncover teachers' hidden beliefs, estimate the degree of internalization and effective belief usage, measure the internal consistency between factors and teachers' belief systems, and explore the interaction between belief and practice.

Third, taking a look at the overall picture, the research on Chinese mathematics teachers' beliefs is more inclined to be macroscopic. They mostly propose their own ideas and insights. The interconnection of empirical research and ideas still needs to be strengthened. In short, there is a need to connect different research results into a more unified whole.

References

Ball, D. L. (1990). Prospective elementary and secondary teachers' understanding of division. *Journal for Research in Mathematics Education, 21*(2), 132–144.

Barkatsas, A., & Malone, J. (2005). A typology of mathematics teachers' beliefs about teaching and learning mathematics and instructional practices. *Mathematics Education Research Journal, 17*(2), 69–90.

Cai, B., & Che, W. (2013). 课堂有效教学:内涵、特征及构成要素 [Effective classroom teaching: the connotation, features and constituent elements] [in Chinese]. *Educational Science Research, 1,* 12–17.

Cai, J. (2004). Why do U.S. and Chinese students think differently in mathematical problem solving? Impact of early algebra learning and teachers' beliefs. *Journal of Mathematical Behavior, 23,* 135–167.

Cai, J. (2005). U.S. and Chinese teachers' knowing, evaluating, and constructing representations in mathematics instruction. *Mathematical Thinking and Learning, 7,* 135–169.

Cai, J., Ding, M., & Wang, T. (2014). How do exemplary Chinese and U.S. mathematics teachers view instructional coherence? *Educational Studies in Mathematics, 85*(2), 265–280.

Cai, J., Kaiser, G., Perry, B., & Wong, N. Y. (Eds.). (2009). *Effective mathematics teaching from teachers' perspectives: National and cross-national studies*. Rotterdam: Sense Publishing.

Cai, J., & Wang, T. (2010). Conceptions of effective mathematics teaching within a cultural context: perspectives of teachers from China and the United States. *Journal of Mathematics Teacher Education, 13*(3), 265–287.

Chen, L., & Jiang, S. (2009). 数学有效教学中的四个"发生性"特征 [Four characteristics of effective teaching in mathematics] [in Chinese]. *Theory and Practice of Education, 4,* 54–55.

Chen, S. W. (2008). 关于小学数学教师数学观的调查分析 [An investigation on elementary mathematics teachers' mathematics beliefs] [in Chinese]. *Shanghai Educational Research, 7,* 37–38.

Correa, C. A., Perry, M., Sims, L. M., Miller, K. F., & Fang, G. (2008). Connected and culturally embedded beliefs: Chinese and US teachers talk about how their students' best learn mathematics. *Teaching and Teacher Education, 24*(1), 140–153.

Ding, F. (2008). 中学数学教学信念与教学行为的相关研究 *[A study of the relationship between teaching belief and teaching behaviors of mathematics teacher in middle school] [in Chinese] (Unpublished master dissertation).* Nanning, China: Guangxi Normal University.

Ding, F., & Jia, Y. (2008). 中学数学教师教学信念的现状调查 [The status investigation on high school mathematics teachers' teaching beliefs] [in Chinese]. *Journal of Hunan Agricultural University (Social Science), 3,* 34–35.

Ernest, P. (1989). The impact of beliefs on the teaching of mathematics. In P. Ernest (Ed.), *Mathematics teaching: The state of the art* (pp. 249–254). London: Falmer Press.

Haser, C. (2006). *Investigation of pre-service and in-service teachers mathematics related beliefs in Turkey and the perceived effect of middle school mathematics education program and the school contexts on these beliefs (Unpublished doctoral dissertation)*. Michigan, U.S.: Michigan State University.

Huang, Q. (2004). 数学教师的数学观和数学教育观 [The mathematics belies and mathematics educational beliefs of mathematics teachers] [in Chinese]. *Journal of Mathematics Education, 4,* 24–27.

Jin, M., Guo, Y., & Dai, F. (2009). 数学教师信念研究综述 [Mathematics teachers' beliefs: A synthesis of the research] [in Chinese]. *Journal of Mathematics Education, 1,* 25–30.

Lerman, S. (1983). Problem-solving or knowledge-centred: The influence of philosophy on mathematics teaching. *Journal of Mathematical Education in Science and Technology, 1,* 59–66.

Leung, F. K. S. (1995). The mathematics classroom in Beijing, Hong Kong and London. *Educational Studies in Mathematics, 29,* 297–325.

Leung, F. K. S. (2001). 中国传统的数学观和教育观对新世纪数学教育的启示 [The traditional Chinese views of mathematics and education: implications for mathematics education in the new millennium] [in Chinese]. *Journal of Mathematics Education, 10*(3), 5–7.

Leung, F. K. S., Graf, G. D., & Lopez-Real, F. J. (Eds.). (2006). *Mathematics education in different cultural traditions: The 13th ICMI study*. New York: Springer.

Li, H. (2015). 基于层次分析法的数学教师信息技术信念评估指标权重的确定 [Determination on the index weight of mathematics teachers' information technology beliefs assessments based on AHP]. *Educational Science Tribune, 2,* 53–55.

Li, M. (2013). 高中数学教师的教师信念研究 *[Study on senior high school mathematics teachers' beliefs] [in Chinese] (Unpublished master dissertation).* Changchun, China: Northeast Normal University.

Li, Q., Ni, Y., & Xiao, N. (2007). 教师变量对小学生数学学习观影响的多层线性分析 [Multilevel analysis of effects of teachers' variables on primary school students' conceptions of mathematical learning] [in Chinese]. *Psychological Development and Education, 2,* 93–99.

Liu, C. (2005). 古代中国与古希腊的数学观之比较研究 [Comparative study of ancient Chinese and Greek mathematical view] [in Chinese]. *Journal of Harbin University, 26*(8), 20–23.

Liu, Z., & Zhang, S. (2014). 数学教师信念与学生数学信念系统对学生数学学习兴趣影响的路径研究 [Research on the effects of mathematics teachers' beliefs and students' mathematics belief system on students' mathematics learning interest] [in Chinese]. *Journal of Nanchang Normal College, 6,* 5–9.

Long, B., & Chen, X. (2005). 有效教学的概念重构和理论思考 [Conceptual reconstruction and theoretical thinking of effective teaching] [in Chinese]. *Journal of Educational Science of Hunan Normal University, 4*(4), 39–43.

Ma, L. P. (1999). *Knowing and teaching elementary mathematics: Teachers' understanding of fundamental mathematics in China and the United States*. Hillsdale, NJ: Erlbaum.

Ma, Y. P., Lam, C. C., & Wong, N. Y. (2006). Chinese primary school mathematics teachers working in a centralized curriculum system: A case study of two primary schools in North-East China. *Compare: A Journal of Comparative and International Education, 36*(2), 197–212.

Peng, Y. (2007). 中学数学教师数学教学观的现状调查研究 *[The research and analysis about the actuality of instructive concept of mathematics in middle school] [in Chinese] (Unpublished master dissertation).* Changchun, China: Northeast Normal University.

Schmidt, W. H., Wang, H. C., & McKnight, C. C. (2005). Curriculum coherence: An examination of US mathematics and science content standards from an international perspective. *Journal of Curriculum Studies, 37,* 525–559.

Shahvarani, A., & Savizi, B. (2007). Analyzing some Iranian high school teachers' beliefs on mathematics, mathematics learning and mathematics teaching. *Journal of Environmental and Science Education, 2*(2), 54–59.

Thompson, A. G. (1992). Teachers' Beliefs and conceptions: A Synthesis of the Research. In D. A. Grouws (Ed.), *Handbook of research on mathematics teaching and learning: A project of the national council of teachers of mathematics* (pp. 127–146). New York: Macmillan Press.

Törner, G., & Grigutsch, S. (1994). Mathematische Weltbilder bei Studeienanfängern - eine Erhebung. *Journal für Mathematik-Didaktik, 15*(3–4), 211–251.

Tsailexthim, D. A. (2007). *Cross-national comparative study of teaching philosophies and classroom practices: The first-grade classrooms in Thailand and The United States (Unpublished doctoral dissertation).* California, U.S.: California State University.

Tuo, Z. (2014). 小学数学教师信念结构及特征的个案研究 *[A case study of the structure and characteristics of primary mathematics teachers' beliefs] [in Chinese] (Unpublished master dissertation).* Changchun, China: Northeast Normal University.

Van Dijk, T. A. (1997). The study of discourse. In T. A. van Dijk (Ed.), *Discourse as structure and process* (Vol. 1, pp. 1–34). Thousand Oaks, CA: SAGE Publications.

Wang, X. (2009). 初中数学教师的教师信念的研究 *[Research on mathematics teachers' beliefs in junior high school] [in Chinese] (Unpublished master dissertation).* Changchun, China: Northeast Normal University.

Wang, X. F. (2014). 中学数学教师数学认识信念对教学行为的影响研究 *[A study of the impact of teaching beliefs on teaching behaviors of mathematics teachers in middle school] [in Chinese] (Unpublished doctoral dissertation).* Nanjing, China: Nanjing Normal University.

Wang, X. P., Deng, P., & Wei, Y. (2008). 几种数学观下的数学教学 [Mathematics teaching in some different views of mathematics] [in Chinese]. *Curriculum, Teaching Material and Method, 28*(2), 53–57.

Wong, N. Y. (2002). 数学观研究综述 [State-of-art in conception of mathematics research] [in Chinese]. *Journal of Mathematics Education, 11*(1), 1–8.

Wong, N. Y., Lam, C. C., Huang, K. M., Ma, Y. P., & Han, J. W. (2002). 中国内地中学教师的数学观 [The mathematics views of Chinese Mainland high school teachers]. *Curriculum, Teaching Material and Method, 1,* 68–73.

Wu, D. W. (2014). 高中数学教师对有效教学相关变量的认识及其意义——以吉林省为例 [High school mathematics teachers' perceptions of effective teaching on relevant variables and their implications] [in Chinese]. *Curriculum, Teaching Material and Method, 9,* 68–73.

Wu, W. L. (2006). 小学教师数学观的调查研究 *[An investigation of elementary school teachers' mathematics beliefs] [in Chinese] (Unpublished master dissertation).* Beijing: Capital Normal University.

Wu, Y. J. (2011). 关于有效教学的再认识 [About the re-recognition of effective teaching] [in Chinese]. *Curriculum, Teaching Material and Method, 7,* 9–14.

Xie, S. Y. (2014). 中学数学教师认识信念系统量表的编制与信效度检验 [Compilation of epistemological beliefs system scale for high school mathematics teacher and testing it for reliability and validity] [in Chinese]. *Journal of Mathematics Education, 4,* 47–54.

Xu, F. J. (2010). 对数学有效教学的几点思考 [Some thoughts about effective mathematics teaching] [in Chinese]. *Journal of Curriculum Teaching and Research, Z2,* 74–75.

Xu, L. (2001). 中学数学教师教学信念的个案研究 [A case study of high school mathematics teachers' teaching beliefs] [in Chinese]. *Maths Teaching In Middle School, 4,* 32–34.

Yu, P. (2004). 数学教育心理学 *[Mathematics education psychology] [in Chinese].* Nanning: Guangxi Educational Press.

Yu, P. (2007). 教师的认识信念系统及其对教学的影响 [Teachers' epistemological belief systems and their impacts on teaching] [in Chinese]. *Teacher Education Research, 19*(4), 18–22.

Yu, P. (2013). 数学教师认识信念的一个理论框架与量表设计 [Theory framework and scale design of mathematics teachers' epistemological beliefs] [in Chinese]. *Journal of Mathematics Education, 4,* 34–38.

Zhang, D. Z. (2007). 中华文化对今日数学教育之影响 [The Chinese culture's influence on mathematics education today] [in Chinese]. *Journal of Basic Education, 16*(1), 45–54.

Zhang, G. F., & Yu, Y. H. (2010). 新课程理念下如何追求有效教学 [How to seek effective teaching under the new curriculum ideas] [in Chinese]. *Mathematics Learning and Research, 2* (68), 70.

Zhang, M. J. (2006). 教师、家长的数学信念和学习观与学生的数学焦虑的相关研究 *[Research on the relationship between secondary school students' mathematics anxiety and teachers' beliefs, parents' beliefs and students' learning beliefs] [in Chinese] (Unpublished master dissertation).* Shanghai, China: East China Normal University.

Zhang, Q. P., Wong, N. Y., & Lim, C. C. (2009). 中国内地数学信念研究的综述 [Comprehensive analysis of beliefs about mathematics in Mainland China] [in Chinese]. *Journal of Mathematics Education, 6,* 16–22.

Zhao, Y. (2014). 不同学程的数学教师对平面几何的教学信念调查 [Investigation of teachers beliefs about teaching plane geometry in different programs] [in Chinese]. *Journal of Guizhou Normal University, 3,* 51–55.

Zhou, S. (2007). 师范生数学教学信念的发展研究 *[Developmental research on student teachers' mathematics teaching beliefs] [in Chinese] (Unpublished doctoral dissertation).* Shanghai, China: East China Normal University.

Zhou, Z. (2001). 关于高师院校数学教师的信念的调查研究 *[An investigation of the beliefs of normal university mathematics teachers] [in Chinese] (Unpublished doctoral dissertation).* Shanghai, China: East China Normal University.

Part VI
The Evaluation of Mathematics Education

Chapter 21
The Assessment of Mathematics Classroom Instruction in Primary and Secondary Schools

Lianhua Ning

Studies of the assessment of mathematics classroom instruction in primary and secondary schools in China can date back to the lesson observation and evaluation of the 1950s. Under the teaching and research systems formed at different levels, provincial, municipal, district (county) and school, lesson observation and evaluations have become the main way of assessing mathematics classroom instruction and are regular, daily activities. Along with the reform of basic education this century, the ideas, content, and formats for assessing mathematics classroom instruction have changed a lot.

First of all, this chapter introduces the research status of mathematics classroom instruction assessment in primary and secondary schools with respect to the content and the functions and standards generally, and then it introduces the elements, index system, assessment forms, and developing trends.

21.1 The Research Profile of Assessment of Mathematics Classroom Instruction

Since the 1990s, the research on basic mathematics education in China has flourished gradually. There is a richness of theoretical and practical studies concerning the assessment of mathematics classroom instruction. Typically, these studies have focused on the definitions, functions, standards, and forms of assessment (Cao, Li, & Qin, 2011).

L. Ning (✉)
Nanjing Normal University, Nanjing, China
e-mail: ninglh@126.com

Y. Cao and F.K.S. Leung (eds.), *The 21st Century Mathematics Education in China*, New Frontiers of Educational Research,
https://doi.org/10.1007/978-3-662-55781-5_21

21.1.1 The Definition of Assessment of Mathematics Classroom Instruction

Many scholars have defined the assessment of mathematics classroom instruction from different perspectives, based on the characteristics of mathematics classroom instruction in primary and secondary schools in China. Some representative views are discussed in the next section.

In a mathematics class, the teaching is assessed through a process of qualitative and quantitative evaluation of the actual levels of teaching and learning. This is seen as a way to improve the quality of classroom teaching (Wang & Li, 1990). The evaluation process is based on value judgments of the effects of teaching and learning by means of evaluation standards and scientific methods. It is vital, therefore, to establish scientific evaluation standards and form a corresponding index system, which not only relates to the implementation of evaluation principles and the function of evaluation, but also affects the improvement of teaching quality directly (Li, 1994).

In the classroom, in order to achieve the teaching purpose, teachers need first to understand the requirements of the syllabus and textbooks, then to utilize effective teaching methods to promote the students' mastery of mathematics knowledge, develop their mathematical abilities and form proper mathematical ideas. Therefore, the classroom teaching reflects the teachers' thoughts and skills, and the students' participation and learning levels. The assessment of mathematics classroom instruction in primary and secondary schools takes these elements and their interactions into account, so as to improve the effectiveness of teaching. Through the assessment, we can have a deep understanding of the teaching rules, promote the communication between teachers and their self-reflections, and motivate teachers to improve their teaching skills, abilities, and enthusiasm. At the same time, the assessment should take the students' learning attitudes and abilities into consideration. In former assessment models, we only paid attention to the teachers, but now we know teaching is a bilateral activity, and only by taking the learning activities as one element of the assessment we can reach the purpose of improving the students' learning and thinking abilities and building up positive views of their learning and of mathematics (Weng, 2002).

The teaching evaluation is generally a qualitative and quantitative process in the sequence of identification, diagnosis, guidance, adjustment, encouragement, and improvement (Lu & Fu, 2004). Based on the observation of the teaching process and results, we make scientific judgments about the teaching effect, the learning quality and personality development of students, identify the problems that exist, and then adjust and optimize the teaching process.

The objectives of teaching evaluation vary as different people may have different focuses. They can be classified into several groups, the evaluation of teachers, students, the teaching process and teaching effect, the teacher's teaching and the students' learning, and the classroom activities as a whole (Cao et al., ibid).

The assessment of mathematics classroom instruction in primary and secondary schools generally involves the following links. The first step, according to the overall objectives proposed by the National Curriculum Standards for Mathematics (or the Syllabus), is to set the teaching and learning purposes relating to specific teaching content as the reference standards for the assessment. In order to fully reflect the key role of students in the evaluation, they can participate in the development of standards. The second step is to create some learning situations to achieve the pre-set teaching goals and check the extent to which the goals are reached in light of the actual situation. The third is to collect all information about each link of the mathematics teaching by quantitative or qualitative methods. This information can include the students' mastery of basic knowledge and basic skills, developments in their ability to cooperate, explore and innovate, and the affective attributes such as attitudes and values. The fourth step is to process and analyze the information scientifically, then give feedback and make the necessary adjustments. Thus teachers and students can carry out self-evaluation, self-regulation, and self-control objectively, which can encourage the students' study as well as improve the teachers' teaching.

21.1.2 The Function of Assessment of Mathematics Classroom Instruction

The evaluation process has become an important and indispensable link of classroom teaching. Different scholars have different views about this. Some think that the assessment of mathematics classroom instruction has the function of identification, guidance, diagnosis, encouragement, feedback, and regulation decision-making (Chen, 2004).

In general, the function of the assessment of mathematics instruction can be summed up as being for the purposes of guidance, encouragement, identification, diagnosis, improvement, examination, selection, feedback, and regulation. Prior to 2000, assessment processes placed more emphasis on the function of examination, identification, and selection, while now more attention is paid to the functions of diagnosis, improvement, and regulation (Tu & Ji, 2007).

21.1.3 The Standards for Assessing Mathematics Classroom Instruction

The process of developing evaluation standards for classroom instruction in China starts with the teaching theory and then the standards of a good class.

Almost all of the core literature about the assessment of mathematics instruction is concerned with the establishment of standards. Since the assessment objects are different, so the emphasis of the evaluation standards is also different. Research on the strategy of establishing standards for the assessment of mathematics classroom instruction can be divided roughly into two categories (Cao et al., ibid). The first is to divide the mathematics classroom instruction into teaching goal, teaching content, teaching method, teaching process, and teaching effect. Another approach is to divide the mathematics classroom instruction into teachers' teaching and students' learning, based on the specific behaviors observed during the teaching process.

21.2 The Elements of Assessing Mathematics Instruction

The assessment of mathematics instruction involves a great deal of content, and the evaluation elements vary according to different focuses. Some research has indicated that the most important elements are the teaching goal, teaching content, teaching process, teaching method, and teaching effects.

21.2.1 Assessment of the Mathematics Teaching Goal

The assessment of the mathematics classroom teaching goal is mainly stated from the following aspects:

1. Whether the teaching purposes are explicit and concrete. What mathematics knowledge and skills should be mastered, what abilities, feelings and attitudes should be developed and which level should be reached all need to be articulated clearly.
2. Whether the setting of the teaching goal is reasonable. First, we should consider whether the teaching goal can be understood and accepted by the students, whether it is good for their mathematics learning, and whether it is within their zone of proximal development. Second, we should consider whether the teaching goal can be implemented smoothly and whether it enables the students' best development.
3. Whether the foothold of the teaching purpose is science. The teaching purpose should not only be knowledge acquisition, but also the development of process knowledge. The mathematical processes that the students need to experience, the activities that they need to take part in, and the realizations that they need to obtain should all be included.

21.2.2 Assessment of the Mathematics Teaching Content

Not only is the mathematics teaching content an important resource for teachers, it is also the main tool for identifying students' learning. The quality and efficiency of the mathematics teaching content are usually discussed from the following viewpoints.

1. Whether the teachers present and explain the mathematics teaching content correctly and the students understand it accurately. The scientificity and veracity of the mathematics teaching content are essential; in other words, teachers must know and understand the teaching content and other content related to it, and their understanding should be deeper than the students'. The knowledge conveyed should be correct and accurate.
2. Whether the background material of the mathematics knowledge has been explored fully. The mathematics learning content should be realistic, meaningful, and challenging. We should confirm whether it is presented in a way that engages the students actively in observation, experimenting, guessing, reasoning, and communication. A proper exploration of the background material not only enhances students' understanding in depth and breadth, but also makes the mathematics classroom teaching become rich and colorful.
3. Whether the design of the teaching content is appropriate. We should confirm that the lesson design stresses the key points and clarifies the difficult ones, that the level of difficulty is in line with the students' current development levels and within their zone of proximal development, that the students are encouraged to 'rediscover' the knowledge, and that the necessary scaffolding is provided. These are important indicators of the extent to which the mathematics teaching content is implemented.

21.2.3 Assessment of the Mathematics Teaching Process

The mathematics teaching process is based on the interactions that occur between teacher and students. The focus should be not only on obtaining mathematics knowledge, but also on developing mathematics abilities and shaping the student's personality. The assessment of the mathematics teaching process focuses mainly on the following aspects.

1. Whether the steps of the teaching process are sequential and linked appropriately. For example, we should confirm whether the teacher and students have a good relationship, whether the teaching purpose, teaching content, and teaching method are addressed fully, and whether the time allocation for each step is

reasonable. It is also important to consider the transitions between each teaching link.

2. Whether the organization of the teaching process is helpful for the students to construct their knowledge structure by themselves. We should confirm whether the teaching process provides the students with the situation, time, and space for self-study and discovery. It is also important to observe the students' participation, whether they are engaged actively in their learning, and the nature of the teacher's guidance.
3. The specific evaluation indicators include the effectiveness of the interactions between teachers and students, and between students and students, the flow of the exchange of information, the timeliness and appropriateness of feedback, the teacher's flexibility in adjusting the lesson flow according to the feedback, and the teacher's overall ability to handle the whole teaching process. Combined, these indicators can reflect a clear understanding of the mathematics teaching efficiency.

21.2.4 Assessment of the Mathematics Teaching Method

Mathematics teaching method is an important component of mathematics classroom teaching. The evaluation of the mathematics classroom teaching methods mainly involves the following aspects.

1. Whether the chosen teaching method is effective. The chosen teaching method needs to be adapted to the specific teaching content and to allow for the smooth achievement of the lesson's goals. The teaching method needs to take account of the students' learning styles and guide independent, cooperative and inquiry-based learning.
2. Whether the teaching method is suitable for students' ages and current development levels. The teaching method needs to address the psychological needs of the students' cognitive development and to challenge this development.
3. Whether the teaching method is instructive. The teaching method needs to be able to arouse the students' desire to learn, inspire them to think actively, establish specific learning aims and strong learning motivation and explore knowledge actively.
4. Whether the chosen teaching method is integrated with modern teaching approaches and is varied appropriately. This is the key to realizing the effectiveness of the mathematics classroom teaching method. We should evaluate the quality of the various teaching methods in the overall context and avoid evaluating a single method.

21.2.5 Assessment of Mathematics Teaching Effect

The teaching effect itself is a basic element of the assessment of the mathematics classroom teaching. The assessment of teaching effect of a mathematics class can be stated from the following aspects.

1. Whether the teaching tasks and requirements have been completed and whether the teaching purposes and goals have been achieved. To measure the degree of the teaching goals, we need to take into account an overall consideration of the following aspects: the mastery of knowledge, the cultivation of ability, the levels of mathematics thinking and the development of individual mental and emotional qualities.
2. Whether the students are obtaining process knowledge in addition to content knowledge. Try to observe the students' performances to identify whether they are involved actively in the process of mathematics study. For example, do they think positively? To what extent can they answer the teachers' questions? Can they solve the problems by themselves? How about the performance of the underachievers? What roles are the teacher and students playing in the process of teaching? Can they complete the basic tasks smoothly?
3. Whether the cognitive and affective loads on the students are appropriate and whether they are learning happily and feel comfortable. Teachers and students should utilize the minimum amount of time and energies to achieve optimal teaching effects. It is also important to consider the progress of the whole class, not only to focus on a few students.

21.3 The Form of Assessment of Mathematics Classroom Instruction

It can be seen from the various descriptions that the criteria and approaches for assessing mathematics classroom instruction in primary and secondary schools in China are diverse. Almost each city, district (county), and school has its own system. However, they have two features in common, teachers' teaching and students' learning, and the key elements are the teaching goal, teaching content, teaching process, teaching methods and teaching effect.

21.3.1 The Form of Assessment of Mathematics Classroom Instruction: in View of the Teacher's Teaching

Evaluating a mathematics class from the perspective of the teacher's teaching is one of the main ways to assess mathematics classroom instruction. This includes the effectiveness of the teaching, whether the teaching goals and requirements are explicit and concrete, whether the teaching content is correct, whether the teaching methods are reasonable, whether the language is vivid, lively, accurate and concise, whether the key points are highlighted, and whether the teaching goals are attained.

The first step in evaluating classroom teaching is to determine the first-level criteria, the main factors that need to be studied. The next step is to analyze the content of each main factor and determine the second-level system of evaluation, the sub-factors. These sub-factors reflect the actual situation in the class. Each of these is given a numerical rating, from which an overall score is calculated.

Below is an example of a widely used form for the evaluation of the teacher's teaching. Evaluators record the scores for all factors through classroom observation or watching videos, and then calculate the integrated score according to the setting weights, thus to evaluate the class Table 21.1.

The process of scoring this evaluation form is as follows. First, find the sum of the multiple of the score of each level and its corresponding 2-level weight, and that is the score for the main factor of the corresponding column. Second, find the sum of the multiple of the score of the main factor and its corresponding 1-level weight, and then we get the total points. Generally, points from 86 to 99 will be graded as level A and regarded as outstanding. Points from 71 to 85 will be graded as level B and regarded as good. Points from 60 to 70 will be graded as level C and regarded as general. Points below 60 will be graded as level D and regarded as a failure.

There are a few aspects that require attention when using this evaluation form.

1. The evaluation should focus on the mathematics classroom teaching process as a whole, with more importance attached to the teaching process than the result.
2. The assessment of mathematics instruction should focus on how the teachers guide the students to participate actively in the mathematics activities; for example, the quality of the initial mathematics situation, the design and effectiveness of the activities, the handling of scaffolding and hints, and the extent to which children are encouraged to be enquiring and reflective. It is especially important to pay attention to the students' experiences in the mathematics learning process and not to focus on the teachers' teaching of mathematics knowledge or the training of basic skills.
3. Emphasis should be placed on nurturing mathematical talent. It is important to look at students' performances on a range of content in order to get an overall picture of their quality.
4. The ultimate purpose of evaluation should be to improve the quality of mathematics teaching in line with the goals of the Mathematics Curriculum Reform. It should also encourage the teachers to think constantly about their teaching practice and try to improve it and inspire their students to participate actively in mathematics activities. By doing this, the effect of mathematics teaching can be improved considerably.

Table 21.1 Sample evaluation form for mathematics classroom instruction (from the point of teacher's teaching) time

Teacher		Subject	School	Grade				Score
Evaluation factors				Evaluation grade				
Main factor	Weight of the first-level index	Sub-factor	Weight of the second-level index	A	B	C	D	
Teaching purpose (A_1)	0.15 (B_1)	1. Reflect the function of education and pay attention to students' overall development	0.20					
		2. Highlight the process-oriented targets and focus on students' learning process	0.30					
		3. Aim at students' zone of proximal development	0.35					
		4. Guidance, stimulation and regulation	0.15					
Teaching content (A_2)	0.15 (B_2)	1. Accuracy of content taught	0.35					
		2. Links to students' daily lives and existing knowledge and experience.	0.35					
		3. Fun element	0.30					
Teaching process (A_3)	0.25 (B_3)	1. Suitable linkage between aspects of the lesson.	0.20					
		2. Time distribution.	0.20					
		3. Students encouraged to construct their own knowledge develop mathematical abilities	0.35					
		4. Flow of information exchange and flexibility in response to feedback	0.25					

(continued)

Table 21.1 (continued)

Teacher		Subject	School	Grade				Score
Evaluation factors				Evaluation grade				
Main factor	Weight of the first-level index	Sub-factor	Weight of the second-level index	A	B	C	D	
Teaching method (A_4)	0.25 (B_4)	1. Appropriate choice of methods according to content and students' age groups	0.15					
		2. Stimulation of students' interest in study	0.25					
		3. Promotion of independent, cooperative and inquiry-based learning	0.25					
		4. Teacher as an inspirer and guide, allowing space for students to explore	0.20					
		5. Varied range of teaching methods as appropriate	0.15					
Teaching effect (A_5)	0.20 (B_5)	1. Fulfilment of lesson goals	0.30					
		2. Active engagement and thinking by students	0.25					
		3. Reasonable cognitive and affective loads on teachers and students	0.25					
		4. Promotion of mutual development of teachers and students	0.20					

21.3.2 The Form of Assessment of Mathematics Classroom Instruction: in View of the Students' Learning

To evaluate a class from the perspective of the students' learning is to make a judgment about their learning processes and results. It involves testing their mastery of the basic knowledge and basic skills, the development of their capabilities and the analysis and evaluation of their learning behaviors, attitudes, emotional factors, and other elements. The main purpose of evaluating the learning situation of all students is to understand their learning processes and promote their further development. As well, the feedback can help them to identify their own weaknesses, then be used to improve the teachers' teaching and students' learning.

The students' learning situation is usually evaluated through classroom observation, mathematics tests, and other qualitative performance assessment such as interviews, mathematics diaries, and portfolios.

21.3.2.1 Classroom Observation

The classroom is the major context for students' learning, so classroom observation is the most important way for teachers to find out about their students. By carrying out purposeful, systematic observation of what students do and their emotional responses, we can get the most direct and real information about them. For example, we should observe whether the students take part in the class activities actively (speak actively, ask and answer questions, etc.). We should also observe whether they can communicate with partners actively to assess their awareness of cooperation and communication. We should observe the students' confidence and interest as indicators of their emotions and attitudes.

There are many entry points for classroom observation, so different researchers have different perspectives. For example, in Yiming Cao's book *The Microanalysis of Mathematics Classrooms in Middle Schools in China from International Perspectives*, the teachers' teaching behavior and students' learning behavior are coded in detail through classroom observation, so as to evaluate both (Cao, 2011). In their book *Classroom Observation—Leading to a Professional Classroom Observation and Comment,* Shen Yi and Yunhao Cui developed a framework for classroom observation consisting of 4 dimensions, 20 Perspectives and 68 observation points; of these, 17 points are related to students' learning (Shen & Cui, 2011).

The following are some reference indicators for observing a mathematics class from the angle of students' learning Table 21.2.

This sheet examines seven aspects of student learning; these basically reflect the students' mastery of knowledge and skills, their involvement in classroom activities, the quality of the interaction between teachers and students, the condition of their mathematical thinking and the state of the students' emotions, attitudes, and

Table 21.2 Sample evaluation sheet for classroom observation (students' learning)

Items	Factors	A	B	C	Remarks
Mastery of knowledge and skills	Understand knowledge				A = fully understanding B = preliminary understanding C = just take part in the activities
	Card relationships				
	Basic exercises				
	Solve problems				
Attitude	Listen				A = serious B = not so serious C = not serious at all
	Practice				
Participation and interaction	Speak with initiative				A = active B = not so active C = not active at all
	Ask questions and show a spirit of inquiry				
	Discuss and communicate				
Self-efficacy	Put forward some special views.				A = often B = not so often C = very few
	Dare to try and to question				
Listen and express	Listen to others				A = can B = not so often C = very few
	Express their views				
Organization of thought	Express themselves in logical order				A = strong B = not so strong C = not at all
	Use a logical, clear process to solve problems				
Creation of thought	Find out another way to solve the question				A = can B = not so strong C = not at all
	Use nonstandard thinking				

values. It must be noted, however, that it requires the observers to have abundant experience and good decision-making abilities.

21.3.2.2 Mathematics Tests

Generally, teachers create oral questions, in-class practice, written assignments or formative tests for the students to answer orally or in writing. From these tests, we can identify the students' levels of knowledge, skill, ability, and other aspects. Mathematics tests are important ways to evaluate the students' learning and also relatively objective and accurate methods for reflecting their mastery of knowledge and the development of their abilities.

Actually, some of the oral questions teachers ask in class are also a kind of oral test. Aiming to have an understanding of the student' mastery of the new learned knowledge, these questions can provide a foundation for the teachers to determine the teaching process and subsequent tasks. This is referred to as the questioning–answering evaluation pattern.

In-class practice and written assignment are also a form of testing, helping the teachers to know the students. It is a characteristics of Chinese classes that teachers assign practice exercises and written homework in class to learn about the students' knowledge and skills. In the past, students had separate exercise books for classwork and homework. Now, almost every school uses the learning material published by the education press or has its own school-based textbooks. The teachers are required to mark the assignments carefully every day and give scores or grades. They are also encouraged to give criticism or praise and suggest further requirements or advice. But the most common requirement is to revise the questions on which the students are incorrect. This kind of regular evaluation helps the teachers to have a grasp of the students' learning and provides essential material for subsequent teaching. It is an effective way to assess mathematics classroom instruction in primary and secondary schools.

Individual, timed mathematics tests are the main form of assessment in Chinese mathematics classes. These are a necessary supplement for in-class practice and written assignments. The main purpose is to find out the students' levels of attainment at the end of a project or unit of work, so it is often called a 'periodical achievement test'. These tests can be diagnostic or formative, so as to evaluate the situation and existing problems regarding the development of students' knowledge and ability and then take timely corrective action. According to the types of questions, these tests can be subjective or objective. For a mathematics teacher, the first challenge is to design test questions of high quality. The questions should be representative and target different levels. The teacher needs to do a scientific analysis and summary of the results and to take timely measures to improve the teaching. That is to say, the reliability, validity, and purposefulness of the testing result are of great importance in evaluating mathematics teaching.

21.3.2.3 Qualitative Performance Assessment

Reflecting the achievement of knowledge and skill through actual tasks, qualitative performance assessment combines the assessments derived from teachers' and students' self-evaluation. This can reflect the students' development and progress and increase their confidence to learn mathematics well.

Qualitative performance assessment can help teachers to collect information about their students in many ways, and ensure that the assessment is comprehensive and scientific and, to some extent, that it makes up for the insufficiencies of the paper-and-pencil test. Some students are not able to perform well in mathematics tests due to anxiety. Some like to think deeply about questions and cannot solve all questions smoothly within the stipulated time. Some are good at operations and experiments. Hence the paper-and-pencil test cannot reflect mathematics ability completely. Recording students' performances through interviews, mathematics diaries and portfolios can help us to have an overall view of their development.

Interviews are generally conducted in the form of questionnaires and open questions designed to gain an in-depth understanding of students' performances, learning situations, learning experiences or achievements. We can identify what is going on in students' inner worlds, which fills the gaps left by some external assessment methods such as classroom observation and mathematics tests.

Mathematics diaries provide a way for students to self-evaluate and reflect on their mathematics knowledge and methods. By guiding the students to write down their real feelings, difficulties and interests freely, first-hand information can be gathered to assess their real learning situation.

The portfolio records important data that reflect the students' progress. It provides a good formative assessment of their growth. A diverse range of materials can be collected, including the most satisfactory homework, the most significant achievements from inquiry activities, the most impressive questions, reflection after solving a question, the experience of reading mathematics books, or short mathematics papers.

21.4 Developing Trends in Assessing Mathematics Instruction

Along with the new curriculum reform, the ideas, content, form and other aspects of the assessment of mathematics instruction in primary and secondary schools are undergoing major changes.

21.4.1 Ideas About Evaluation as a Component of the Mathematics Curriculum Reform

Changes to the assessment of mathematics instruction are significant in the mathematics curriculum reform. Some guiding suggestions and ideas are provided by *The Outline of Curriculum Reform of Fundamental Education (trial)*. It is emphasized in this guideline that a quality education evaluation system should promote students' all-round development, and encourage teachers to upgrade and develop curricula. There are five key recommendations in the guideline. The first is to focus on development and to be aware of the changing nature of assessment. The second is to pay close attention to the overall evaluation and individual differences, using a varied range of evaluation criteria. The third is to emphasize qualitative evaluation and use multiple assessment methods combining both qualitative and quantitative evaluations. The fourth is to use both self-assessment and peer-assessment. The fifth is to combine summative and formative assessment techniques.

The 2011 version of The Mathematics Curriculum Standards for Full-time Compulsory Education advises that the main purpose of assessment is to understand students' learning processes fully, to motivate students' learning and teachers' teaching. Assessment systems need to have multiple assessment aims and methods, with attention paid to both the results and the processes. The focus should be not only on the level of mathematics learning but also on affective variables (Ministry of Education, 2011).

The Mathematics Curriculum Standard for Senior High School (Trial) (The Research Group for Mathematical Curriculum Standards, 2004) pointed out that the evaluation system had changed a lot due to the requirements imposed by modern society. The requirements are similar to those described in the previous paragraph: a reasonable and scientific evaluation system with a range of ideas, content, formats, and policies; a focus on both the result and the process of the learning; and a focus on affective outcomes as well as on mathematics learning.

It has been explained here that the new mathematics curriculum standard puts expansion first. It points out that the goal of assessing mathematics teaching and learning is to promote the students' all-roundedness and active development, not just to discriminate their learning and intellectual levels. Second, the evaluation embodies diversity and a focus on the process. The evaluation should permeate the whole teaching and learning process and take all situations, not just the learning result, into consideration.

As a consequence of changes to the assessment of the new mathematics curriculum, researchers have put forward some new ideas about classroom teaching evaluation. For example, Zhou, Lei, and Yan (2003) suggested that evaluation should pay more attention to students' attitudes, temperaments, mathematics emotions, problem solving, and other non-intelligence aspects. It is important to emphasize the process of the discovery and exploration of mathematics concepts and rules. Activities should be related to the students' real lives. Written reports, investigation reports, and other nonstandard questions should be used to evaluate the students' thought processes and cognitive characteristics. Evaluation methods should be a combination of formal and informal, closed and open tests, written and oral exams, etc.

The scope of the evaluation should be expanded beyond the classroom. Combined assessment by teachers, other students, students themselves and parents can help students to develop the idea of assessing themselves and others, to help teachers and parents have a full grasp of the students' learning situations, to mobilize all kinds of resources and enhance students' enthusiasm to study mathematics and develop their self-study abilities.

In mathematics teaching, teachers should be clearly aware of the importance of evaluation. They should help students to know more about themselves and build up self-confidence, display initiative and raise their creativity and divergent thinking, with the overall aim of improving mathematics ability in students of different ability levels (Kuang, 2006).

In general, there are three key aspects to the new approach to evaluating mathematics instruction. The first one is encouraging students to build self-confidence by stimulating evaluation. The second one is to help their self-realization through the educational assessment. The third is to promote the development of the students' individuality through developmental assessment (Zhou, 2006).

21.4.2 New Trend in Assessing Students' Learning Processes

The process is a better reflection of the developments and changes that occur in students than the results are. Therefore, the assessment of mathematics instruction should focus on both the outcome and process. The evaluation process should include indicators of the students' interests and attitudes, their self-confidence to learn mathematics, independent thinking, awareness of cooperation and communication, and the level of mathematics cognition.

21.4.2.1 Improving Students' Understanding of the Value of Mathematics

Knowledge of the value of mathematics is important to cultivate students' mathematical thinking and promote their general development. As society develops, mathematics not only plays an important role in natural science but also in technology, social science, and people's daily lives. Abstract content and precise reasoning can help to develop their scientific spirit and rational thinking. In addition, appreciating the beauty of mathematics is an important part of understanding the value of mathematics.

21.4.2.2 Formation of Students' Thinking Processes and Habits

In the past there has been a tendency to pay too much attention to mathematics results—whether students remember certain formulae, or whether a mathematics problem has been solved—but not enough to the process of their thinking and understanding. This may lead to a decline in their enthusiasm to explore and think independently and even the development of their thinking skills.

Independent thinking is a basic characteristic of mathematics learning. We should focus on whether students are willing to think, are good at thinking and can insist on thinking in the process of assessment, then guide them to improve their thinking methods and processes constantly. Teachers should help students to develop critical-thinking habits that pursue truth instead of books, encourage them to think carefully instead of blindly following, and teach them the way to judge things and solve problems.

21.4.2.3 Students' Participation, Communication, and Cooperation with Others

By doing mathematics and taking part in mathematics activities, students enrich their experiences and build mathematics knowledge in their own ways. This kind of knowledge can be really understood and owned. The assessment of mathematics classroom instruction should pay attention to whether the students are willing to and really do participate in the mathematics activities, particularly those that engage them in thinking.

In addition, the assessment of mathematics classroom instruction should focus on whether the students are willing to communicate and cooperate with others in the activities. In the process of communication and cooperation, students will not only learn to listen to and understand others, but also learn to express themselves correctly and to recognize that they can achieve more effective results by combining their strengths rather than working alone.

21.4.2.4 Reflection

Reflection is self-monitoring, self-regulation, and self-evaluation of one's own learning activities and cognition. This process can mainly be understood from three aspects. One is the orientation and planning of students' learning activities in mathematics. The second point is conscious examination and feedback in mathematics activities. The third point is that students adjust, correct and manage their mathematics learning activities consciously.

The objects and content of reflection are varied. For example, the review of mathematics concepts, the application of mathematics formulae, reviews of problem-solving processes and the rethinking of mathematics methods are all important. Reflection can improve the efficiency of learning and enhance the students' understanding and mastery of mathematics knowledge, which helps to develop higher-level skills such as critical and creative thinking.

Teachers should train their students to reflect consciously on their mathematics learning. Specifically, teachers can teach students to check and adjust their mathematics thinking processes in such ways as writing about their attainments, writing mathematics diaries, engaging them in research on some typical problems and the exchange of experiences. Questions can be asked such as: What are you doing? (Can it be presented clearly?) Why are you doing so? (Can it reach the goal?) What are the benefits? (If we get the outcome, then what will be done next?) By asking these questions, we ask ourselves how we find and solve problems, which basic methods and skills are used, what setbacks we meet and what mistakes have been made, where difficulties are likely to appear, why we have made similar mistakes, and what lessons we can learn. This self-review turns our experiences and feelings to process knowledge.

21.4.3 *Evaluation of Students' Mathematics Abilities*

Mathematics ability is an important part of mathematics quality as well as the key point for students' independent learning and sustainable development. Since ability level is a part of the mastery and application of mathematics knowledge, the evaluation of students' mathematical ability runs through the whole process of knowledge construction and problem solving.

21.4.3.1 Evaluation of the Ability to Find and Pose Problems

The ability to recognize problems in related materials and pose problems through abstraction and generalization has become an important standard to measure students' mathematics quality. Compared with solving a problem, finding and posing a problem is much more important, because it can reflect the extent of students' exploration, practice, and creative thinking. The mathematics exploration and mathematics modeling activities emphasized in *The Mathematics Curriculum Standard for Senior High School (Trial)* reflect this idea to some extent.

To put this idea into practice effectively in the classroom, teachers should encourage their students to think actively and make brave guesses and inspire them to find mathematics problems and models from experiments and daily life. Guide them to appreciate others' findings and gradually develop the habits of questioning and doubting. Of course, these abilities cannot be developed in a short period of time. Therefore, the evaluation of this ability should first address the students' enthusiasm and self-confidence and aim at their problem consciousness, instead of just pursuing the quality of the problem.

21.4.3.2 Ability to Collect Information Actively and Solve Problems

Mathematics classroom instruction aims to develop students' abilities to collect information and analyze and solve problems, so as to adapt to the requirements of the rapidly changing information society. Evaluation should focus on whether students can choose effective methods to collect and sort information according to a problem, whether they can put forward some available suggestions based on some relative knowledge and factors, and whether they can build appropriate mathematics models and solve problems independently or in cooperation with others.

The evaluation of mathematics problem-solving ability should always pay attention to the process of analyzing and processing data, the rationality of the mathematics model and the whole process of using mathematics knowledge to analyze and solve problems. Instead of just focusing on one stage or one certain outcome, the evaluation should also pay attention to the students' attitudes and methods. At the same time, we should notice whether the students can consciously question, adjust and improve the collected information and the solutions to problems in the process of evaluation.

21.4.3.3 Mathematics Expression and Communication Skills

Since mathematics language is accurate, simple and formal, it is important that students can express things and communicate with others using both mathematics and natural language. This includes whether they can express a problem accurately, and whether they can express their minds and opinions using various languages such as text, symbols, graphics, and actions, both orally and in writing. Of course, the purpose of mathematics expression and communication skills is not just to tell, but also to make judgments, question, and reply.

With regard to mathematics expression and communication skills, teachers should have patience and understanding of students' immature opinions or poor expressions. As well, they should encourage them and listen to their expressions and views carefully, appreciate their communications and give them timely observations and instruction about the thoughts, opinions and accuracy and form of expression.

21.4.4 Pluralism

In order to promote the overall development of students, the assessment of mathematics instruction should be pluralistic, including plurality of subjects, methods, contents, and standards.

21.4.4.1 Plurality of Subjects

Plurality of subjects refers to the combination of all assessments from teachers, students themselves, other students, parents and others in society. Their involvement in the mathematics classroom teaching process fully reflects the main idea that we should give each student an all-round and unbiased evaluation.

As always, when the assessment of students' mathematics learning is made mainly by teachers it is a one-way process between educator and educatee. It is hard to avoid one-sided and biased conclusions, which is no good for the overall understanding and judgment of the students. Plurality makes the evaluation much more objective and brings encouragement and feedback into play.

21.4.4.2 Pluralism in Evaluation Methods

Pluralism in evaluation methods refers to the combination of qualitative and quantitative methods, written and oral approaches, and results and processes. The assessment of mathematics classroom teaching cannot depend only on one of these ways, but should take various methods into consideration to get a more objective and scientific conclusion.

Students' mathematics learning has long been evaluated by just one paper or one score. We have to admit that education is complicated and students are also vastly different. Their attitudes to mathematics learning, their abilities to cooperate and communicate, the sensitivity of problems, and their habits and depth of thinking cannot just be measured by one paper or one score. Therefore, evaluation based on the combination of qualitative and quantitative, written and oral, and results and process is necessary and can fully reflect the real situation.

In the form of observation, recording, interchanging and discussing, the qualitative evaluation of students' mathematics learning can be carried out by means of portfolio, comments, reviews, and assessment of research projects. In particular, using the portfolio to keep track of a student's study has been praised widely in recent years. It is an effective way to record the growth paths of all students.

However, the encouragement of pluralism in evaluation methods does not mean that the function of quantitative evaluation methods should be ignored. In fact, tests or quizzes should always be the basic evaluation methods, but we should reform and adjust the specific means of quantitative evaluation, so as to make it more scientific, reasonable and adapted to the needs of the new curriculum. For example, the content and forms of the question can be changed, or we can create a good problem situation and broaden the links between different aspects of knowledge. The idea is to combine the questions organically combined with the real experience and test the students' understanding of the mathematics nature.

21.4.4.3 Pluralism in Evaluation Content

Pluralism in evaluation content means that the evaluation includes not only knowledge, skills and abilities, but also the processes and methods, emotional attitudes and values and other related content. The previous assessment of mathematics classroom teaching stressed the evaluation of knowledge and skills while ignoring these other aspects. Such a system leads directly to situations such as "the exercise-stuffed teaching method", "the high-grades but low ability" and "nice memory but poor creativity". The encouragement of pluralism in evaluation content referred to in *The Mathematics Curriculum Standard for Senior High School* (*Trial*) shows that our country is trying to change this situation and enables students to develop mathematics accomplishments and physical and mental qualities.

Evaluations of mathematics classroom teaching should focus on knowledge and skills, processes and methods, emotional attitudes and values. Since the evaluation of the latter two points is weak, we should enhance them on purpose.

21.4.4.4 Pluralism in Evaluation Standards

Pluralism in evaluation standards means different students have different standards and that we should evaluate the content from different angles. On one hand, the evaluation should respect students' individual differences and their choices instead of expecting a uniform standard. On the other hand, the evaluation of certain mathematics content should not only depend on whether the targeted result is reached, but also on the experiences during the learning process.

As for evaluating students' mathematics learning through testing, the standards usually focus on the result and each step is strictly graded. Such a pattern leads to the emphasis on problem-solving method, specification and accuracy, and the students gradually lose their thought and character, even forget their innovative skills. To change this situation, we must reform the standard of evaluation to make it open and plural. Not only the answers should be considered, but also the students' experience, their mathematics comprehension and consideration, their ability to solve problems, and their intelligence and personality development.

To realize the diversification of evaluation standards, the presentation of the evaluation results should be varied. Based on the principle of respecting and promoting the overall development of every student, we can choose different ways to display the evaluation results. For example, let students themselves summarize their successes and failures in a test. Let students introduce their experiences of mathematics attainment and teachers make corresponding grades. Have students present short papers and research summaries and appraise these performances.

Above all, it is important to be aware that pluralistic assessment can evaluate and encourage students from all angles and directions, which helps each student obtain successful experience and develop much better.

References

Cao, Y. (2011). 国际视野下的中国中学数学课堂微观分析 *[Micro analysis of Chinese Middle School Mathematics Classroom in International Perspective]*. Beijing: Beijing Normal University Press.

Cao, Y., Li, J., & Qin, H. (2011). 我国数学课堂教学评价研究综述 [Literature review of assessment for mathematics teaching and learning in Chinese classroom]. Bulletin des Sciences Mathematics, 8.

Chen, Y. (2004). 数学教学评价的内容和方法 [The Content and Methods of Mathematics Teaching Assessment]. *Journal of Qiannan Normal University of Nationalities, 6,* 66–70.

Kuang, S. (2006). 初探小学数学教学评价 [Preliminary Investigation of Assessment of Primary School Mathematics Teaching and Learning]. *Scientific Consult, 12,* 44–45.

Li, Y. (1994). 数学教育概论 *[An Introduction to Mathematics Education]* (p. 146). Beijing: Science and technology of China Press.

Lu, S., & Fu, L. (2004). 数学教学论 *[Mathematics teaching theory]* (p. 186). Beijing: Science Press Co.Ltd.

Ministry of Education. (2011). 全日制义务教育数学课程标准 *(试验稿) [Mathematics curriculum standards for full-time compulsory education (trial)]*. Beijing: Beijing Normal University Press.

Research Group for Mathematics Curriculum Standard. (2004). 普通高中数学课程标准*(实验稿)* 解读 *[Study of high school mathematics curriculum standard (Trial)]* (p. 328). Nanjing: Jiangsu Education Press.

Shen, Y., & Cui, Y. (2011). 课堂观察——走向专业的听评课 *[Classroom observation: A way towards professional lesson observation and evaluation]*. Shanghai: East China Normal University Press.

The Research Group for Mathematical Curriculum Standards. (2004). 普通高中课程标准 *[National mathematical curriculum standards for high school]*. People's Education Press.

Tu, R., & Ji, S. (2007). 数学课程与教学论新编 *[New version of mathematics curriculum and teaching theory]* (p. 147). Nanjing: Jiangsu Ecuation Press.

Wang, D., & Li, M. (1990). 数学教学论 *[Mathematics Teaching Theory]* (p. 292). Haikou: Nan Hai Publishing Co.

Weng, K. (2002). 数学教育学教程 *[Handbook of mathematics education]* (p. 209). Chengdu: Si Chuan University Press.

Zhou, K., Lei, K., & Yan, X. (2003). 新课程视野中的数学教育 *[Mathematics education from new curriculum perspective]* (pp. 312–313). Chengdu: Sichuan University Press.

Zhou, X. (2006). 新课程中的数学教学评价 [Assessment of mathematics teaching and learning in new curriculum]. Education Guiding, 2–28(3).

Chapter 22
Primary and Secondary Mathematics Selective Examinations

Rongbao Tu

Abstract It is well known that China has a large population but relatively scarce high-quality educational resources. Under such conditions, primary and secondary mathematics selection examinations play an increasingly important role. They have been a prerequisite for almost all students to advance to the next stages of learning; the examination process has been described vividly as "crossing a single-plank bridge."

It is well known that China has a large population but relatively scarce high-quality educational resources. Under such conditions, primary and secondary mathematics selection examinations play an increasingly important role. They have been a prerequisite for almost all students to advance to the next stages of learning; the examination process has been described vividly as "crossing a single-plank bridge."

The primary and secondary school selective examinations are mainly divided into two stages. The first of these is the summative examination at the end of nine-year compulsory education, the *Zhong Kao*. It targets junior middle school graduates, and according to the examination results, they are able to enter full-time ordinary senior middle school or secondary vocational–technical school. The other one is the National University Entrance Exam (*Gao Kao*). Its targets are full-time senior middle school graduates or candidates with equivalent education backgrounds. They can enter different universities or colleges according to the examination results. The two examination systems are introduced separately in the following sections.

R. Tu (✉)
Nanjing Normal University, Nanjing, China
e-mail: rbtu304@163.com

Y. Cao and F.K.S. Leung (eds.), *The 21st Century Mathematics Education in China*, New Frontiers of Educational Research,
https://doi.org/10.1007/978-3-662-55781-5_22

22.1 An Overview of the Ninth-Grade Mathematics Graduation Examination

The Ninth-Grade Mathematics Graduation Examination, the *Zhong Kao,* is one of the summative tests administered during compulsory education for the purpose of evaluating fully and accurately the extent to which junior high school graduates achieve the graduate level set by the *Mathematics Curriculum Standards for Compulsory Education (2011 Edition)* (Ministry of Education, 2011). The results of this exam not only determine whether the students can graduate, but are also an important basis for high school enrollment. The *Zhong Kao* has the functions of screening, guidance, feedback, regulation, and incentive.

22.1.1 The Management System of the Zhong Kao

Under the unified management by the Administrative Department of Education, the specific work of the *Zhong Kao* is entrusted to relevant departments, including the teaching research departments, the examination bodies, and the admission departments. In order to ensure the quality of the exam, the education administrative departments at or above the prefecture-level are responsible for it.

The education administrative departments of the provinces (autonomous regions and municipalities) determine or put forward instructive ideas about the topics and the organization of this examination.

Each province or city has its own personnel qualification system of test formulation, checking and evaluation. Only those who have been trained properly and have obtained the corresponding qualifications are eligible for these jobs.

Each province or city also has its own management system that includes the monitoring and checking of the test formulation and supervision of the test evaluation. In recent years, all provinces and cities have adopted modern methods to improve the quality of test marking.

To improve the science of the test formulation and management of this examination, the ministries of education and their administrative departments in provinces, autonomous regions, and municipalities entrust relevant departments to operate examination evaluation research groups. Besides providing guidance for the various institutes developing the test, the examination evaluation research group also evaluates the formulation and management of the examination regularly, makes evaluation reports, and guides the reforming of the exam. Each province or city must analyze and summarize the examination process and outcomes after the test marking, and submit papers and relevant materials as required.

22.1.2 The Mathematics Proposition of the Zhong Kao

22.1.2.1 The Objectives

The *Zhong Kao* has a number of objectives. First, it is intended to promote rudimentary knowledge, basic skills, basic methods, and comprehensive abilities which are necessary for lifelong study and to improve the quality of teaching in compulsory education. As well, it aims to promote quality-oriented education, strengthen the curriculum reform, and encourage the development of innovative thinking and practical abilities in students, in active ways. Further, the purpose of the examination is to identify qualified, excellent students for all kinds of mid-level schools, but particularly to enhance the development of senior secondary education.

22.1.2.2 The Principles

1. Guidance

The *Zhong Kao* is required to be student-based, with emphasis on competence, applicability, exploration, and a comprehensive grasp of mathematics. Thus, the test-formulation process should serve as a guide to schools to strengthen their mathematics programmes, enable mathematics teachers to improve their instruction, and guide students to learn how to learn.

2. Relevance

As the junior middle school is an important stage to lay students' mathematical foundations, it is necessary for the test formulation to be in strict accordance with the *Mathematics Curriculum Standards for Compulsory Education (2011 Edition)* (Ministry of Education, 2011) and the reality of mathematics teaching.

3. Comprehensive coverage

The examination is required to be based on a comprehensive examination of students' basic knowledge and skills. Great importance is attached to examining students' abilities to draw on their existing knowledge to analyze and solve problems, along with a range of basic mathematics qualities and abilities.

4. Scientificity

The test-formation process needs to ensure that all examination questions are correct and scientific, using standard expressions, concise language and clear questions, information, and graphs.

5. Appropriateness

The mathematics questions should have a suitable difficulty ratio and enable students of all levels to perform well. The setting of the questions should be at different levels, with an appropriate starting point and increase in difficulty, thus reflecting the examination's function as a tool for selection.

22.1.2.3 The Content

The content of the *Zhong Kao* consists of "Number and Algebra," "Space and Shape," "Statistics and Probability," and "Task-based Learning."

1. Number and Algebra

This category covers the topics of real number, fractional and integral expression, equation and equations, inequality and inequalities, functions, and other knowledge. It evaluates whether students can find the relationships and laws among numbers, graphs, and real-world problems, and use appropriate tools to represent and process quantity relationships and change rules. It also tests the students' ability to apply mathematics concepts and to solve problems using algebraic knowledge and methods.

2. Space and Shape

Items in this category test the fundamental proprieties of basic graphs (straight line, circle) and their relationships. They also test the fundamental proprieties and application of translational and rotational symmetry, how to use the coordinate system to determine an object's location, and students' spatial sense and their ability to prove some mathematical propositions relative to the triangle and quadrangle, based on some basic facts.

3. Statistics and Probability

This category tests the methods used by students to describe data and the idea of estimating a population from a sample. It also tests simple calculations of the probability of events, and the ability to make informed predictions and inferences by collecting, sorting, describing, and analyzing the data as well as describing the possibilities of events.

4. Task-based Learning

Based on some of the content related to the three categories described above, this aspect refers to basic methods for studying problems, students' awareness and abilities to understand and apply mathematical knowledge in a variety of contexts, their thinking abilities, and their understanding of related mathematical knowledge.

In addition to the above, the examination covers some basic mathematical thinking methods, for example, representing numbers with letters, equations and functions, the whole-part method, the graph transformation method, the coordinates method, and statistical and stochastic thought.

22.1.2.4 Operation

Initially, about 3–5 people are selected to work on developing a proposed examination paper; their selection is based on the recommendations of schools or research departments and through the auditing and filtering of the Education Administrative Departments. Since the high school entrance examination is a large-scale social educational test with high risk and high stakes, these people receive some training and then go to an isolated place to begin the process.

1. The paper structure

The test-design group's first task is to determine the proportions of the four content areas to be included in the examination paper. This is based on the proportions appearing in the curricula for grades seven to nine. They also need to decide the numbers of questions, the point allocation, and the time allocation for each part, then the structure of the paper and the desired difficulty coefficients.

2. A detailed catalog

After the question types, question numbers and the test structure have been confirmed, the committee is required to construct a bilateral table as a detailed catalog of all the test requirements.

3. Design and adjustment

The test-design group sets questions according to the bilateral table. When all members of the test-design group have finished their assigned preliminary work, they discuss the items one-by-one and modify, adjust, process, and polish them. Then the group considers and adjusts the examination questions and papers systematically. Finally, they are required to predict the difficulty coefficients of the questions and papers.

22.1.2.5 Checking the Proposed Paper

To ensure the quality of the examination papers, provinces and cities all have test-review systems. While checking each test question, the reviewers should first read the key points of the paper (including the standard answers and scoring criteria), proofread and solve each item, and consider the words and expressions. They also need to communicate with the test designers to gain an understanding of the guiding objectives, the basic idea, the arrangement of questions, the predictive grades, and so on. Finally, the reviewers propose amendments and fill in a review feedback form.

22.1.2.6 The Paper Structure

The structure of the examination papers includes both explicit and implicit structures. The former refers to the paper's framework structure and question structure, while the latter refers to the knowledge structure, ability structure, and difficulty structure of the paper.

The framework structure generally includes the name, number of pages and full marks of the paper, the examination time, the instructions and other notes, the test questions, the answer sheet, standard answers, the scoring criteria, and detailed scoring regulations.

Almost all of the papers developed across the country use three kinds of questions: multiple-choice, fill-in-the-blanks, and calculation questions. The allocation of marks for the first two of these, combined, cannot exceed 40% of the total score.

The knowledge structure includes the test content, knowledge coverage, knowledge focus and the proportion of each knowledge element, the cognitive levels of items required to examine, and the proportion each cognitive level accounts for. The ratio of marks assigned to "Number and Algebra," "Space and Shape," and "Statistics and Probability," respectively, is usually close to 4:4:2.

When it comes to ability structure, the *Zhong Kao* examination papers basically cover number and symbol sense, spatial ideas and sense, reasoning abilities, application, presenting problems from a mathematical perspective, and solving practical problems by drawing on prior knowledge.

The difficulty structure refers to the difficulty of the paper on the whole, the difficulty of each question, and the proportion of all difficult questions at different levels. Although different areas may have different difficulty structures due to the local education resources, the difficulty coefficients in all areas are approximately from 0.60 to 0.75. Similarly, the proportions of all difficult questions at different levels are not the same, with a majority proportion being 7:2:1 for easy, medium, and hard, respectively.

22.1.3 Test Evaluation

Generally, the test-evaluation work is organized uniformly by the local Education Administrative Departments. Each one sets up a leading group and chooses excellent mathematics teachers from local junior high schools to undertake the task in a centralized way. They arrange the place and time for this work. In addition establishing the systems and rules for arranging the papers and ensuring confidentiality, the inspection system involves ensuring the quality of the marking.

22.1.4 Assessment

22.1.4.1 Analysis and Assessment of the Paper

The Ministry of Education requires all provinces and cities to set up evaluation teams to undertake the analysis and assessment of their papers. The tasks carried out by most test-assessment teams include as follows: sample, count, draw the difficulty curves for questions and histograms of the scores, analyze the data collected and the graphs drawn, and then make comments on each item. In addition, they convene workshops for teachers, to discuss the advantages and disadvantages of the paper and the questions one-by-one, and record the common errors and various solutions for each question. They conduct questionnaire surveys about the teachers' perceptions of the mathematics examination paper of the year and finally write the assessment report and the paper-analysis report.

22.1.4.2 Assessment by Higher-Level Departments

The Ministry of Education and Administrative Departments in the provinces, autonomous regions, and municipalities entrust relevant departments to set up examination evaluation research groups. These groups evaluate the test formulation and management of the exam and make regular evaluation reports.

22.1.5 Sample of Zhong Kao *Mathematics Papers*

Since the *Zhong Kao* mathematics papers are designed and organized independently by counties or cities, there are hundreds of different sets of papers around the country each year. However, given that they are all designed according to the *Full-time Compulsory Education Mathematics Curriculum Standards*, the contents and formats of these exam papers are substantially the same. The following is a sample of a mathematics *Zhong Kao* paper designed in Nanjing, Jiangsu Province.

The Graduation Examination for Ninth Grade in Nanjing, 2015
MATHEMATICS

Section A. Multiple Choice Questions (Question amount: 6; Full marks: 12; 2 marks for each question. Only one item meets the requirements of the question. Please choose the right answer and fill in the corresponding circle on the answer sheet.)

1. What is the value of $|-5+3|$?

A. -2 B. 2 C. -8 D. 8

2. What is the value of $(-xy^3)^2$?

A. x^2y^6 B. $-x^2y^6$ C. x^2y^9 D. $-x^2y^9$

3. In $\triangle ABC$, DE // BC and $\frac{AD}{DB}=\frac{1}{2}$ (see the diagram below), which of the follow conclusions could be right?

A. $\frac{AE}{EC}=\frac{1}{2}$ B. $\frac{DE}{BC}=\frac{1}{2}$

C. $\frac{\text{the circumference of }\triangle ADE}{\text{the circumference of }\triangle ABC}=\frac{1}{3}$ D. $\frac{\text{the square of }\triangle ADE}{\text{the square of}\triangle ABC}=\frac{1}{3}$

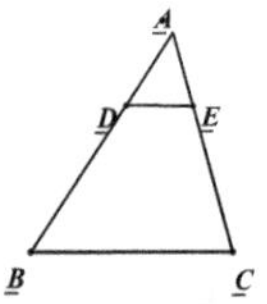

4. The number of motor vehicles in a city was 2×10^6 at the end of 2013, which increased by 3×10^5 in 2014. How can you represent the quantity of the motor vehicles of the city at the end of 2014 in scientific notation?

A. 2.3×10^5 B. 3.2×10^5 C. 2.3×10^6 D. 3.2×10^6

5. Estimate the value of $\frac{\sqrt{5}-1}{2}$; which of the following could be right?

A. between 0.4 and 0.5 B. between 0.5 and 0.6

C. between 0.6 and 0.7 D. between 0.7 and 0.8

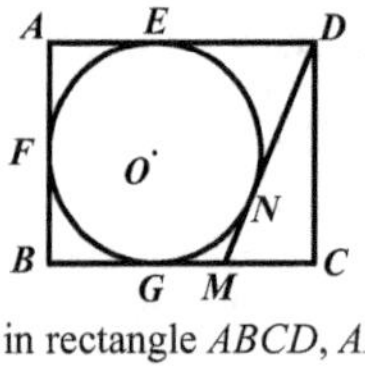

6. In the figure above, in rectangle $ABCD$, AB=4 and AD=5. AD、AB、BC are tangent to $\odot O$ at E, F, G respectively. Draw a tangent line to $\odot O$ passing through D and intersecting with BC at M.

The point N is the point of tangency. What is the length of DM?

A. $\frac{13}{3}$ B. $\frac{9}{2}$ C. $\frac{4}{3}\sqrt{13}$ D. $2\sqrt{5}$

Section B. Fill in the blanks (Question amount: 10; Full marks: 20; 2 marks for each question. Please write the answers directly on the answer sheet.)

7. The square roots of 4 are______; the arithmetic square root of 4 is______.
8. If $\sqrt{x+1}$ is meaningful in real texts, then the range of x is ______.
9. The value of $\frac{\sqrt{5}\times\sqrt{15}}{\sqrt{3}}$ is______.

10. $(a-b)(a-4b)+ab$ can be factored as ______.

11. The solution of the inequalities $\begin{cases}2x+1>-1\\2x+1<3\end{cases}$ is ______.

12. It is given that 1 is one root of the function $x^2+mx+3=0$, then another root is ______ and the value of m *is* ______.

13. In a rectangular coordinate system, the coordinate of the point A is $(2, -3)$, A' is the symmetric point of A about the x-axis. A'' is the symmetric point of A' about the y-axis. So the coordinates of the point A'' are (______, ______).

14. An engineering team has 14 employees. Their types of work and corresponding monthly salaries are shown in the following table.

type of work	Number of people	the average monthly salary
electrician	5	7000
carpenter	4	6000
brick layer	5	5000

Now the engineering team has an employees' turnover. To be specific, they removed two carpenters and added an electrician and brick layer. Then the variance of the monthly salary of the engineering team staff is______ (fill in the "smaller", "the same" or "bigger") compared with the before variance.

15. In the figure above, the pentagon ABCDE is inscribed in the circle with center O. If ∠CAD=35°, then ∠B+∠E=______°.

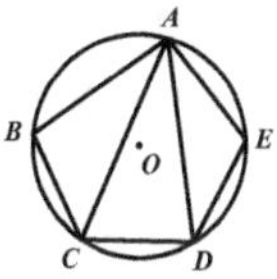

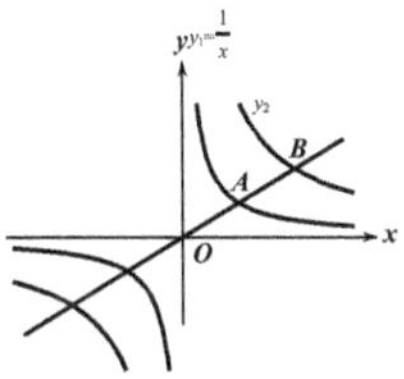

16. In the figure above, a line passing through the point O intercepts with the graphs of the inverse proportional function y_1, y_2 at the points A, B respectively. A is the midpoint of OB. If $y_1=$

$\frac{1}{x}$, then the function expression of y_2 about x is ______.

Section C. Essay Questions (Question amount: 11; Full marks: 88. Give your answers in the designated areas. Explanations, full proofs, or steps of mathematical calculations should be included.)

17. (6 marks) Solve the inequality $2(x+1)-1 \geq 3x+2$ and express its solution on the number axis.

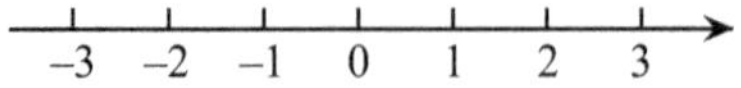

18. (7 marks) Solve the equation $\frac{2}{x-3} = \frac{3}{x}$.

19. (7 marks) Calculate $\left(\frac{2}{a^2-b^2} - \frac{1}{a^2-ab}\right) \div \frac{a}{a+b}$

20. (8 marks) In the figure below, in triangle ABC, CD is the height on side AB and $\frac{AD}{CD} = \frac{CD}{BD}$.

(1) Prove that $\triangle ACD \backsim \triangle CBD$;

(2) Find $\angle ACB$.

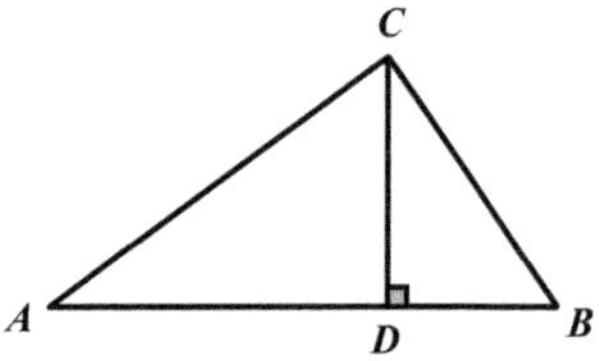

21. (8 marks) In order to understand the scores that 100000 primary, secondary and university students get in the 50 meters running in a given area in 2014, about 10% of each of these groups of students are chosen by the department of education to take the measurement. By arranging the sample data and combining with the sampling results in 2010 we get the following chart.

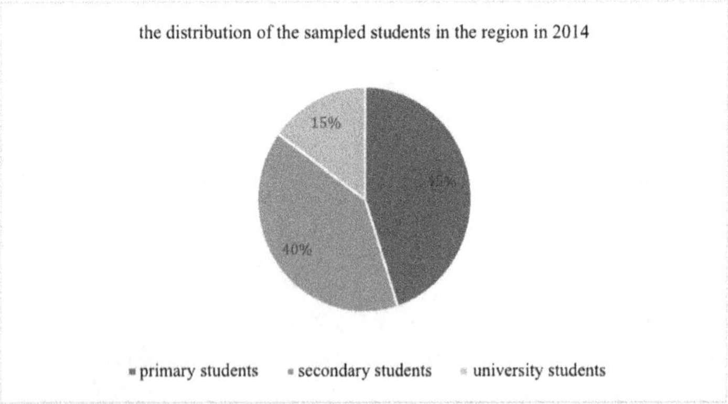

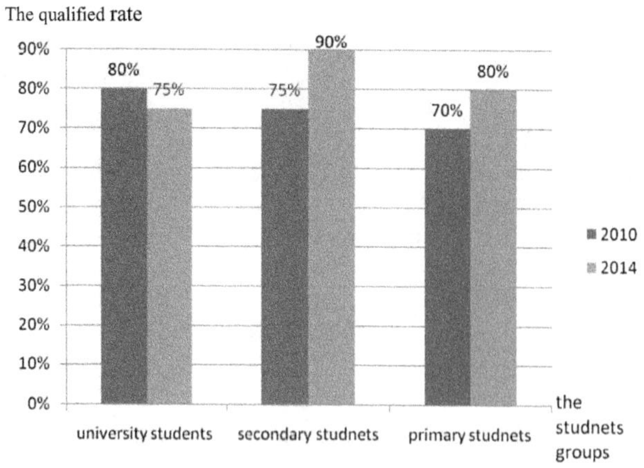

(1) The total number of the primary, secondary and university students chosen in this test is____, and the number of primary students is ____.

(2) According to the results, we can estimate that the number of secondary students competing in the 50 meters running among the 100000 students in this region in 2014 is ____.

(3) Compare the qualifying rates of the students in 50 meters running in 2010 and 2014 and write your conclusion.

22. (8 marks) There are one 10-yuan note, one 20-yuan note and one 50-yuan note in someone's purse. Take out two notes from the purse randomly.

(1) Find the probability that the total value of the notes taken out is 30-yuan.

(2) Find the probability that the total value of the notes taken out can buy an item of 51- yuan.

23. (8 marks) As in the figure below, a blue ship in location A is to the west of the wharf O, a red ship in location C is to the north of the wharf O and $\angle CAO=45°$. If the blue ship is sailing at a steady speed of 45km/h from west to east, the red ship is sailing at a steady speed of 36 km/h straight to the north. 0.1 hour later, the blue ship and the red ship arrive at B and D respectively and $\angle DBO=58°$. Find the distance from B to O.

(reference information: $\sin 58° \approx 0.85$, $\cos 58° \approx 0.53$, $\tan 58° \approx 1.60$)

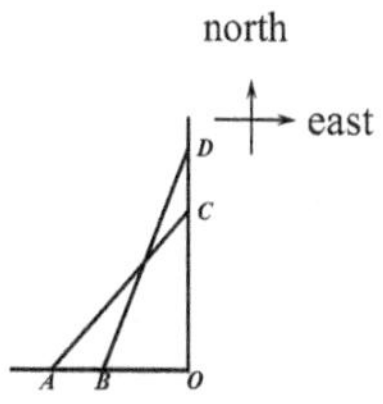

24. (8 marks) As in the figure, $AB \parallel CD$ and the point E、F is in the line AB and CD respectively. Combine the points E and F. The angular bisectors of $\angle AEF$ and $\angle CFE$ intercept at the point G, the angular bisectors of $\angle BEF$ and $\angle DFE$ intercept at the point H.

(1) Prove that the quadrangle $EGFH$ is a rectangle;

(2) Sara investigates deeper after finishing the proof of question (1). Draw a line MN through the point G such that MN // EF and it intercepts with AB and CD at P and Q respectively. Then a quadrangle $MNQP$ is produced. Sara guesses that the quadrangle $MNQP$ is a rhombus. Please fill in the following diagram to complete her idea.

For AB//CD, MN//EF, PQ//EF, we know that the quadrangle $MNQP$ is a parallelogram. To prove that the quadrangle $MNQP$ is a rhombus we just need to prove $NM=NQ$.

From the given information________, MN // EF,

we can get $NG = NF$, so we need to prove $GM = FQ$, that is to prove $\triangle MGE \cong \triangle QFH$. We can easily get ________ , ________,

So we just need to prove $\angle MGE = \angle QFH$, $\angle QFH = \angle GEF$, $\angle QFH = \angle EFH$, ________, that is the proof.

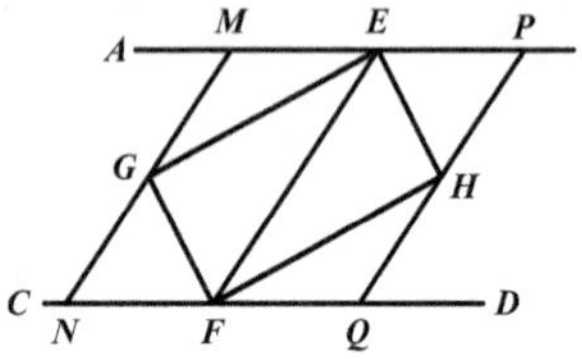

25. (10 marks) In the figure below, $ABCD$ is a square with sides of length 4. Please find all different isosceles triangles having a length of 3 and taking A as a vertex and other two vertices are in the side of the square $ABCD$. (requests: only the sketches are needed and mark 3 on the sides whose length are 3.)

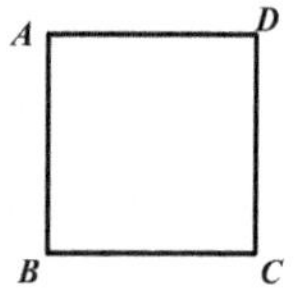

26. (8 marks) See the diagram below, the quadrangle $ABCD$ is inscribed in the circle with center O. The extension cord of BC and AD intercept at the point E and $DC=DE$.

(1) Prove that $\angle A = \angle AEB$;

(2) Combine the point O and E and OE intercepts with CD at the point F. If $OE \perp CD$, prove: $\triangle ABE$ is an equilateral triangle.

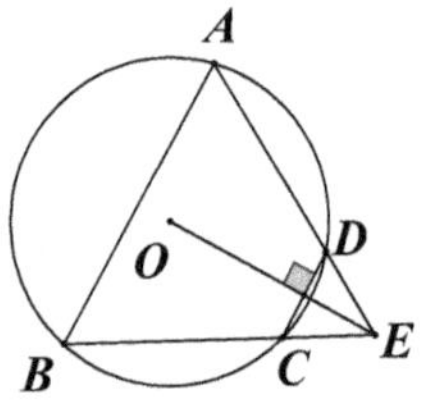

27. (10 marks) One enterprise produces and sells a product. Assume the sales are equal to the output. In the diagram below, the line ABD represents the function relationship between the

production cost per kilogram (noted by y_1) and the output (noted by x). The segment CD represents the function relationship between the sale price (noted by y_2) and the output (noted by x).

(1) Please explain the real meaning of the abscissa and ordinate of the point D.

(2) Find the function expression of y_1 and x.

(3) When can the enterprise get the biggest profit and what is the biggest profit?

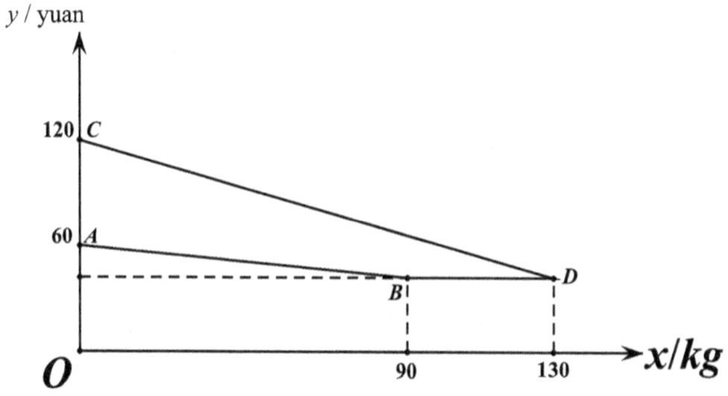

22.2 The University Entrance Examination

The University Entrance Examination (*Gao Kao*) is one kind of selective examination held for graduates from ordinary high schools to gain university or college admission. It is the most influential educational and social test in China. Full-time ordinary high school graduates and students with equivalent educational levels can take this test. The results of the *Gao Kao* are central to university recruitment.

22.2.1 *The Purpose and Function of the* Gao Kao

Broadly speaking, the main purpose of the *Gao Kao* is to provide effective results for university recruitment. As an important means of selecting talent, this examination not only has the function of screening, selecting, orientating, and evaluating, but also the functions of providing feedback, diagnosing, regulating, and stimulating; however, of all of these, the functions of selecting, orientating, and evaluating matter the most.

22.2.1.1 *Gao Kao* as a Selection Examination for Colleges or Universities

Full-time ordinary high school graduates and students with equivalent educational levels take this selective and large-scale test. The purpose is to select high-quality students for admission to colleges or universities. For economical and historical reasons, the percentage of senior middle school students able to attend universities is relatively low in China; therefore, the *Gao Kao* examination becomes fiercely competitive.

22.2.1.2 *Gao Kao* as a Guide for Mathematics Teaching in Senior Middle Schools

The *Gao Kao* examination also has a strong effect on mathematics teaching. On one hand, the proposal, test contents, and ability requirements are decided according to the mathematics syllabus for senior middle schools; at the same time, the objectives and format of the exam and even its outcomes are used to provide direct guidance for mathematics teaching in the middle schools.

22.2.1.3 *Gao Kao as* a Way to Evaluate Teaching Effectiveness in Senior Middle Schools

There are various methods of evaluating the effectiveness of mathematics teaching, but the *Gao Kao* examination is undoubtedly the most important one. According to the scores, we can know whether or to what extent the students' knowledge, skills, and abilities have reached the secondary schools' education targets.

22.2.2 *Characteristics of the* Gao Kao

The *Gao* Kao has a number of carefully designed and implemented characteristics which are discussed in detail in this section.

22.2.2.1 *Gao Kao* as a Norm-Referenced Test

The college entrance examination not only determines whether the candidates have reached a certain level, but also compares their scores with those of the group, thus singling out the best to receive a higher education. In this way, the *Gao Kao* is considered to be a kind of norm-referenced test, and therefore, it should have high reliability, validity, and the necessary discrimination.

22.2.2.2 *Gao Kao* as a Test of Mathematics Basics

To ensure that all university entrants will be able to continue with their mathematics learning, the *Gao Kao* needs to be able to examine their acquisition of the necessary basic mathematics. The test incorporates four key aspects of basic mathematics. The first one is basic knowledge, which includes all of the concepts, principles, properties, formulae, axioms, and theorems involved in the middle school mathematics curriculum. The second one is basic skills, namely computation, drawing, and reasoning according to certain procedures or steps. The third one is mathematics thinking methods. Specifically, the substitution method, comparative method, mathematical induction, the methods for completing the square, undetermined coefficients, transformation, and the unified method are the general mathematical methods tested. The methods of analysis, integration, induction, enumeration, and proof by contradiction often appear in the test of logical method. The mathematics thinking methods incorporate number–shape combination, functions and equations, classified discussion, and equivalent transformation.

22.2.2.3 The *Gao Kao* as a Test of Mathematical Abilities

The *Gao Kao* is a kind of academic aptitude test. Not only does it investigate candidates' grasp of mathematics knowledge, it also examines the ability and general mental capacity shown when students are using related theory and methods based on this knowledge. In line with this idea, the examination tends to test students' competence.

22.2.2.4 The *Gao Kao* as a Test of Both Difficulty and Speed

Because of its high abstraction, systematic structure, and logic, there are very few questions that can be answered just by mechanical memory or intuitive impression. In fact, the questions that involve speculation are generally somewhat difficult and require candidates to have a certain level of ability to observe, analyze, and reason. In order to distinguish students of different levels and select satisfactory students for university entrance, it is inevitable that some questions are difficult and require higher ability levels. Therefore, an important characteristic of the examination is its ability to discriminate according to difficulty. As well, there are more than 130 knowledge points in the middle school. To investigate the students' understanding of these points comprehensively, the number of questions is so large that the candidates should be able to solve problems at a fast speed. It is believed that high proficiency, simplicity, and speed in solving problems are tests of flexibility and agility, another important aspect of the examination.

22.2.3 The Content and Requirements of the Gao Kao

Before 2007, the content of the *Gao Kao* examination was derived from the national teaching material, the teaching plan for full-time senior middle schools, and the mathematics teaching syllabus for full-time senior middle schools. Since 2007, some provinces have begun to design their tests automatically based on the national mathematics curriculum standards for senior middle schools (trial) and the teaching content and requirements decided in the corresponding teaching materials (Editor groups, 2004).

22.2.3.1 The Content and Requirements for the Inspection of Knowledge

The examination divides the assessment knowledge into three levels, from low to high; these are knowing, understanding and mastery, and flexible and synthetic application.

Knowing involves having a preliminary and perceptual awareness of the meaning and relevant background of the knowledge listed, knowing some relevant content and being able to apply this directly to related questions.

Understanding and mastery refer to having a deep and rational understanding of the listed content, being able to explain, exemplify, transform and deduce the knowledge, and apply it directly to related questions.

Flexible and synthetic application involves mastering the inner relationships of the contents systematically, and applying the listed knowledge to analyze and solve complicated or comprehensive problems.

The content of the examination includes plane vectors, sets and simple logic, functions, inequalities, trigonometric functions, sequences, the functions of lines and circles, the functions of conic curves, lines, planes and simple geometries, permutations, combinations and the binomial theorem, probability and statistics, limits, derivatives, and extension of the number system from real numbers to complex numbers. Under the background of the new curriculum reform, the content has been adjusted slightly; for example, algorithms, statistical cases, geometric probability models, matrixes, coordinates, and parametric equations have been added recently, while complex calculations and reasoning of the angles and distances in three-dimensional geometry have been reduced.

22.2.3.2 Examination of Mathematical Thinking and Methods

The examination of mathematical thinking and methods runs through the whole paper, in both general and more focused senses, and in a hierarchical way. The *Gao Kao* places an emphasis on testing common properties and methods, breaking down particular skills and examining the students' grasp of mathematical thinking and methods. These can be divided into three categories: mathematical ideas, mathematical thinking, and mathematical methods. Specifically, mathematical ideas include functions and equations, number–shape combinations, classification and integration, transformation, specialization and generalization, finite and infinite, and probability and necessity. Mathematical thinking includes the methods of analysis, integration, induction, deduction, observation, testing, and specialization. Mathematical methods include, among others, the methods of completing the square, reduction, and undetermined coefficients.

22.2.3.3 Examination of Mathematical Abilities

The examination of mathematical abilities mainly includes thinking ability, computing power, spatial ability, practical ability, and innovation, with thinking ability being the core point. The *Gao Kao* tends to test students' competence, which fully reflects the importance of an ability test. In recent years, with the deepening of concepts brought about by the curriculum reform, research study courses have been introduced. The *Gao Kao* highlights students' abilities to pose, analyze, and solve problems, and to engage in mathematics enquiry, mathematics modeling, mathematics communication, and mathematics practice.

22.2.4 *Format and Design Features of the* Gao Kao *Papers*

22.2.4.1 Format

The *Gao Kao* can be designed in two ways: one is the national uniform design and the other is the province-based autonomous design. Some provinces choose the former and others the latter. In 2015, for example, there were 16 college entrance mathematics examination papers in total, and two of them were at the national level, namely National Volume I and National Volume II. Four provinces using the National Volume I were Henan, Hebei, Shanxi, and Jiangxi, while 13 used the National Volume II, the Guangxi Zhuang Autonomous Region, Guizhou, Gansu, Qinghai, Ningxia Autonomous Region, Hainan, Heilongjiang, Jilin, Liaoning, Xinjiang Uygur Autonomous Region, Yunnan, the Tibetan Autonomous Region, the Inner Mongolian Autonomous Region. Guangdong, Shandong, Jiangsu, Fujian, Zhejiang, Anhui, Tianjin, Beijing, Shaanxi, Hunan, Hubei, Sichuan, Chongqing, Shanghai, and another 14 provinces and cities use their self-made examination papers.

Even though these 16 college entrance mathematics examination papers from 2015 were all based on the *Curriculum Standard of Mathematics in High Schools* (*Experimental*), there were some differences in focus and formats. Most papers had three kinds of question types: multiple-choice (choose from 1 to 4), blank-filling, and calculation questions, with about 10, 6, and 6, respectively, of each type.

Generally, the total score was 150, and the test time was 120 min. Some papers, for example, Jiangsu's and Shanghai's, only had two question types, blank-filling and calculation questions. The total values may also have been different. For instance, the total value of the paper for Jiangsu province was 160 marks.

22.2.4.2 Design Features

A high-quality *Gao Kao* paper should have a certain difficulty coefficient, good discrimination validity, and be reliable and valid (Tu, Wang, & Ning, 2006).

1. The difficulty coefficient
   The difficulty coefficient is an index reflecting the difficulty of a question.
   The computational formula is $P = \frac{\overline{X}}{M}$

More specifically, P represents the difficulty coefficient. M indicates the total score of a question, and $\overline{X}$ indicates the average score for all students who participate in the examination. For example, if the total score of a question is 20 and the average score is 16, then the difficulty coefficient is 0.8.

It is easy to see that the higher the value of P, the easier the question is, and on the contrary, is the smaller the value of P, the more difficult the question is.

There are three difficulty levels for the *Gao Kao* questions, with a ratio of 3:5:2, respectively, for simple questions (the difficulty coefficient is above 0.7), medium–difficult (the difficulty coefficient is from 0.4 to 0.7), and difficult (the difficulty coefficient is below 0.4). The difficulty coefficient for the overall test paper should typically be kept at about 0.60. All the mentioned 16 papers in 2015 reached this requirement.

2. Discrimination validity

The discrimination validity means the extent to which a question can tell the students' actual ability levels. The scores of questions with high discrimination validity may have a large range, while those with low discrimination validity can make the scores very close, and thus not discriminate between different levels of student ability.

To calculate the discrimination validity, we can use the formula $D = \frac{\mu_h - \mu_l}{\mu}$, where D represents the discrimination validity of a question, μ stands for the total score of the question, μ_h represents the average score of the high-scoring group, and μ_l represents the average score of the low-scoring group (respectively the top and bottom 27% of students).

In general, questions with discrimination validity from 0.4 to 1.0 are the best, then those from 0.3 to 0.4; those from 0.2 to 0.3 are barely acceptable, and those from 0 to 0.2 are unsuitable. All questions with discrimination validity lower than 0.3 should be eliminated or improved.

The difficulty coefficient has a close relationship with the discrimination validity. When the difficulty coefficient is too big or too small, the discrimination validity will be low in response. If the difficulty coefficient is around 0.5, the distribution of the scores of the students will be discrete and normal, thus this result helps to make comparisons of each student's position relative to the total group.

In addition, the discrimination validity is relative. We can get different discrimination validity values with different formulae. Therefore, when comparing the discrimination validity of multiple questions, we must use the same formula.

Since the *Gao Kao* is one kind of selective examination, it is most important for it to have a high level of discrimination validity. Given that the computation of the discrimination validity is based on the numbers and the scores of the examinees, which are difficult to obtain, it is not possible to present specific examples here.

3. Reliability

Reliability is an index describing the stability and reliability of the test results, namely how closely the score reflects the students' actual ability. A reliability coefficient varies between 0 and 1; the greater the value is, the higher the reliability is, and the more reliable the test is. For regular and influential tests like *Zhong Kao* and *Gao Kao*, the reliability coefficient is generally above 0.9.

There are many methods to calculate reliability, including the test–retest, parallel forms, and split-half methods. In the test–retest method, we test the same group of students twice with the same test paper and then calculate the correlation coefficient between the scores of the two tests, which is called the reliability coefficient. The parallel approach uses two test papers similar in testing content and requirements, the style and quantity of the questions, and the difficulty and the discrimination validity of corresponding questions; the correlation coefficient between the two tests is the equivalent reliability coefficient. In the split-half method, the questions are sequenced according to difficulty from easy to hard, and then divided into two equivalent parts and calculate the correlation coefficient.

Cronbach developed a basic formula to calculate the reliability coefficient as $r = \frac{n}{n-1} \cdot \frac{S^2 - \sum_{i=1}^{n} S_i^2}{S^2}$. In this formula, n stands for the total number of the questions, S^2 represents the variance of the scores of all examinees, and S_i^2 represents the variance of the scores of all examinees in by question.

For example, the first part of a test paper is multiple choice (accounting for 60 points), and the rest four calculation questions account for 40 points (each makes up for 10 points). The outcome of the test is as follows: $S = 12.7$, $S_1 = 6.2$, $S_2 = 3.8$, $S_3 = 3.1$, $S_4 = 3.3$, $S_5 = 4.2$. So the reliability coefficient of the test is

$$r = \frac{n}{n-1} \cdot \frac{S^2 - \sum_{i=1}^{n} S_i^2}{S^2} = \frac{5}{5-1} \cdot \frac{12.7^2 - (6.2^2 + 3.8^2 + 3.1^2 + 3.3^2 + 4.2^2)}{12.7^2}$$
$$\approx 0.54.$$

There are several ways to improve the reliability coefficient of a test. For instance, improve the number of questions and expand the coverage of the test so as to reduce the possibility of accidental correct answers. Use questions with proper difficulty and discrimination validity to the greatest extent. Arrange the questions from easy to hard to keep the students in calm, which helps they play at their normal level in the examination.

Since the development of the *Gao* Kao papers has a long history, the reliability coefficients of the papers are generally strong. For example, in 2015, all 16 papers had coefficients of around 0.9. Specially, the reliability coefficient for Jiangsu was 0.91.

4. Validity

The validity is an index reflecting the effectiveness and veracity of an examination, reflecting the extent to which it has reached its established goals. For a mathematics examination, the validity generally refers to the content validity, which is the extent to which the knowledge and abilities stated in the objectives have been reflected in the content.

To determine the validity, we should establish validity criteria; generally the students' performance of the grasping of knowledge and abilities in daily study and the evaluation of experienced teachers are used here. Then we calculate the correlation coefficient between the scores of the examination and the scores under the validity criteria and take this value as the validity.

For example, if there are n students take the exam and their scores in the exam and under the validity criteria are x_i and y_i respectively, then the validity is

$$r = \frac{\sum_{i=1}^{n}(x_i-\bar{x})(y_i-\bar{y})}{\sqrt{\sum_{i=1}^{n}(x_i-\bar{x})^2 \cdot \sum (y_i-\bar{y})^2}} \quad (i = 1, 2, 3\ldots, n),$$

In the formula, $\bar{x}$ and $\bar{y}$ represent the corresponding average value. Ordinarily, a reasonable validity value should between 0.4 and 0.7.

The key to improve the validity of an exam is to plan a good test paper. So before the test programming, we should analyze the content and the ratios of knowledge and ability in each part.

It is hard to calculate the validity of *Gao Kao*. First, teachers estimate the validity of the paper according to the scores of the students in their own classes, and then the testing and measurement organization collects information describing the basic situation in each school and each area. This understanding is used as a basis for further development of the papers.

22.2.5 Sample of Gao Kao *Mathematics Papers*

The design and establishment of the test paper is a complex cyclical process (Ren & Kong, 2010), as shown in the following figure.

Examination paper design process

The following is a sample *Gao Kao paper*:

National University Entrance Examination 2015

Mathematics for Science (New Curriculum Ⅱ)

This paper consists of two parts: Part A (Multiple Choice Questions) and Part B (Non-multiple Choice Questions)

Part A (Full marks: 150 points. Given time: 120 minutes)

I. Multiple Choice Questions:

(Altogether 12 items, 5 points for each; for each item, only one out of the four choices satisfies the requirement of the item.)

1. Given Set A={−2, −1, 0, 2}, B={x|(x−1)(x+2)<0}, then A∩B=

(A) {−1, 0} (B) {0, 1} (C) {−1, 0, 1} (D) {0, 1, 2}

2. Given that a is a real number and $(2+ai)(a-2i)=-4i$, then a =

(A) 1 (B) 0 (C) 1 (D) 2

3. The column diagram below shows the emission loads of sulfur dioxide from 2004 to 2013 (unit: million tons), which of the following conclusions is incorrect?

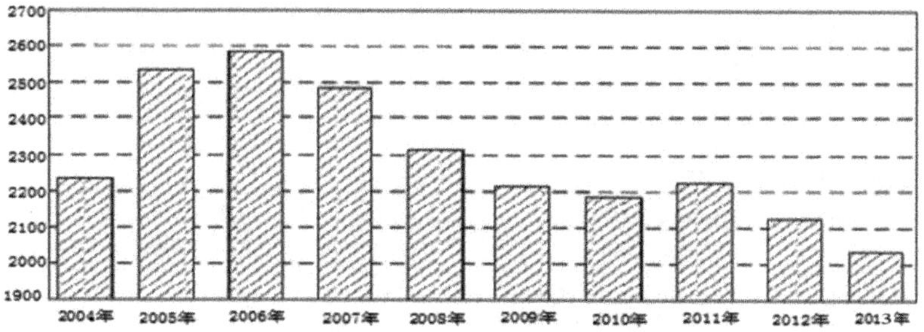

(A) Compared year-by-year, the effect of reducing the emission of sulfur dioxide in Year 2008 is the most significant.

(B) The sulfur dioxide emission control has shown effectiveness in Year 2007.

(C) The sulfur dioxide emission has shown a decreasing tendency from Year 2006.

(D) Since Year 2006, the sulfur dioxide emission has a positive correlation with the year.

4. Given geometric sequence $\{a_n\}$ and $a_1=3$, $a_1+a_3+a_5=21$, then $a_3+a_5+a_7=$

(A) 21 (B) 42 (C) 63 (D) 84

5. Given function $f(x)=\begin{cases}1+\log_2(2-x), x<1,\\ 2^{x-1}, x\geq 1\end{cases}$, then $f(-2)+f(\log_2 12)=$

(A) 3 (B) 6 (C) 9 (D) 12

6. A cube is truncated by a plane, the three views of the remainder is shown on the right, then the ratio of the volume of the separated partto the volume of the remainder is

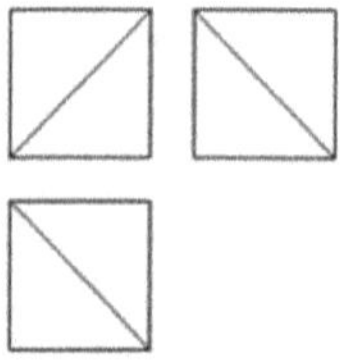

(A) $\frac{1}{8}$ (B) $\frac{1}{7}$ (C) $\frac{1}{6}$ (D) $\frac{1}{5}$

7. A circle passes through three points A (1, 3), B (4, 2), C (1, −7) and intersects y-axis at M and N, then $|MN|=$

(A) $2\sqrt{6}$ (B) 8 (C) $4\sqrt{6}$ (D) 10

8. The flow chart on the right shows an algorithm derived from the "decrease technique" in the Chinese ancient mathematical masterpiece *Nine Chapter Arithmetic*. Execute the program; if the inputs a, b are respectively 14 and 18, then the output $a=$

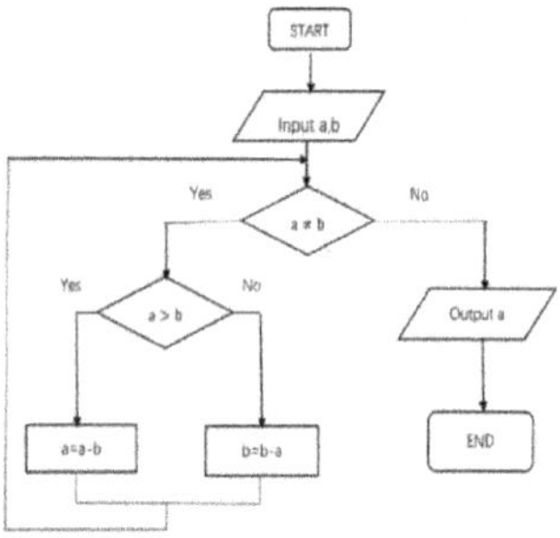

(A) 0

(B) 2

(C) 4

(D) 14

9. Given that A, B are two points on the sphere centered at O, ∠AOB=90°, C is a moving point on the sphere. If the maximum volume of triangular pyramid O-ABC is 36, then the surface area of the sphere is

(A) 36π (B) 64π (C) 144π (D) 256π

10. As shown in the figure, in rectangle ABCD: AB=2, BC=1, O is the midpoint of AB, point P moves along the sides BC, CD and DA, ∠BOP=x. Expressed the sum of the distances from the moving point P to A and B as function $f(x)$, then the image of $y=f(x)$ is roughly like

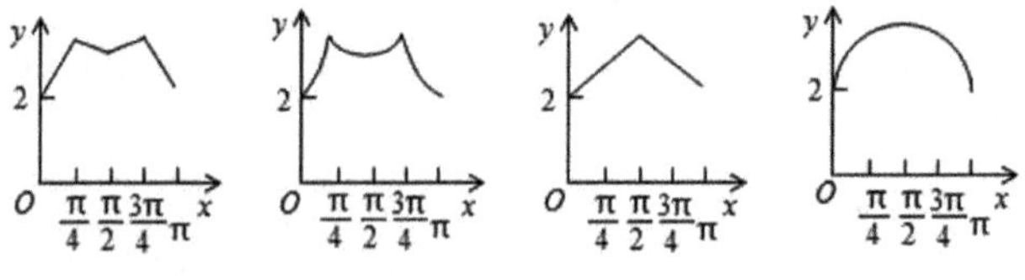

(A) (B) (C) (D)

11. Given that A, B are respectively the left and right vertices of hyperbola E. M is a

point on E, ΔABM is an isosceles triangle and its vertex angle is 120°, then the eccentricity of E is

(A) $\sqrt{5}$ (B) 2 (C) $\sqrt{3}$ (D) $\sqrt{2}$

12. Given that $f'(x)$ is the derived function of odd function $f(x)(x\in\mathbf{R})$, $f(-1)=0$, when $x>0$, $xf'(x)-f(x)<0$, then the range of x that makes $f(x)>0$ is

(A) $(-\infty,-1)\cup(0,1)$ (B) $(-1,0)\cup(1,+\infty)$

(C) $(-\infty,-1)\cup(-1,0)$ (D) $(0,1)\cup(1,+\infty)$

Part B

This part includes two types of questions: required ones and optional ones. Questions 13 to 21 are required; candidates should answer each of them. Questions 22 to 24 are optional.

II. Fill in the Blanks: altogether 4 items, 5 points for each.

13. Given that $\boldsymbol{a}$ and $\boldsymbol{b}$ are unparalleled vectors, vectors $\lambda\boldsymbol{a}+\boldsymbol{b}$ and $\boldsymbol{a}+2\boldsymbol{b}$ are parallel, then real number λ=________. (Fill with digital number)

14. If x, y satisfy the constraints $\begin{cases} x-y+1\geq 0, \\ x-2y\leq 0, \\ x+2y-2\leq 0, \end{cases}$, then the maximum value of $z=x+y$ is ____________.

15. Given that, in the expansion of $(a+x)(1+x)^4$, the sum of the coefficients of the odd power terms of x is 32, then a =__________.

16. Given that S_n is the sum of the first n terms of sequence $\{a_n\}$ and $a_1=-1$, $a_{n+1}=S_n S_{n+1}$, then S_n=________.

III. Calculation Questions: Answers should provide captions, proofs and/or calculation steps.

17. (Full mark: 12 points)

In ΔABC, D is a point on BC, AD bisects ∠BAC, the area of ΔABD is two times the area of ΔADC.

(1) Find $\dfrac{\sin\angle B}{\sin\angle C}$;

(2) If AD=1, DC=$\dfrac{\sqrt{2}}{2}$, find the lengths of BD and AC.

18. (Full mark: 12 points)

A company did a survey to assess customers' satisfaction with its products. It selected 20 customers randomly from each of District A and B. The scores assigned by the customers were as follows:

District A: 62 73 81 92 95 85 74 64 53 76
78 86 95 66 97 78 88 82 76 89

District B: 73 83 62 51 91 46 53 73 64 82
93 48 65 81 74 56 54 76 65 79

(1) Based on the above two sets of data, complete the stem-leaf plot of customer satisfaction scores below and compare the average satisfaction scores and the degree of dispersion of the two districts (accurate values are not required, only conclusions are needed);

(2) According to scores assigned by the customers, the customers' satisfaction was divided into three levels, from low to high:

Satisfaction score	Lower than 70 points	70 points to 89 points	No less than 90 points
Level of Satisfaction	Not satisfied	Satisfied	Very satisfied

Note event C as "the level of customers' satisfaction of District A is higher than that of District B". Supposing the customers' satisfaction results of the two districts are independent, take the frequency of the events as the probability of the corresponding event occurring, and find the probability of event C with the given data.

A地区		B地区
	4	
	5	
	6	
	7	
	8	
	9	

19. (Full mark: 12 points)

As shown in the Figure, in Cuboid ABCD-$A_1B_1C_1D_1$, AB=16, BC=10, AA_1=8, E and F are points on A_1B_1 and D_1C_1 respectively, $A_1E=D_1F=4$. A plane α that passes through E and F intersects with the surfaces of the cuboid. The segments of interaction form a square.

(1) Draw the square in the figure (the drawings and reasons are not required);

(2) Find the sine value of the angle formed by AF and Plane α.

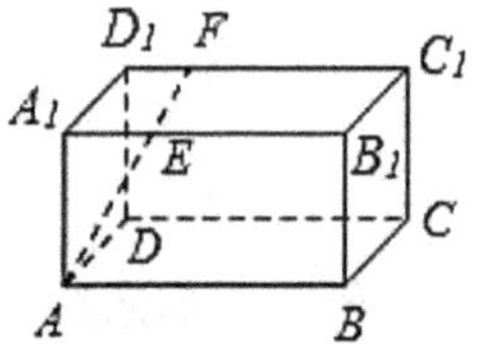

20. (Full mark: 12 points)

Given Ellipse C: $9x^2+ y^2 = m^2$ (m>0), l is a line that does not pass through the origin O and is not parallel to the coordinate axes. l and C have two points of intersection: A and B, M is the midpoint of segment AB.

(1) Prove that the product of the gradient of OM and the gradient of l is constant

value;

(2) If l passes through point ($\frac{m}{3}$, m), extend the segment OM intersects C at point P, can quadrilateral OAPB be a parallelogram? If yes, find the gradient of l at the time; if it cannot, explain the reason.

21. (Full mark: 12 points)

Given function $f(x)=e^{mx}+x^2-mx$.

(1) Prove that $f(x)$ is monotonically decreasing on $(-\infty, 0)$ and monotonically increasing on $(0, +\infty)$;

(2) If for any $x_1, x_2\in[-1,1]$, there is $|f(x_1)-f(x_2)|\leq e-1$, find the range of m.

Please choose one from questions 22, 23, and 24 to answer. If more than one question is answered, the score will only be assigned to the first question answered. Please write down clearly the question number beside your answer.

22. (Full mark: 10 points) Elective 4—1: Selective lecture of geometrical proof

As shown in the figure, O is a point inside isosceles triangle ABC, ⊙O intersects the base BC of $\triangle ABC$ at points M and N and the height AD of the base at point G, it is also tangent to AB and AC at points E and F respectively.

(1) Prove: $EF // BC$;

(2) If AG equals to the radius of ⊙O and $AE=MN=2\sqrt{3}$, find the area of quadrilateral $EBCF$.

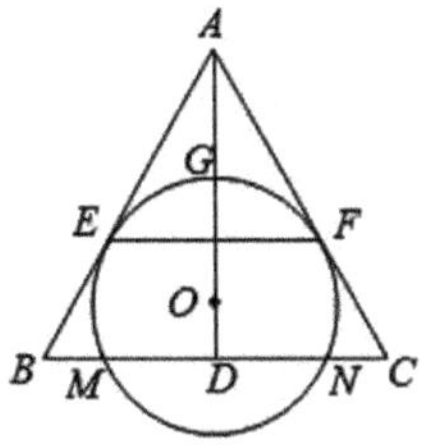

23. (Full mark: 10 points) Elective 4-4: Coordinates and parametric equations

In rectangular coordinate xOy, curve C_1: $\begin{cases} x = t\cos\alpha, \\ y = t\sin\alpha, \end{cases}$ (t is the parameter, $t\neq0$) , where $0\leqslant\alpha<\pi$. In the polar coordinate with O as the origin and x-axis as its positive axis, curve C2: $\rho = 2\sin\theta$, curve C3: $\rho = \cos\theta$.

(1) Find the rectangular coordinates of the intersection of C_2 and C_3;

(2) If C_1 and C_2 intersect at point A, C_1 and C_3 intersect at point B, find the maximum value of $|AB|$.

24. (Full mark: 10 points) Elective 4-5: Selection lecture of inequality

Given that a, b, c, d are positive numbers and $a+b=c+d$, prove that:

(1) If $ab>cd$, then $\sqrt{a}+\sqrt{b}>\sqrt{c}+\sqrt{d}$;

(2) $\sqrt{a}+\sqrt{b}>\sqrt{c}+\sqrt{d}$ is the necessary and sufficient condition of $|a-b|<|c-d|$.

22.2.6 Marking and Evaluation of the Gao Kao *Paper*

Each province designates staff to get together and mark the papers. The markers can be university teachers, middle school teachers, researchers, or graduate mathematics students. The multiple-choice questions are checked by computers, and the blank-filling questions and calculation questions are distributed to specific groups. Each problem should be marked by two persons to ensure that the error rate is less than 0.01%. In recent years, network grading has been taken up by almost all provinces, which improves the efficiency and correctness greatly by using the inherent strengths of searching, recognition, and statistics of the computer.

The marks provide measurable criteria for individual and group assessment that enable colleges and universities to select talented students. As well, each province should evaluate the results of the exam to see whether they are valid, reliable, and achieve the desired results. Several evaluation indexes are proposed, and based on these criteria, we gather and statistically analyze the marks, and they are used in the process of evaluating the examination.

References

Editor groups. (2004). 《普通高中数学课程标准(实验)》解读 [*Interpretation of the full-time compulsory education mathematics curriculum standard*. Nanjing: Jiangsu Education Publishing Group.

Ministry of Education, P. R. C. (2011). 全日制义务教育数学课程标准 *(2011) [Full-time compulsory education mathematics curriculum standard (2011)]*. Beijing: Beijing Normal University Publishing Group.

Ren, Z., & Kong, F. (Eds.). (2010). 数学教育评价新论 *[New arguments on the mathematics education evaluation]*. Beijing: Beijing Normal University Publishing Group.

Tu, R., Wang, G., & Ning, L. (2006). 新编数学教学论 *[New mathematics teaching theory]*. Shanghai: East China Normal University Press.

Chapter 23
Practice and Theoretical Thinking in Constructing a Developmental Assessment System for Mathematics Ability

Chunli Zhang

Mathematics ability, also called mathematics quality, is considered to be a special mental ability, which means "a steady individual mental characteristic required by completing mathematics activities (study, research, practice) smoothly efficiently; it is a relatively steady mental characteristic formed, developed, and mainly presented during the teaching activities" (Zhao, 1987).

In recent years, international mathematics assessment projects have paid more and more attention to tests of mathematics ability or mathematics quality. Since the new course reform in our country, a lot of changes have occurred in the ideas, content methods and evaluators of mathematics learning assessment: from traditional identification assessment to developmental assessment based on modern teaching concepts; from paying attention to basic knowledge and abilities to paying attention to the learning processes and methods; from the single assessment method of the written examination to diversified methods including class observation, homework analysis, and student portfolios; and from teachers as the main evaluators to diversification which encourages student to assess each other and themselves (Gong, 2003). Influenced by international and domestic trends in assessment reform, we recognize the need for a mathematics ability assessment framework that is geared to international standards and adaptable to the new course reform in our country.

23.1 Analysis of Mathematics Ability Structure

Many psychologists and mathematics educators have done research on the basic structure of mathematics ability. For example, Revesz, an American psychologist, maintained that mathematics ability has two basic forms: applicable ability and

C. Zhang (✉)
Faculty of Education, Beijing Normal University, Beijing, China
e-mail: 97070@bnu.edu.cn

Y. Cao and F.K.S. Leung (eds.), *The 21st Century Mathematics Education in China*, New Frontiers of Educational Research,
https://doi.org/10.1007/978-3-662-55781-5_23

creative ability. Weidling, a Swedish psychologist, suggested that it is composed of four elements: understanding mathematical problems, learning mathematical problems, the ability to combine them with other problems, signs, methods, and demonstrations, and the ability to apply them to solve mathematical problems (Shi & Hua, 2008). One of the most influential researchers in this field was Krutetsky of the former Soviet Union; through extensive experiment and research with various students, he systematically studied the nature and structure of mathematics ability and maintained that it is composed of nine elements: the ability to generalize mathematics materials; the ability to formalize mathematics materials and to use structure of the form, i.e., the structure of relation and connection, to conduct operations; the ability to use numbers and other signs to conduct operations; the ability to conduct consecutive and rhythmic logical reasoning; the ability to use simplified thinking structures; the ability to reverse the psychological process, i.e., from the positive to the negative thinking series; flexibility of thinking, i.e., the ability to transit from one mental operation to another; mathematics memory ability; and the ability to form a concept of space (Krutetsky, 1988).

In this era of knowledge-based economies and information-based societies, the fierce international competition and challenges require members of society to be capable of knowledge innovation and scientific innovation, to be able to collect, sort and express information using mathematical methods, establish models, solve problems, and create value for society. In China, therefore, the standards for students' mathematical abilities have been developed to reflect these needs.

First, elements of the structure of mathematics ability have been expanded. While paying attention to the traditional three mathematics abilities (ability to calculate, logical thinking ability, and spatial concepts), the current mathematics course standards have suggested some further abilities. The first of these are the spirit of innovation and ability to practice; second, based on the increasing importance of the information society, is the ability to obtain information and process data; third is knowledge of real-life spatial concepts and graphs, and the ability to think in images; fourth is the ability to use statistics to observe and analyze problems in real life.

Second, the classifications of ability have been given less emphasis in the new standards and the comprehensiveness of mathematics ability has been emphasized more. The fundamental objective of mathematics education is to develop a spirit of innovation and ability to use mathematics knowledge to solve problems in real life. However, in real life, being able to find the solution to any problem is not the effect of a single ability; it usually requires an interaction of multiple abilities that include being able to apply relevant prior experience and knowledge, hence comprehensive mathematical ability is essential. In fact, our country's mathematics course standards propose the development of students' thinking abilities, the effective use of statistics, spatial concepts, a spirit of innovation, and the ability to apply knowledge to solve problems, but these abilities are not described clearly or in detail.

Third, the standards emphasize the development of general ability. This includes observational ability, memory, attention, and imagination. These are important to the development of mathematical abilities, just as they are in other subject areas.

23.2 Mathematics Ability Assessment Framework

Assessment of mathematics ability should be based on both content and behavioral standards. Content standards refer to descriptive knowledge that we think valuable and expect students to know, the procedural knowledge that can be used to work things out and strategic knowledge that involves knowing why and when to use the other types of knowledge. Only if we transfer the content standards into corresponding behavioral standards will we be able to provide a basis for assessment. Clearly, different types of test questions need to be used to examine different levels of students' abilities involved in solving mathematics problems (Ma & Zhang, 2003). According to students' understanding and thinking levels, levels of mathematics ability are divided into three levels: simple skills, conceptual understanding and application, and problem solving. As well, according to the idea of SOLO classification, students' responses to a mathematics problem can vary greatly; even if they are at the same level of thinking, they will present different degrees of proficiency and accuracy according to the complexity and novelty of the problem situation (Biggs & Collis, 1982).

To illustrate this complexity, we can build a preliminary three-dimensional model for the assessment of mathematics ability. The content domain is the first dimension of the model, and students' understanding and thinking levels are the second. The students' actual responses, from the simple to the complicated, constitute the third dimension of the model. These three dimensions together constitute a three-dimensional theoretical framework for mathematics ability assessment, as shown in the following figure:

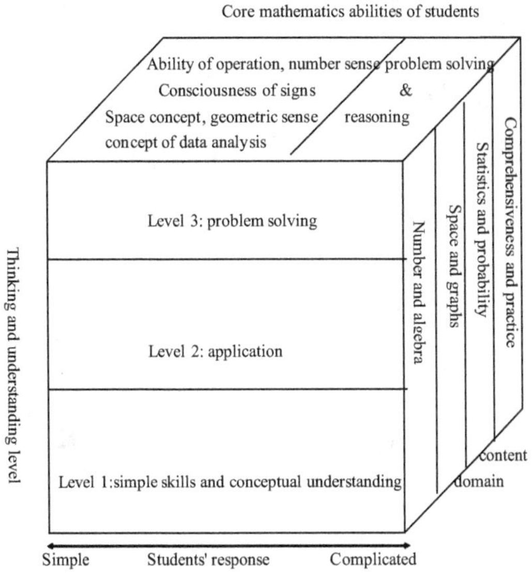

23.3 Behavioral Standards for Mathematics Ability Assessment

Nowadays, many teachers are implementing the course reform in their mathematics classes according to the content standards in the mathematics course standards. However, teachers are finding that these content standards cannot be transferred directly to assessment standards. In fact, when assessing the difficulty of a question, it is usually discovered that both objective difficulty and subjective difficulty are involved and are related to multiple factors, including the complexity of the knowledge and thinking ability required by the question, the novelty of the problem situation, the type of the question, and the value of the question for students (Lei, 2006). For this reason, we need to make the descriptors of different ability levels more specific, operable, and systematic. This is actually a kind of analytical method to solve the problem, by developing a set of scientific behavioral standards for the assessment of mathematics ability by exploring the thinking characteristics and main behavioral expressions related to different ability levels.

23.3.1 Level One: Simple Skills and Conceptual Understanding

Simple skills mainly include arithmetic operations or algebraic operations and the use of mathematics and measuring tools to carry out calculations, construction, and measurement. Conceptual understanding means that students can express, illustrate, and judge under specific conditions according to the definition, nature, and characteristics of the concept. Understanding of a concept also includes identifying the question that can be solved according to the given information (such as data) and knowing the differences and connections between a given object and relevant others in order to carry out comparisons, classifications and ranking, etc. The main behavioral expressions of conceptual understanding are as follows:

Memorization/recognition	Be able to recognize an object in a specific situation according to its characteristics
Simple operation	Know adding, subtracting, multiplying, and dividing and the operation rules of their mixed operation; evaluate expression and formula, simplify an algebraic or numeric expression; combine like terms, solve equations, etc.
Simple measurement and construction	Be able to use simple measuring tools, and construct using a straight edge and compass according to the given conditions
Classification and ranking	Classify objects, graphs, numbers, expressions, and concepts according to common properties; be able to classify a certain object correctly and rank according to a certain property

(continued)

(continued)

Expression and extraction	Use models to express numbers; use graphs, tables, charts, coordinate graphs to present mathematics information or data; be able to use equivalent representation notation to express certain given mathematics nature or relationship. Be able to identify and extract useful information (e.g., data) that will help solve problems
Illustration	Be able to describe relevant characteristics of an object from illustrations of it; be able to explain a given equation or expression using a problem or situation
Judgment	Be able to judge the differences and connections between a given object and relevant others according to meaning, nature, and characteristics of the concept

23.3.2 Level Two: Application

When students reach the level of application, it means that they can accurately choose and use proper procedures, apply rules to various changing situations and solve problems using the methods of specific models or signs. Students mastering procedural knowledge usually connect a calculation method with the given problem situation, correctly use algorithms and can communicate the result of an operation in the problem setting. The application level also involves non-computational skills such as reading ability, ability to produce charts, conducting geometric construction, and explaining meanings of the model. The main behavioral expressions of application are as follows:

Choosing	Choose proper algorithm, model, rule, formula, and unit to solve problems, in which the algorithm, rule, or method to solve the problems is known by students
Modeling	Use proper models, such as equivalent expressions and charts, to solve conventional problems
Explaining	Explain a given mathematics model (equality, charts, etc.).
Using tools	According to requirements, use a series of mathematics tools and steps to complete graphs with given requirements
Solving conventional problems	Apply facts, procedures, concepts, and other knowledge to solve conventional problems (including problems in real life), which means that the problems are similar to those that students may encounter in the class
Verification	Verify/check whether the solution or the result is correct; evaluate the rationality of the solution to the problem or the result

23.3.3 Level Three: Problem Solving

When students reach this ability level, it means that they can make reasonable assumptions, guess, predict, and deduce from the given information, apply reasoning in the new environment to carry out analysis and assessment, connect all of their prior mathematics concepts, procedures, reasoning, and skills of information exchange to solve problems and testify or refute the truth of a certain statement using mathematics methods and reasoning. The main behavioral expressions of problem solving are as follows:

Assumption/guess/prediction	Make reasonable assumptions, guesses, and effective predictions and deductions from the given information
Analysis	Determine, describe, and use the relationships of variables or objects in the mathematics situation; analyze univariate statistical data; use geometric graphs to simplify the problem; to narrate the nature of an unfamiliar given cube
Generalization and extension	Use a more general, broader, or wider-applied term to expand the thinking and results of problem solving
Connection/ composition/ integration	Connect new knowledge with existing knowledge; connect the different elements of knowledge and relevant expression; connect relevant mathematics thoughts or objects; integrate the mathematics process to obtain results; merge the results again to obtain further results
Solving unconventional problems	Be able to apply the mathematics process to unfamiliar situations to solve a problem which has not been encountered in class but is similar to those that have been encountered
Testifying	According to mathematics results or properties, test the truth of a certain conclusion or refute a certain statement using mathematical reasoning
Assessment	Discuss and conduct critical evaluations of mathematics thoughts, guesses, solutions to problems, and proofs

23.4 Monitoring Students' Thinking and Understanding Through Assessment

The core of assessing mathematics ability is to monitor students' thinking and understanding. The method of assessing mathematics ability needs to be consistent with the teaching, as one part of the daily mathematics teaching practice. During the assessment, it is not enough to figure out how many concepts a student can identify or how many skills he/she can master; instead, the assessment should focus on how students identify the concepts and master the skills and processes used to apply these concepts and skills in understanding problems, establishing models, solving problems, and demonstrating thinking in increasingly complicated and novel tasks and situations; that is, assessing what the students have understood rather than what they know.

The first step to change the assessment practice is to develop procedure-oriented assessment tasks. Through procedural assessment tasks, students can link the mathematics concepts, mathematics skills, mathematics thoughts and methods, strategies for problem solving and various experiences (including experiences of mathematics activities and experiences in life), and use them to solve some unconventional problems. In fact, the relevant content of the procedural assessment tasks and the traditional problems that students are required to solve according to fixed procedures they have learnt are quite similar, but the relevant thinking processes are different. The new approach is to examine the abilities of exploring, discovering, and creating instead of memorizing and imitating. By encouraging students to study the procedural assessment tasks in various situations of reality or virtual reality, we expect them to demonstrate the habits and abilities to think mathematically.

The second step to change the assessment practice is to enhance procedural assessment, especially to monitor students' progress in various aspects of their mathematics ability, which is a key aspect of assessment practice reform. Traditional mathematics teachers use quizzes, chapter tests, scores and numbers of correct answers, and regular statistics describing students' performance rankings in the class to monitor their mathematics learning. With the increasing introduction of procedural assessment tasks, teachers are desperately in need of another method to assess students' mathematics ability performances. Then, the scores and correct answers will only be a part of the examination indexes; the processes and methods of solving problems will become a more important indicator of different mathematics ability levels.

The third step in changing the assessment practice is to use the information from the assessment to improve teaching. The purpose of this step is to expand the view of class assessment for teachers. Teachers should consider the assessment as an activity integrated in the daily teaching practice, reflect the idea and guiding function of the assessment in daily teaching and promote the development of students' mathematics ability levels by giving targeted feedback and guidance on students' performances.

In short, in the framework system of mathematics ability assessment for students, assessment cannot be separated intentionally from teaching; assessment provides feedback information for teaching; information from assessment, including ability level, thinking characteristics and behavioral performance, can be used as evidence to assess the students' achievement and progress as well as the basis for teaching decisions. Only in this way can the assessment fully play a significant role in promoting the development of students' mathematics abilities.

23.5 Framework for a Developmental Assessment System for Mathematics Learning

In the twenty-first century, educators in China have been making efforts to design learner-centered educational environments and to promote students' construction of knowledge by developing seminars, cooperative group activities and research studies; at the same time, they also face an inevitable problem, which is how to develop proper assessment tools to assess students' understanding and achievements in mathematics.

Educational assessment theory integrates many viewpoints of cognitive psychology, especially cognitive constructivism psychology; as explained above, it is not appropriate just to measure students' abilities to memorize and imitate, it is necessary to assess their understanding of mathematics knowledge, ability to solve problems, practical abilities, emotions, and attitudes (Hargreaves, 1989). Formative assessment, summative assessment, and performance assessment are three parts of a complete mathematics learning developmental assessment system. Formative assessment has the function of diagnosis, with learning content and the specific behavioral objectives as references; the main daily assessment methods and means include asking questions in the class, blackboard-writing, class practice, formal examinations, and homework examinations. Summative assessment has the function of identification, with course objectives (course standard) as references, adopting paper-and-pencil testing as the main assessment method and means (Tang, 2000, pp. 40–41). Performance assessment has the functions of diagnosis and promotion, with individual development as a reference, adopting performance tasks which specifically include open tasks using paper-and-pencil, and performance tests using tools other than paper-and-pencil, mathematics investigations and experiments, mathematics diaries, and learning portfolios (Tang, 2000, pp. 111–117). An all-round assessment platform can be built by collecting a wide variety of artifacts through all three of formative, summative, and performance learning.

As shown in Fig. 23.1, modern mathematics assessment requires the establishment of a developmental assessment system with diversified objectives and various methods. The assessment not only focuses on knowledge and skills, but also covers processes and methods, especially the abilities of solving problems, cooperation, and communication and practice, as well as emotional factors and attitudes toward learning mathematics. The assessment methods are varied, including written tests, oral tests and activity reports, class observations, after-class interviews, homework analysis, and the establishment of best work portfolios and learning portfolios. The presentation of evaluation results can be both quantitative and qualitative. As well, attention should be paid to the students' development processes, with the focus on vertical assessment that is monitoring each student's progression across time,

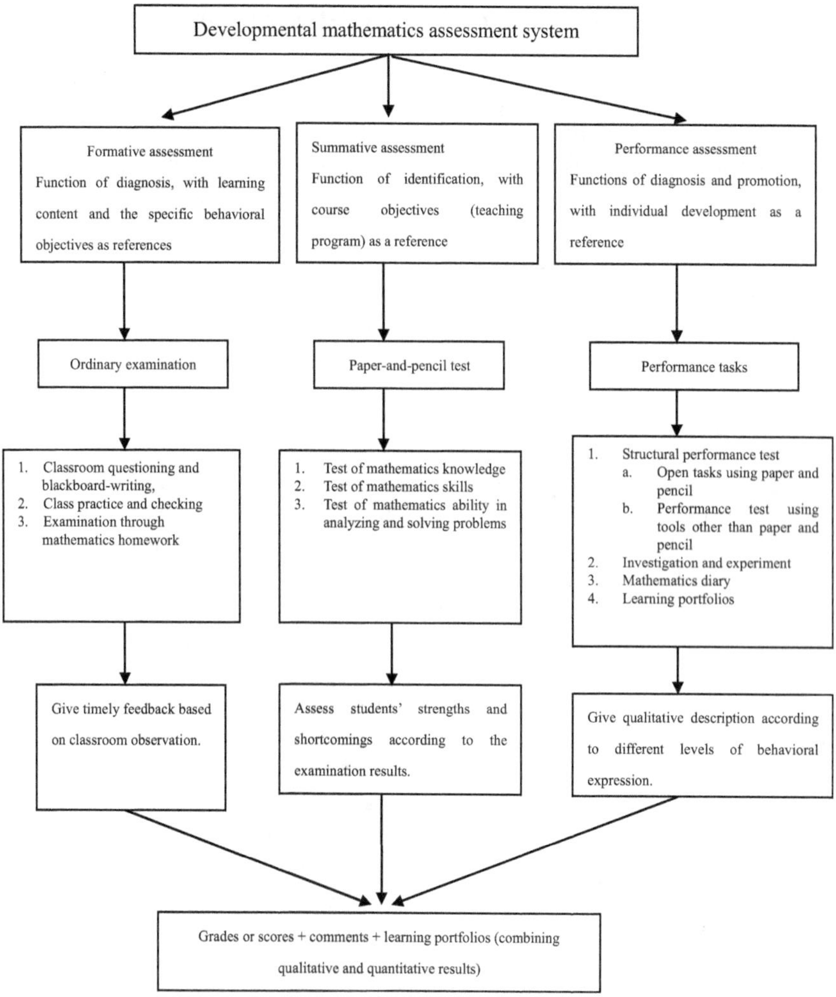

Fig. 23.1 Developmental mathematics learning assessment system

focusing on added value in the grade and quality of students rather than simply grading and ordering so that students can experience their own progress. In addition, developmental assessment involves more people in the assessment process, encouraging self-assessment and mutual assessment of students and strengthening communication and cooperation between parents and teachers so as to ensure the smooth implementation of developmental assessment.

23.6 Diversified Assessment Tools in the Developmental Assessment System

One of the important tasks of the developmental assessment system is developing diversified assessment tools and methods. In our research project with experimental teachers from fourteen schools, commenced in 2002, we have focused mainly on seven aspects of the development and implementation of assessment tools.

The first method is observing students' performances in the classroom. We provided students with a dynamic focusing process, the effect of which was to strengthen teachers' observations and the assessment of students' daily behaviors in the class. The teachers first developed a class observation table and conducted an assessment from five aspects: participation in learning, interaction during learning, thinking occurring during learning, achievement, and emotional condition during learning. In the experiment, the teachers realized that assessment was everywhere, especially that the ongoing in-class assessment played an important role in encouraging the learning initiative and correcting students' mistakes. When the teachers changed former words of praise such as "Hey, hey, you are so great!" to vivid and specific words like "You brought up a question that I had not thought of as a teacher!" or "You found these mathematical problems in daily life, you are so good at observing and thinking," these comments enable the students to figure out the learning behaviors expected by their teachers, and they could also experience the careful observations being made by their teachers and the encouragement from the teachers' hearts.

Second, through measuring levels or aspects of mathematics ability by tests (written test, oral test, presentation etc.), diversified assessment tools enabled the teachers to break through limitations of traditional assessment ideas and tools, even "manmade" barriers, so that they could view the changes in students' academic performance and social growth in a certain period of time from a more comprehensive and objective standard. We brought up six principles involved in making up mathematics tests: focusing on the development of each student as the objective; focusing on the formation of knowledge on the basis of knowledge and basic skills development; connecting with reality, focusing on the application of knowledge in daily life; designing the test to promote teaching and improve learning; paying attention to the diversification of tests; and expressing new people-oriented ideas. The question types should be diversified, with reading, writing, drawing, circling, dotting, linking, and other forms included so as to stimulate and enlighten students' minds and add fun in answering the questions. During test assessments, teachers should fully encourage "diversified answers" and lead their students to think deeply from different points of view while respecting their unique experiences and understanding.

Third, through the learning portfolios, diversified assessment tools provided a chance for teachers to learn about minor changes occurring in their students and for students to present their talents, which could not be achieved through traditional assessment tools. First, we prepared *Student Growth Records* and required that the collected student work should show four characteristics—clear discipline, intuition and operability, students' personalities, and diversification of assessment. In the growth record, we set such topics as "The world of graphs," "Look, my calculation is accurate and fast," "Pleasure of learning," "Mathematics in life," and "My new discoveries." During implementation, we advocated that students should have the right to select their own portfolio content so that the effect of the assessment could be utilized fully to promote their development; we also actively sought support from parents so that this work would be implemented thoroughly.

Fourth, through developing performance tasks, diversified assessment tools also provided a chance for the teachers to assess the students' behaviors of cooperating and communicating with others and their practical and innovative behaviors when working on real-world tasks in groups. This not only introduced cooperative learning into the class, but also allowed the teachers to assess students' use of prior knowledge to solve problems, in ways that traditional paper-and-pencil tests could not. The content for this type of evaluation could be the "practical activity" in the textbook, or an activity theme designed from a real-life situation, a current problem of interest to everyone, or a school or community activity. The teachers were not concerned only with the final result. Instead, they encouraged their students to conduct diversified exploration using different methods and from different angles; the focus was on the spirit of exploration, sense of cooperation, and sense of innovation. The practice showed that such evaluation helped to promote students' active participation, cooperation with others, and being bold in practice.

Fifth, through evaluating each student's mathematics learning totally, diversified assessment required the teachers to grasp the most obvious strengths and shortcomings when the students were engaged in diversified learning activities, so as to point out both the achievements and the aspects they needed to work on. We proposed the following principles: (1) Pay attention to the assessment of process and method (ability to analyze and solve problems, sense of cooperation, etc.); (2) Comments should be real, personalized, objective, and offering incentive; (3) Make comments appropriate to the needs and hopes of different students. During implementation, we emphasized strongly that teachers' comments should be real, personalized and easy to understand, and expressed truthfully from the teachers' feelings. Chinese is a language with a special charm, which can help us to communicate abundant information; integrating feelings in their language, teachers can make the students feel confident and relaxed and thus play an important role in the students' development. We emphasized that the teachers should use comments that were personalized and objective as well as rich and touching, and that they should use language, which could be understood by the students; words, which are too exaggerated, flashy or profound, cannot move students and, instead, they will feel that their teachers are false and deliberately misleading.

Sixth, the diversified assessment encouraged the teachers to look at their students' progress from a developmental viewpoint. Two methods were to improve their ways of correcting homework and establishing make-up test systems. The traditional way to correct homework was to mark answers with "√" or "×"; this method was direct and rigid. When students got their homework back, they would be frustrated to see the "×" and be passive about correcting the wrong answers. It was especially difficult for the slower students, who took a long time to get their homework done, only to receive series of "×." As an alternative, we suggested that the teachers mark mistakes with "\" instead of "×," which could easily be turned into a "√" after the corrections had been made. This reform only concerned a small sign change, but the teachers were excited by it. For a long time, even the students with potential had red flowers in their exercise books; they had more enthusiasm to learn and to do their homework better and better. Fewer students were refusing to hand in homework and the grades improved. The "make-up test" system aimed to fully explore students' differences and develop their individual potentials; it helped self-recognition and confidence building, encouraged students to study actively and improved their abilities through deferred assessment of their knowledge and ability. In the examination at the end of a semester or in a unit test, if a student was not satisfied with the grade of a course, he/she could apply for a "make-up test." The "make-up test" was divided into "immediate make-up test" and "deferred make-up test." The application was evaluated by teachers, students, and parents, and once it was accepted, the applicant had the incentive to achieve the objectives that had been set. Students could choose the better grade from the two examinations to be the final assessment result. The implementation of this system was a great improvement on the one-time assessment of the past, with the students being more motivated to learn and having one more chance to succeed. It made the students much more confident and promoted the continuous development of their potentials.

Finally, the diversified assessment offered a communication channel for students and parents to learn about the learning process and the students' results in and out of school. The schools used "parent–school contact cards," which consisted of four parts: presentation of the individual case, teachers' feelings and comments, the feelings of the students themselves and comments by classmates, and the parents' feelings and comments. Focusing on the distinctive behaviors or problems, the teachers regularly recorded anecdotes about a student on the parent–school contact card and communicated with parents in a timely way. At the same time, in order to strengthen communication between students and their parents, according to the problems in the parent–school contact card, each class made a wall newspaper for each discipline, to encourage the students to evaluate each other; parents were welcomed to look at the wall newspaper. At the end of each semester, a special parent symposium was organized; the parents revisited the problems described in the "parent–school contact card" to make further improvements. Well-equipped schools also used the advantages of the internet to construct assessment platforms to make the parents–school communication channel faster and more efficient. As well, for parents who did not go online regularly, the teachers used text messages to remind them to communicate more with the teachers so as to promote and care more about the students' growth.

23.7 Some Thoughts About the Assessment Practice and Future Theoretical Research

Throughout the years of the project research, we have been surprised at the large amount of practical experience we have gained in relation to assessment. At the same time, we are also thinking more about the new questions for future theoretical research.

23.7.1 How to Better Integrate Assessment and Teaching to Inform the Teaching

One important future direction is to integrate mathematics learning assessment with mathematics teaching in order to use the assessment outcomes to enhance the quality of the learning. For example, open questions or performance tasks can play a role in tests or examinations and can also be introduced in the class as a teaching activity. Another example is that, as well as utilizing students and parents along with teachers to carry out the assessment, this can become a part of the learning process, with students encouraged to assess each other and the teacher. In this way, the teacher can change from the authority and provider of standard answers to promoter, leader, and learning collaborator. Furthermore, making mistakes in learning is inevitable for students; due to differences in family and cultural backgrounds, students are diverse in many ways, including intelligence, their ways of thinking, learning habits, and learning motivation. The use of encouragement and honest feedback can enable students to see their shortcomings so that they can take measures to address them.

23.7.2 How to Make Assessment Standard Fair, Open, and Impartial

Assessment depends a lot on the assessment standard being fair, open, and impartial for all students. Specifically, to achieve a fair assessment standard, mathematics experts and teachers need to take into account individual student differences and the diversity of learning content to ensure that the standard (including assessment content and differentiation requirements) does not favor some students over others. The standard needs to be open and transparent; it can even be developed by both students and teachers together. Teachers need to be impartial about the assessment, with no prejudices toward any students, and to provide comments objectively. An important component of future mathematics learning assessment theory is how to keep the assessment fair, open, and impartial and develop a set of scientific measurement scales and proper assessment language. In the future, we need to improve the operability of the mathematics learning assessment scale so that teachers can easily master and operate it and use it to better observe students' mathematics learning behavior. We also need to improve the ability of teachers to assess themselves, including rethinking assessment behavior and language. There is a need to develop the language of assessment scale so that students can understand it better, then the assessment will become a mirror for students' mathematics learning behavior and will be able to learn about, discover, and perfect themselves.

23.7.3 How to Coordinate Various Assessment Tools and Use Them in Practice

The purpose of developing diversified assessment tools is not for identification or screening; instead, it is about learning comprehensively about students' developing mathematics learning abilities. Hence, the framework for the developmental mathematics learning assessment system needs to be quite different from that of the traditional assessment. But we also have to realize that developmental assessment is not a denial of the traditional assessment; instead, it is an inheritance from and development of the traditional assessment as well as a beneficial supplement to it. It is not intended to abandon the traditional testing system, but rather to add the new and diversified assessment methods to the traditional assessment framework. In the future, what we need to study is not how to get a final score or grade by combining the assessment results obtained through various assessment methods by some kind of weighted average method, but how to select proper assessment methods according to different teaching content and requirements and how to coordinate various assessment tools to work so that the outcomes can be used to provide more feedback and suggestions for students and more varied and real references for mathematics teaching.

Assessment reform is a huge but delicate task. We need to rethink it during the practice so that we can adopt the characteristics of various assessment methods to create a scientific and fair mathematics learning assessment environment for students.

References

Biggs, J. B., & Collis, K. F. (1982). *Evaluating the quality of learning: The SOLO taxonomy* (pp. 125–167). New York: Academic Press.

Gong, L. (2003). 数学教育评价的回顾与展望 [Retrospect and prospect of mathematics education assessment]. *Journal of Mathematics Education, 2,* 31–34.

Hargreaves, A. (1989). *Curriculum and assessment reform* (pp. 1–4). Milton Keynes, Philadelphia: Open University Press.

Krutetskii, V. A. (1988). 中小学生数学能力心理学 [Psychology of mathematics ability of primary school and middle school students] (B. Li, Trans). Shanghai: Shanghai Educational Publishing House.

Lei, X. (2006). 大规模教育考试:命题与评价 [Large-scale educational examination: Proposition and assessment] (pp. 100–134). China: East China Normal University Press.

Ma, Y., & Zhang, C. (2003). 数学教育评价 [Mathematics education assessment]. Beijing: Higher Education Press.

Shi, Y., & Hua, G. (2008). 中小学生数学能力的结构及其培养 [Structure and cultivation of mathematics ability of primary school and middle school students]. *Journal of Educational Studies, 3,* 36–40.

Tang, X. (2000). 课堂教学与学习成效评价 [Classroom teaching and learning evaluation]. Nanning: Guangxi Education Press.

Zhao, Y. (1987). 小学生数学能力的测查与评价 [Examination and assessment of mathematics ability of primary school students] (pp. 120–143). Beijing: Educational Science Publishing House.